Essentials of Hydrogeology

Essentials of Hydrogeology

Isabel Munoz
Amanda Schmidt

Kruger Brentt
Publishers
2025

Kruger Brentt Publishers UK. LTD.
Company Number 9728962

Regd. Office: 68 St Margarets Road, Edgware, Middlesex HA8 9UU

ISBN: 978-1-78715-307-3

For information on all our publications visit our website at http://krugerbrentt.com/

PREFACE

Water is one of the most vital resources on our planet, essential for sustaining life, supporting ecosystems, and driving economic activities. Understanding the behavior and distribution of groundwater—the largest reservoir of fresh water on Earth—is fundamental to managing water resources, mitigating environmental risks, and addressing the challenges of water scarcity and pollution. Hydrogeology, the scientific study of groundwater and its interactions with the earth and the environment, provides valuable insights into the dynamics of subsurface water systems and their implications for society.

This book, titled "Essentials of Hydrogeology," is a comprehensive introduction to the principles, concepts, and applications of hydrogeology. Drawing upon the expertise of geologists, hydrologists, and environmental scientists, this volume offers insights, methodologies, and case studies for understanding and managing groundwater resources.

From the basics of aquifer properties and groundwater flow to the principles of groundwater contamination and remediation, "Essentials of Hydrogeology" covers a wide range of topics essential for mastering the fundamentals of hydrogeology. Through clear explanations, diagrams, and real-world examples, readers will gain a deep understanding of the processes governing groundwater movement, storage, and quality.

Moreover, this book explores the practical applications of hydrogeology in various fields, including water supply, environmental management, and geotechnical engineering. By examining case studies and best practices from around the world, we highlight the role of hydrogeology in addressing complex water-related challenges and promoting sustainable water management practices.

As we confront the growing pressures on our water resources, it is our hope that "Essentials of Hydrogeology" will serve as a valuable resource for students, professionals, policymakers, and anyone interested in understanding the dynamics of groundwater and its significance for society and the environment. Whether you are studying geology, hydrology, environmental science, or working in water resources management, this book invites you to delve into the fascinating world of hydrogeology and explore the solutions it offers to the water challenges of the 21st century.

Isabel Munoz

Amanda Schmidt

CONTENTS

CHAPTER 1

GEOLOGY: AN OVERVIEW

1.1 AN OVERVIEW

Geology is a branch of natural science concerned with the Earth and other astronomical objects, the rocks of which they are composed, and the processes by which they change over time. Modern geology significantly overlaps all other Earth sciences, including hydrology. It is integrated with Earth system science and planetary science.

Geology describes the structure of the Earth on and beneath its surface and the processes that have shaped that structure. Geologists study the mineralogical composition of rocks in order to get insight into their history of formation.

Geology determines the relative ages of rocks found at a given location; geochemistry (a branch of geology) determines their absolute ages. By combining various petrological, crystallographic, and paleontological tools, geologists are able to chronicle the geological history of the Earth as a whole. One aspect is to demonstrate the age of the Earth. Geology provides evidence for plate tectonics, the evolutionary history of life, and the Earth's past climates.

Geologists broadly study the properties and processes of Earth and other terrestrial planets. Geologists use a wide variety of methods to understand the Earth's structure and evolution, including fieldwork, rock description, geophysical techniques, chemical analysis, physical experiments, and numerical modelling.

In practical terms, geology is important for mineral and hydrocarbon exploration and exploitation, evaluating water resources, understanding natural hazards, remediating environmental problems, and providing insights into past climate change.

Geology is a major academic discipline, and it is central to geological engineering and plays an important role in geotechnical engineering.

1.2 HISTORY

The study of the physical material of the Earth dates back at least to ancient Greece when Theophrastus (372–287 BCE) wrote the work Peri Lithon (On Stones). During the Roman period, Pliny the Elder wrote in detail of the many minerals and metals, then in practical use – even correctly noting the origin of amber. Additionally, in the

4th century BCE Aristotle made critical observations of the slow rate of geological change. He observed the composition of the land and formulated a theory where the Earth changes at a slow rate and that these changes cannot be observed during one person's lifetime. Aristotle developed one of the first evidence-based concepts connected to the geological realm regarding the rate at which the Earth physically changes.

Abu al-Rayhan al-Biruni (973–1048 CE) was one of the earliest Persian geologists, whose works included the earliest writings on the geology of India, hypothesizing that the Indian subcontinent was once a sea. Drawing from Greek and Indian scientific literature that were not destroyed by the Muslim conquests, the Persian scholar Ibn Sina (Avicenna, 981–1037) proposed detailed explanations for the formation of mountains, the origin of earthquakes, and other topics central to modern geology, which provided an essential foundation for the later development of the science. In China, the polymath Shen Kuo (1031–1095) formulated a hypothesis for the process of land formation: based on his observation of fossil animal shells in a geological stratum in a mountain hundreds of miles from the ocean, he inferred that the land was formed by the erosion of the mountains and by deposition of silt.

Georgius Agricola (1494–1555) published his groundbreaking work De Natura Fossilium in 1546 and is seen as the founder of geology as a scientific discipline.

Nicolas Steno (1638–1686) is credited with the law of superposition, the principle of original horizontality, and the principle of lateral continuity: three defining principles of stratigraphy.

The word geology was first used by Ulisse Aldrovandi in 1603, then by Jean-André Deluc in 1778 and introduced as a fixed term by Horace-Bénédict de Saussure in 1779. But according to another source, the word "geology" comes from a Norwegian, Mikkel Pedersøn Escholt (1600–1669), who was a priest and scholar. Escholt first used the definition in his book titled, Geologia Norvegica (1657).

William Smith (1769–1839) drew some of the first geological maps and began the process of ordering rock strata (layers) by examining the fossils contained in them.

In 1763, Mikhail Lomonosov published his treatise On the Strata of Earth. His work was the first narrative of modern geology, based on the unity of processes in time and explanation of the Earth's past from the present.

James Hutton (1726–1797) is often viewed as the first modern geologist. In 1785 he presented a paper entitled Theory of the Earth to the Royal Society of Edinburgh. In his paper, he explained his theory that the Earth must be much older than had previously been supposed to allow enough time for mountains to be eroded and for sediments to form new rocks at the bottom of the sea, which in turn were raised up to become dry land. Hutton published a two-volume version of his ideas in 1795.

Followers of Hutton were known as Plutonists because they believed that some rocks were formed by vulcanism, which is the deposition of lava from volcanoes, as opposed to the Neptunists, led by Abraham Werner, who believed that all rocks had settled out of a large ocean whose level gradually dropped over time.

The first geological map of the U.S. was produced in 1809 by William Maclure. In 1807, Maclure commenced the self-imposed task of making a geological survey of the United States. Almost every state in the Union was traversed and mapped by him, the Allegheny Mountains being crossed and recrossed some 50 times. The results of his unaided labours were submitted to the American Philosophical Society in a memoir entitled Observations on the Geology of the United States explanatory of a Geological Map, and published in the Society's Transactions, together with the nation's first geological map. This antedates William Smith's geological map of England by six years, although it was constructed using a different classification of rocks.

Sir Charles Lyell (1797–1875) first published his famous book, Principles of Geology, in 1830. This book, which influenced the thought of Charles Darwin, successfully promoted the doctrine of uniformitarianism. This theory states that slow geological processes have occurred throughout the Earth's history and are still occurring today. In contrast, catastrophism is the theory that Earth's features formed in single, catastrophic events and remained unchanged thereafter. Though Hutton believed in uniformitarianism, the idea was not widely accepted at the time.

Much of 19th-century geology revolved around the question of the Earth's exact age. Estimates varied from a few hundred thousand to billions of years. By the early 20th century, radiometric dating allowed the Earth's age to be estimated at two billion years. The awareness of this vast amount of time opened the door to new theories about the processes that shaped the planet.

Some of the most significant advances in 20th-century geology have been the development of the theory of plate tectonics in the 1960s and the refinement of estimates of the planet's age. Plate tectonics theory arose from two separate geological observations: seafloor spreading and continental drift. The theory revolutionized the Earth sciences. Today the Earth is known to be approximately 4.5 billion years old.

1.3 GEOLOGICAL MATERIAL

The majority of geological data comes from research on solid Earth materials. Meteorites and other extraterrestrial natural materials are also studied by geological methods.

1.3.1 Mineral

Minerals are naturally occurring elements and compounds with a definite homogeneous chemical composition and ordered atomic composition.

Each mineral has distinct physical properties, and there are many tests to determine each of them. Minerals are often identified through these tests. The specimens can be tested for:

1. Luster: Quality of light reflected from the surface of a mineral. Examples are metallic, pearly, waxy, dull.
2. Color: Minerals are grouped by their color. Mostly diagnostic but impurities can change a mineral's color.
3. Streak: Performed by scratching the sample on a porcelain plate. The color of the streak can help name the mineral.
4. Hardness: The resistance of a mineral to scratching.
5. Breakage pattern: A mineral can either show fracture or cleavage, the former being breakage of uneven surfaces, and the latter a breakage along closely spaced parallel planes.
6. Specific gravity: the weight of a specific volume of a mineral.
7. Effervescence: Involves dripping hydrochloric acid on the mineral to test for fizzing.
8. Magnetism: Involves using a magnet to test for magnetism.
9. Taste: Minerals can have a distinctive taste such as halite (which tastes like table salt).

1.3.2 Rock

A rock is any naturally occurring solid mass or aggregate of minerals or mineraloids. Most research in geology is associated with the study of rocks, as they provide the primary record of the majority of the geological history of the Earth. There are three major types of rock: igneous, sedimentary, and metamorphic. The rock cycle illustrates the relationships among them. When a rock solidifies or crystallizes from melt (magma or lava), it is an igneous rock. This rock can be weathered and eroded, then redeposited and lithified into a sedimentary rock. It can then be turned into a metamorphic rock by heat and pressure that change its mineral content, resulting in a characteristic fabric. All three types may melt again, and when this happens, new magma is formed, from which an igneous rock may once again solidify. Organic matter, such as coal, bitumen, oil, and natural gas, is linked mainly to organic-rich sedimentary rocks. To study all three types of rock, geologists evaluate the minerals of which they are composed and their other physical properties, such as texture and fabric.

1.3.3 Unlithified Material

Geologists also study unlithified materials (referred to as superficial deposits) that lie above the bedrock. This study is often known as Quaternary geology, after the

Quaternary period of geologic history, which is the most recent period of geologic time.

Magma

Magma is the original unlithified source of all igneous rocks. The active flow of molten rock is closely studied in volcanology, and igneous petrology aims to determine the history of igneous rocks from their original molten source to their final crystallization.

1.4 WHOLE-EARTH STRUCTURE

1.4.1 Plate tectonics

In the 1960s, it was discovered that the Earth's lithosphere, which includes the crust and rigid uppermost portion of the upper mantle, is separated into tectonic plates that move across the plastically deforming, solid, upper mantle, which is called the asthenosphere. This theory is supported by several types of observations, including seafloor spreading and the global disribution of mountain terrain and seismicity.

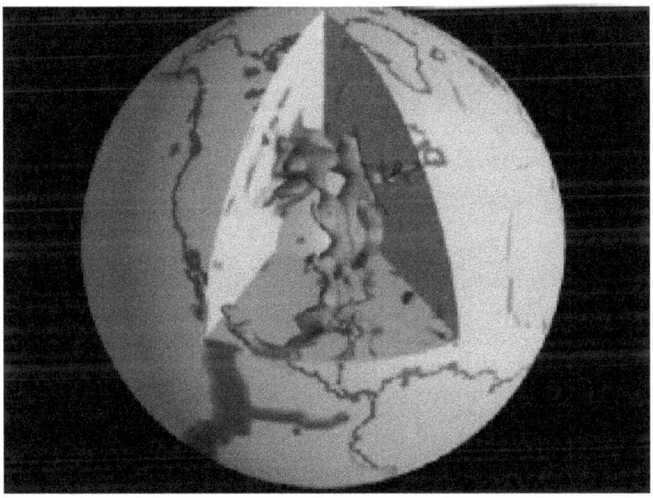

Fig. 1.1: In this Diagram based on Scismic Tomography, Subducting Slabs are in Blue and Continental Margins and a Few Plate Boundaries are in Red

There is an intimate coupling between the movement of the plates on the surface and the convection of the mantle (that is, the heat transfer caused by the slow movement of ductile mantle rock).

Thus, oceanic plates and the adjoining mantle convection currents always move in the same direction – because the oceanic lithosphere is actually the rigid upper thermal boundary layer of the convecting mantle. This coupling between rigid plates moving on the surface of the Earth and the convecting mantle is called plate tectonics.

The development of plate tectonics has provided a physical basis for many observations of the solid Earth. Long linear regions of geological features are explained as plate boundaries.

For example:

1. Mid-ocean ridges, high regions on the seafloor where hydrothermal vents and volcanoes exist, are seen as divergent boundaries, where two plates move apart.
2. Arcs of volcanoes and earthquakes are theorized as convergent boundaries, where one plate subducts, or moves, under another.

Transform boundaries, such as the San Andreas Fault system, resulted in widespread powerful earthquakes. Plate tectonics also has provided a mechanism for Alfred Wegener's theory of continental drift, in which the continents move across the surface of the Earth over geological time. They also provided a driving force for crustal deformation, and a new setting for the observations of structural geology. The power of the theory of plate tectonics lies in its ability to combine all of these observations into a single theory of how the lithosphere moves over the convecting mantle.

1.4.2 Earth Structure

Advances in seismology, computer modeling, and mineralogy and crystallography at high temperatures and pressures give insights into the internal composition and structure of the Earth.

Seismologists can use the arrival times of seismic waves to image the interior of the Earth. Early advances in this field showed the existence of a liquid outer core (where shear waves were not able to propagate) and a dense solid inner core. These advances led to the development of a layered model of the Earth, with a crust and lithosphere on top, the mantle below (separated within itself by seismic discontinuities at 410 and 660 kilometers), and the outer core and inner core below that. More recently, seismologists have been able to create detailed images of wave speeds inside the earth in the same way a doctor images a body in a CT scan. These images have led to a much more detailed view of the interior of the Earth, and have replaced the simplified layered model with a much more dynamic model.

Mineralogists have been able to use the pressure and temperature data from the seismic and modeling studies alongside knowledge of the elemental composition of the Earth to reproduce these conditions in experimental settings and measure changes within the crystal structure. These studies explain the chemical changes associated with the major seismic discontinuities in the mantle and show the crystallographic structures expected in the inner core of the Earth.

1.5 GEOLOGICAL TIME

The geological time scale encompasses the history of the Earth. It is bracketed at the earliest by the dates of the first Solar System material at 4.567 Ga (or 4.567 billion years ago) and the formation of the Earth at 4.54 Ga (4.54 billion years), which is the beginning of the informally recognized Hadean eon – a division of geological time. At the later end of the scale, it is marked by the present day (in the Holocene epoch).

1.5.1 Timescale of the Earth

The following five timelines show the geologic time scale to scale. The first shows the entire time from the formation of the Earth to the present, but this gives little space for the most recent eon. The second timeline shows an expanded view of the most recent eon. In a similar way, the most recent era is expanded in the third timeline, the most recent period is expanded in the fourth timeline, and the most recent epoch is expanded in the fifth timeline.

1.5.2 Important Milestones on Earth

1. 4.567 Ga (gigaannum: billion years ago): Solar system formation
2. 4.54 Ga: Accretion, or formation, of Earth
3. c. 4 Ga: End of Late Heavy Bombardment, the first life
4. c. 3.5 Ga: Start of photosynthesis
5. c. 2.3 Ga: Oxygenated atmosphere, first snowball Earth
6. 730–635 Ma (megaannum: million years ago): second snowball Earth
7. 541 ± 0.3 Ma: Cambrian explosion – vast multiplication of hard-bodied life; first abundant fossils; start of the Paleozoic
8. c. 380 Ma: First vertebrate land animals
9. 250 Ma: Permian-Triassic extinction – 90% of all land animals die; end of Paleozoic and beginning of Mesozoic
10. 66 Ma: Cretaceous–Paleogene extinction – Dinosaurs die; end of Mesozoic and beginning of Cenozoic
11. c. 7 Ma: First hominins appear
12. 3.9 Ma: First Australopithecus, direct ancestor to modern Homo sapiens, appear
13. 200 ka (kiloannum: thousand years ago): First modern Homo sapiens appear in East Africa

1.6 DATING METHODS

1.6.1 Relative Dating

Methods for relative dating were developed when geology first emerged as a natural science. Geologists still use the following principles today as a means to provide information about geological history and the timing of geological events.

The principle of uniformitarianism states that the geological processes observed in operation that modify the Earth's crust at present have worked in much the same way over geological time. A fundamental principle of geology advanced by the 18th-century Scottish physician and geologist James Hutton is that "the present is the key to the past." In Hutton's words: "the past history of our lobe must be explained by what can be seen to be happening now."

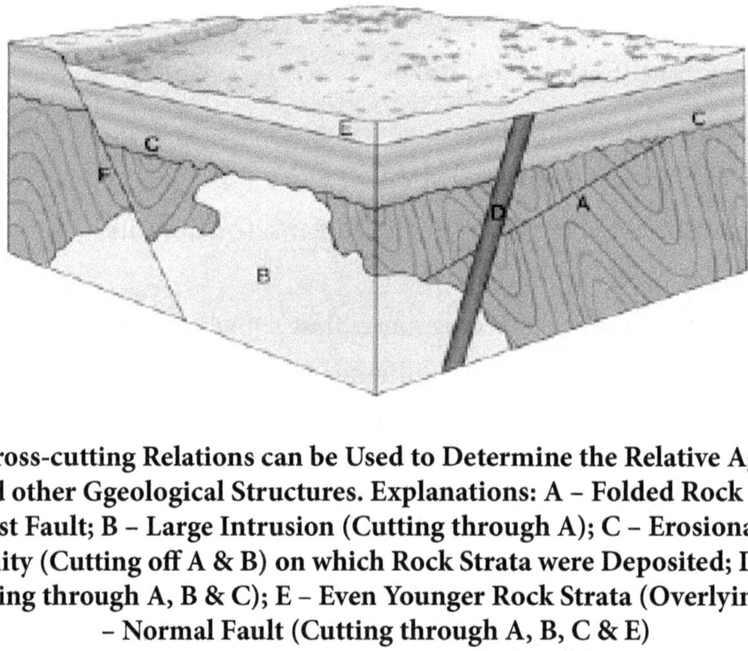

Fig. 1.2: Cross-cutting Relations can be Used to Determine the Relative Ages of Rock Strata and other Ggeological Structures. Explanations: A – Folded Rock Strata Cut by a Thrust Fault; B – Large Intrusion (Cutting through A); C – Erosional Angular Unconformity (Cutting off A & B) on which Rock Strata were Deposited; D – Volcanic Dyke (Cutting through A, B & C); E – Even Younger Rock Strata (Overlying C & D); F – Normal Fault (Cutting through A, B, C & E)

The principle of intrusive relationships concerns crosscutting intrusions. In geology, when an igneous intrusion cuts across a formation of sedimentary rock, it can be determined that the igneous intrusion is younger than the sedimentary rock. Different types of intrusions include stocks, laccoliths, batholiths, sills and dikes.

The principle of cross-cutting relationships pertains to the formation of faults and the age of the sequences through which they cut. Faults are younger than the rocks they cut; accordingly, if a fault is found that penetrates some formations but not those on top of it, then the formations that were cut are older than the fault, and the ones that are not cut must be younger than the fault. Finding the key bed in these situations may help determine whether the fault is a normal fault or a thrust fault.

The principle of inclusions and components states that, with sedimentary rocks, if inclusions (or clasts) are found in a formation, then the inclusions must be older than the formation that contains them. For example, in sedimentary rocks, it is common for gravel from an older formation to be ripped up and included in a newer layer. A similar situation with igneous rocks occurs when xenoliths are found. These foreign bodies are picked up as magma or lava flows, and are incorporated, later to cool in the matrix. As a result, xenoliths are older than the rock that contains them.

The principle of original horizontality states that the deposition of sediments occurs as essentially horizontal beds. Observation of modern marine and non-marine sediments in a wide variety of environments supports this generalization (although cross-bedding is inclined, the overall orientation of cross-bedded units is horizontal).

The principle of superposition states that a sedimentary rock layer in a tectonically undisturbed sequence is younger than the one beneath it and older than the one above it. Logically a younger layer cannot slip beneath a layer previously deposited. This principle allows sedimentary layers to be viewed as a form of the vertical timeline, a partial or complete record of the time elapsed frm deposition of the lowest layer to deposition of the highest bed.

Fig. 1.3: The Permian through Jurassic Stratigraphy of the Colorado Plateau Area of Southeastern Utah is an Example of Both Original Horizontality and the Law of Superposition

The principle of faunal succession is based on the appearance of fossils in sedimentary rocks. As organisms exist during the same period throughout the world, their presence or (sometimes) absence provides a relative age of the formations where they appear. Based on principles that William Smith laid out almost a hundred years before the publication of Charles Darwin's theory of evolution, the principles of succession developed independently of evolutionary thought. The principle becomes quite complex, however, given the uncertainties of fossilization, localization of fossil types due to lateral changes in habitat (facies change in sedimentary strata), and that not all fossils formed globally at the same time.

1.6.2 Absolute Dating

Geologists also use methods to determine the absolute age of rock samples and geological events. These dates are useful on their own and may also be used in conjunction with relative dating methods or to calibrate relative methods.

At the beginning of the 20th century, advancement in geological science was facilitated by the ability to obtain accurate absolute dates to geological events using radioactive isotopes and other methods. This changed the understanding of geological time. Previously, geologists could only use fossils and stratigraphic correlation to date sections of rock relative to one another. With isotopic dates, it became possible to assign absolute ages to rock units, and these absolute dates could be applied to fossil sequences in which there was datable material, converting the old relative ages into new absolute ages.

For many geological applications, isotope ratios of radioactive elements are measured in minerals that give the amount of time that has passed since a rock passed through its particular closure temperature, the point at which different radiometric isotopes stop diffusing into and out of the crystal lattice. These are used in geochronologic and thermochronologic studies. Common methods include uranium–lead dating, potassium–argon dating, argon–argon dating and uranium–thorium dating. These methods are used for a variety of applications. Dating of lava and volcanic ash layers found within a stratigraphic sequence can provide absolute age data for sedimentary rock units that do not contain radioactive isotopes and calibrate relative dating techniques. These methods can also be used to determine ages of pluton emplacement. Thermochemical techniques can be used to determine temperature profiles within the crust, the uplift of mountain ranges, and paleo-topography.

Fractionation of the lanthanide series elements is used to compute ages since rocks were removed from the mantle.

Other methods are used for more recent events. Optically stimulated luminescence and cosmogenic radionuclide dating are used to date surfaces and/or erosion rates. Dendrochronology can also be used for the dating of landscapes. Radiocarbon dating is used for geologically young materials containing organic carbon.

1.7 GEOLOGICAL DEVELOPMENT OF AN AREA

Rock units are first emplaced either by deposition onto the surface or intrusion into the overlying rock. Deposition can occur when sediments settle onto the surface of the Earth and later lithify into sedimentary rock, or when as volcanic material such as volcanic ash or lava flows blanket the surfacc. Igneous intrusions such as batholiths, laccoliths, dike, and sills, push upwards into the overlying rock, and crystallize as they intrude.

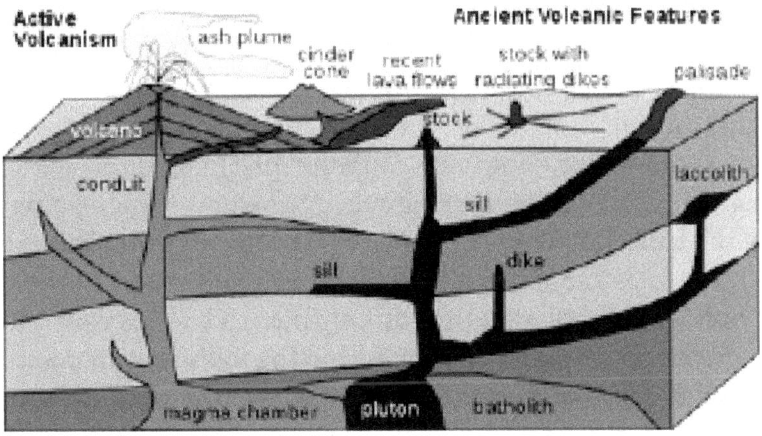

Fig. 1.4: An Originally Horizontal Sequence of Sedimentary Rocks (in Shades of Tan) are Affected by Igneous Activity

After the initial sequence of rocks has been deposited, the rock units can be deformed and/or metamorphosed. Deformation typically occurs as a result of horizontal shortening, horizontal extension, or side-to-side (strike-slip) motion. These structural regimes broadly relate to convergent boundaries, diergent boundaries, and transform boundaries, respectively, between tectonic plates.

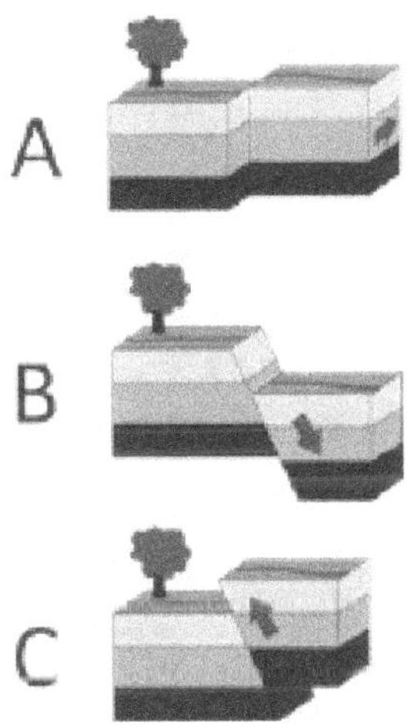

Fig. 1.5: An Illustration of the Three Types of Faults: A. Strike-slip Faults Occur when Rock Units Slide Past one Another. B. Normal Faults Occur when Rocks are Undergoing Horizontal Extension. C. Reverse (or Thrust) Faults Occur when Rocks are Undergoing Horizontal Shortening

The geology of an area changes through time as rock units are deposited and inserted, and deformational processes alter their shapes and locations.

When rock units are placed under horizontal compression, they shorten and become thicker. Because rock units, other than muds, do not significantly change in volume, this is accomplished in two primary ways: through faulting and folding. In the shallow crust, where brittle deformation can occur, thrust faults form, which causes the deeper rock to move on top of the shallower rock. Because deeper rock is often older, as noted by the principle of superposition, this can result in older rocks moving on top of younger ones. Movement along faults can result in folding, either because the faults are not planar or because rock layers are dragged along, forming drag folds as slip occurs along the fault. Deeper in the Earth, rocks behave plastically and fold instead of faulting. These folds can either be those where the material in the center of the fold buckles upwards, creating "antiforms", or where it buckles downwards, creating "synforms". If the tops of the rock units within the folds remain pointing upwards, they are called anticlines and synclines, respectively. If some of the units in the fold are facing downward, the structure is called an overturned anticline or syncline, and if all of the rock units are overturned or the correct up-direction isunknown, they are simply called by the most general terms, antiforms, and synforms.

Fig. 1.6: The San Andreas Fault in California

Even higher pressures and temperatures during horizontal shortening can cause both folding and metamorphism of the rocks. This metamorphism causes changes in the mineral composition of the rocks; creates a foliation, or planar surface, that is related to mineral growth under stress. This can remove signs of the original textures of the rocks, such as bedding in sedimentary rocks, flow features of lavas, and crystal patterns in crystalline rocks.

Extension causes the rock units as a whole to become longer and thinner. This is primarily accomplished through normal faulting and through the ductile stretching and thinning. Normal faults drop rock units that are higher below those that are lower. This typically results in younger units ending up below older units. Stretching of units can result in their thinning. In fact, at one location within the Maria Fold and Thrust Belt, the entire sedimentary sequence of the Grand Canyon appears over a length of less than a meter. Rocks at the depth to be ductilely stretched are often also metamorphosed. These stretched rocks can also pinch into lenses, known as boudins, after the French word for "sausage" because of their visual similarity.

Where rock units slide past one another, strike-slip faults develop in shallow regions, and become shear zones at deeper depths where the rocks deform ductilely.

The addition of new rock units, both depositionally and intrusively, often occurs during deformation. Faulting and other deformational processes result in the creation of topographic gradients, causing material on the rock unit that is increasing in elevation to be eroded by hillslopes and channels. These sediments are deposited on the rock unit that is going down. Continual motion along the fault maintains the topographic gradient in spite of the movement of sediment and continues to create accommodation space for the material to deposit. Deformational events are often also associated with volcanism and igneous activity. Volcanic ashes and lavas accumulate on the surface, and igneous intrusions enter from below. Dikes, long, planar igneous intrusions, enter along cracks, and therefore often form in large numbers in areas that are being actively deformed. This can result in the emplacement of dike swarms, such as those that are observable across the Canadian shield, or rings of dikes around the lava tube of a volcano.

All of these processes do not necessarily occur in a single environment and do not necessarily occur in a single order. The Hawaiian Islands, for example, consist almost entirely of layered basaltic lava flows. The sedimentary sequences of the mid-continental United States and the Grand Canyon in the southwestern United States contain almost undeformed stacks of sedimentary rocks that have remained in place since Cambrian time.

Other areas are much more geologically complex. In the southwestern United States, sedimentary, volcanic, and intrusive rocks have been metamorphosed, faulted, foliated, and folded. Even older rocks, such as the Acasta gneiss of the Slave craton in northwestern Canada, the oldest known rock in the world have been metamorphosed to the point where their origin is indiscernible without laboratory analysis. In addition, these processes can occur in stages. In many places, the

Grand Canyon in the southwestern United States being a very visible example, the lower rock units were metamorphosed and deformed, and then deformation ended and the upper, undeformed units were deposited. Although any amount of rock emplacement and rock deformation can occur, and they can occur any number of times, these concepts provide a guide to understanding the geological history of an area.

1.8 METHODS OF GEOLOGY

Geologists use a number of fields, laboratory, and numerical modeling methods to decipher Earth history and to understand the processes that occur on and inside the Earth. In typical geological investigations, geologists use primary information related to petrology (the study of rocks), stratigraphy (the study of sedimentary layers), and structural geology (the study of positions of rock units and their deformation).

In many cases, geologists also study modern soils, rivers, landscapes, and glaciers; investigate past and current life and biogeochemical pathways, and use geophysical methods to investigate the subsurface.

Sub-specialities of geology may distinguish endogenous and exogenous geology.

1.8.1 Field Methods

Geological field work varies depending on the task at hand. Typical fieldwork could consist of:

1. Geological mapping:
    a. Structural mapping: Identifying the locations of major rock units and the faults and folds that led to their placement there.
    b. Stratigraphic mapping: pinpointing the locations of sedimentary facies (lithofacies and biofacies) or the mapping of isopachs of equal thickness of sedimentary rock.
    c. Surficial mapping: Recording the locations of soils and surficial deposits
2. Surveying of topographic features:
    a. compilation of topographic maps
    b. Work to understand change across landscapes, including:
        i. Patterns of erosion and deposition
        ii. River-channel change through migration and avulsion
        iii. Hillslope processes
3. Subsurface mapping through geophysical methods:
    a. These methods include:

   i. Shallow seismic surveys
   ii. Ground-penetrating radar
   iii. Aeromagnetic surveys
   iv. Electrical resistivity tomography
   b. They aid in:
   i. Hydrocarbon exploration
   ii. Finding groundwater
   iii. Locating buried archaeological artifacts
4. High-resolution stratigraphy:
   a. Measuring and describing stratigraphic sections on the surface
   b. Well drilling and logging
5. Biogeochemistry and geomicrobiology:
   a. Collecting samples to:
   i. determine biochemical pathways
   ii. identify new species of organisms
   iii. identify new chemical compounds
   b. and to use these discoveries to:
   a. understand early life on Earth and how it functioned and metabolized
   b. find important compounds for use in pharmaceuticals
1. Paleontology: excavation of fossil material:
   a. For research into past life and evolution
   b. For museums and education
- Collection of samples for geochronology and thermochronology
- Glaciology: measurement of characteristics of glaciers and their motion

1.8.2 Petrology

In addition to identifying rocks in the field (lithology), petrologists identify rock samples in the laboratory. Two of the primary methods for identifying rocks in the laboratory are through optical microscopy and by using an electron microprobe. In an optical mineralogy analysis, petrologists analyze thin sections of rock samples using a petrographic microscope, where the minerals can be identified through their different properties in plane-polarized and cross-polarized light, including their birefringence, pleochroism, twinning, and interference properties with a conoscopic lens.

In the electron microprobe, individual locations are analyzed for their exact chemical compositions and variation in composition within individual crystals.

Stable and radioactive isotope studies provide insight into the geochemical evolution of rock units.

Petrologists can also use fluid inclusion data and perform high temperature and pressure physical experiments to understand the temperatures and pressures at which different mineral phases appear, and how they change through igneous and metamorphic processes. This research can be extrapolated to the field to understand metamorphic processes and the conditions of crystallization of igneous rocks. This work can also help to explain processes that occur within the Earth, such as subduction and magma chamber evolution.

1.8.3 Structural Geology

Structural geologists use microscopic analysis of oriented thin sections of geological samples to observe the fabric within the rocks, which gives information about strain within the crystalline structure of the rocks. They also plot and combine measurements of geological structures to better understand the orientations of faults and folds to reconstruct the history of rock deformation in the area. In addition, they perform analog and numerical experiments of rock deformation in large and small settings.

The analysis of structures is often accomplished by plotting the orientations of various features onto stereonets. A stereonet is a stereographic projection of a sphere onto a plane, in which planes are projected as lines and lines are projected as points. These can be used to find the locations of fold axes, relationships between faults, and relationships between other geological structures.

Among the most well-known experiments in structural geology are those involving orogenic wedges, which are zones in which mountains are built along convergent tectonic plate boundaries. In the analog versions of these experiments, horizontal layers of sand are pulled along a lower surface into a back stop, which results in realistic-looking patterns of faulting and the growth of a critically tapered (all angles remain the same) orogenic wedge.

Numerical models work in the same way as these analog models, though they are often more sophisticated and can include patterns of erosion and uplift in the mountain belt. This helps to show the relationship between erosion and the shape of a mountain range. These studies can also give useful information about pathways for metamorphism through pressure, temperature, space, and time.

1.8.4 Stratigraphy

In the laboratory, stratigraphers analyze samples of stratigraphic sections that can be returned from the field, such as those from drill cores. Stratigraphers also analyze data from geophysical surveys that show the locations of stratigraphic units in the subsurface.

Geophysical data and well logs can be combined to produce a better view of the subsurface, and stratigraphers often use computer programs to do this in three dimensions. Stratigraphers can then use these data to reconstruct ancient processes occurring on the surface of the Earth, interpret past environments, and locate areas for water, coal, and hydrocarbon extraction.

In the laboratory, biostratigraphers analyze rock samples from outcrop and drill cores for the fossils found in them. These fossils help scientists to date the core and to understand the depositional environment in which the rock units formed. Geochronologists precisely date rocks within the stratigraphic section to provide better absolute bounds on the timing and rates of deposition. Magnetic stratigraphers look for signs of magnetic reversals in igneous rock units within the drill cores. Other scientists perform stable-isotope studies on the rocks to gain information about past climate.

1.9 PLANETARY GEOLOGY

With the advent of space exploration in the twentieth century, geologists have begun to look at other planetary bodies in the same ways that have been developed to study the Earth. This new field of study is called planetary geology (sometimes known as astrogeology) and relies on known geological principles to study other bodies of the solar system. This is a major aspect of planetary science, and largely focuses on the terrestrial planets, icy moons, asteroids, comets, and meteorites. However, some planetary geophysicists study the giant planets and exoplanets.

Although the Greek-language-origin prefix geo refers to Earth, "geology" is often used in conjunction with the names of other planetary bodies when describing their composition and internal processes: examples are "the geology of Mars" and "Lunar geology". Specialized terms such as selenology (studies of the Moon), areology (of Mars), etc., are also in use.

Although planetary geologists are interested in studying all aspects of other planets, a significant focus is to search for evidence of past or present life on other worlds. This has led to many missions whose primary or ancillary purpose is to examine planetary bodies for evidence of life. One of these is the Phoenix lander, which analyzed Martian polar soil for water, chemical, and mineralogical constituents related to biological processes.

1.10 APPLIED GEOLOGY

1.10.1 Economic geology

Economic geology is a branch of geology that deals with aspects of economic minerals that humankind uses to fulfill various needs. Economic minerals are those extracted profitably for various practical uses. Economic geologists help locate and manage the Earth's natural resources, such as petroleum and coal, as well as mineral resources, which include metals such as iron, copper, and uranium.

Mining Geology

Mining geology consists of the extractions of mineral resources from the Earth. Some resources of economic interests include gemstones, metals such as gold and copper, and many minerals such as asbestos, perlite, mica, phosphates, zeolites, clay, pumice, quartz, and silica, as well as elements such as sulfur, chlorine, and helium.

Petroleum Geology

Petroleum geologists study the locations of the subsurface of the Earth that can contain extractable hydrocarbons, especially petroleum and natural gas. Because many of these reservoirs are found in sedimentary basins, they study the formation of these basins, as well as their sedimentary and tectonic evolution and the present-day positions of the rock units.

1.10.2 Engineering Geology

Engineering geology is the application of geological principles to engineering practice for the purpose of assuring that the geological factors affecting the location, design, construction, operation, and maintenance of engineering works are properly addressed. Engineering geology is distinct from geological engineering, particularly in North America.

In the field of civil engineering, geological principles and analyses are used in order to ascertain the mechanical principles of the material on which structures are built. This allows tunnels to be built without collapsing, bridges and skyscrapers to be built with sturdy foundations, and buildings to be built that will not settle in clay and mud.

1.10.3 Hydrology

Geology and geological principles can be applied to various environmental problems such as stream restoration, the restoration of brownfields, and the understanding of the interaction between natural habitat and the geological environment. Groundwater hydrology, or hydrogeology, is used to locate groundwater, which can often provide a ready supply of uncontaminated water and is especially important in arid regions, and to monitor the spread of contaminants in groundwater wells.

1.10.4 Paleoclimatology

Geologists also obtain data through stratigraphy, boreholes, core samples, and ice cores. Ice cores and sediment cores are used for paleoclimate reconstructions, which tell geologists about past and present temperature, precipitation, and sea level across the globe. These datasets are our primary source of information on global climate change outside of instrumental data.

1.10.5 Natural Hazards

Geologists and geophysicists study natural hazards in order to enact safe building codes and warning systems that are used to prevent loss of property and life. Examples of important natural hazards that are pertinent to geology (as opposed those that are mainly or only pertinent to meteorology) are:

1. Avalanches
2. Earthquakes
3. Floods
4. Landslides and debris flows
5. River channel migration and avulsion
6. Rockfalls
7. Sinkholes
8. Soil liquefaction
9. Subsidence
10. Tsunamis
11. Volcanoes

CHAPTER 2

DEFINING HYDROGEOLOGY

2.1 OVERVIEW OF HYDROGEOLOGY

Hydrogeology, the study of the movement, distribution, and properties of groundwater, stands as a critical discipline within the broader field of geology. Its significance lies in unraveling the secrets of Earth's hidden water pathways, understanding the dynamics between surface water and subsurface aquifers, and addressing the challenges of managing and sustaining this invaluable resource. As a science at the intersection of geology, hydrology, and environmental science, hydrogeology delves into the intricate interactions between rocks, soils, and water, playing a pivotal role in the exploration, extraction, and protection of groundwater. Hydrogeology stands as a captivating discipline that unveils the mysteries of Earth's water systems. From deciphering the nuances of the hydrologic cycle to addressing contemporary challenges and pioneering innovative solutions, hydrogeologists play a crucial role in safeguarding one of the planet's most precious resources. Through their expertise, the hidden pathways of groundwater are revealed, guiding humanity towards a more sustainable and harmonious coexistence with the dynamic world beneath our feet.

2.1.1 The Hydrologic Cycle: A Dance of Water on Earth

At the heart of hydrogeology lies the exploration of the hydrologic cycle, a complex and perpetual dance of water across the Earth's surface. This intricate cycle involves the continuous movement of water between the atmosphere, land, and oceans. Precipitation, in the form of rain or snow, replenishes surface water bodies and infiltrates the ground. Infiltrated water may become part of the soil moisture, be taken up by plants, or penetrate deeper into the subsurface to recharge aquifers. The hydrogeologist's lens provides insight into this intricate web, studying the pathways and transformations water undergoes in its journey through the cycle.

2.1.2 Aquifers: Earth's Subsurface Reservoirs

Central to hydrogeology is the exploration of aquifers, the geological formations capable of storing and transmitting groundwater. Aquifers vary in terms of porosity, permeability, and storage capacity. Understanding these properties is essential for characterizing aquifers and predicting how water moves through them. Aquifers may be unconfined, interacting directly with the land surface, or confined, sandwiched between impermeable layers. This nuanced understanding of aquifer properties is fundamental to the sustainable management of groundwater resources.

2.1.3 Groundwater Flow: Unraveling Subsurface Pathways

The movement of groundwater beneath the Earth's surface is a dynamic process governed by Darcy's Law. This fundamental principle outlines the relationship between the hydraulic gradient, permeability, and groundwater flow velocity. Hydrogeologists meticulously map and model groundwater flow patterns, investigating factors influencing the movement of water through aquifers. Recognizing the intricate interplay between geological formations and the hydraulics of subsurface flow is crucial for predicting groundwater availability and addressing issues of contamination and depletion.

2.1.4 Aquifer Testing and Analysis: Probing Subsurface Characteristics

Aquifer testing emerges as a crucial tool in hydrogeology, allowing scientists to probe the characteristics of subsurface formations. Pumping tests, slug tests, and other methods provide data to analyze aquifer parameters such as transmissivity, storativity, and specific yield. The interpretation of these tests aids in assessing the performance and sustainability of aquifer systems, offering insights into how water interacts with geological formations and responds to human activities.

2.1.5 Groundwater Chemistry: The Chemical Fingerprint of Earth's Waters

The chemical composition of groundwater is a key aspect of hydrogeological investigations. Hydrogeologists analyze major and minor ions in groundwater to understand its origin, evolution, and potential for contamination. Water quality parameters are scrutinized to assess the health of groundwater resources. The study of groundwater chemistry extends to investigating natural processes such as mineral dissolution and precipitation, as well as anthropogenic influences such as pollution and industrial activities.

2.1.6 Well Design and Construction: Accessing the Subsurface Veins

Wells serve as windows into the subsurface world, allowing hydrogeologists to access and monitor groundwater. Well design and construction are critical aspects of this process. Factors such as screen selection, casing materials, and proper grouting influence the integrity of the well and the reliability of data collected. Hydrogeologists employ their expertise to design wells that provide accurate representations of subsurface conditions while minimizing the risk of contamination.

2.1.7 Groundwater Exploration Techniques: Mapping the Subsurface

Advancements in technology have empowered hydrogeologists with an array of exploration techniques to map the subsurface. Remote sensing, Geographic Information Systems (GIS), geophysical methods like electrical resistivity and ground-penetrating radar, and borehole logging contribute to the comprehensive understanding of subsurface geology. These tools aid in identifying potential groundwater resources, delineating aquifer boundaries, and assessing the feasibility of well sites.

2.1.8 Managed Aquifer Recharge: Nurturing Earth's Reservoirs

In the face of growing water demands and climate variability, managed aquifer recharge has emerged as a strategic approach to enhance groundwater availability. This process involves intentionally directing surface water into aquifers for storage during times of abundance, creating a buffer against droughts or over-extraction. Hydrogeologists play a pivotal role in designing and implementing managed aquifer recharge projects, ensuring the sustainable utilization of subsurface reservoirs.

2.1.9 Groundwater and Environmental Challenges: Balancing Act for Sustainability

As populations burgeon and industrial activities intensify, hydrogeologists face the challenge of balancing the demand for groundwater with its sustainable management. Over-extraction, saltwater intrusion, land subsidence, and contamination are among the pressing issues that hydrogeologists grapple with. Their expertise becomes paramount in developing strategies to mitigate these challenges, ensuring that groundwater remains a reliable and resilient resource for generations to come.

2.1.10 Emerging Trends and Future Directions: The Evolving Landscape

The landscape of hydrogeology is ever-evolving, driven by technological advancements, interdisciplinary collaborations, and an increasing awareness of environmental sustainability. Hydrogeologists find themselves at the forefront of innovation, integrating their knowledge with fields such as remote sensing, data analytics, and modeling. As the world faces unprecedented challenges, hydrogeology is poised to lead the way in developing solutions and strategies for the sustainable management of Earth's water resources.

2.2 IMPORTANCE OF HYDROGEOLOGY IN EARTH SCIENCES

The importance of hydrogeology in Earth Sciences cannot be overstated. This discipline serves as a cornerstone for understanding the intricate interactions between water and the subsurface, guiding sustainable water resource management practices. From mapping groundwater flow patterns to addressing environmental challenges and pioneering innovative solutions, hydrogeologists play a pivotal role in safeguarding this invaluable resource for current and future generations.

2.2.1 Understanding the Hydrologic Cycle: A Fundamental Earth Process

The hydrologic cycle, often referred to as the water cycle, is a fundamental Earth process that hydrogeologists study meticulously. It involves the continuous movement of water between the atmosphere, land, and oceans, encompassing processes such as precipitation, infiltration, runoff, and evaporation. Hydrogeology plays a crucial role in deciphering this cycle, providing insights into the distribution of water on Earth and the pathways it takes as it navigates through the various reservoirs, both above and below ground.

2.2.2 Aquifers: The Earth's Subsurface Reservoirs

At the heart of hydrogeology lies the exploration of aquifers—geological formations capable of storing and transmitting groundwater. Understanding aquifers is paramount for managing water resources sustainably. These subsurface reservoirs vary in terms of porosity, permeability, and storage capacity. Hydrogeologists employ sophisticated techniques to characterize aquifers, enabling accurate predictions of groundwater availability and guiding prudent decisions regarding extraction rates and conservation measures.

2.2.3 Groundwater Flow: A Dynamic Puzzle Beneath Our Feet

Groundwater flow, governed by principles outlined in Darcy's Law, represents a dynamic puzzle beneath the Earth's surface. Hydrogeologists investigate how water moves through subsurface formations, mapping flow patterns, and examining factors influencing groundwater velocities. This knowledge is crucial for predicting the movement of contaminants, understanding the potential for aquifer recharge, and managing water resources in a sustainable manner.

2.2.4 Aquifer Testing and Analysis: Probing Subsurface Properties

Aquifer testing emerges as a powerful tool in hydrogeology, allowing scientists to probe subsurface properties with precision. Pumping tests and slug tests provide data for the analysis of critical aquifer parameters such as transmissivity, storativity, and specific yield. This information aids in assessing the performance of aquifer systems, contributing to effective resource management and the sustainable use of subsurface water.

2.2.5 Groundwater Chemistry: A Window into Earth's Subsurface

The chemical composition of groundwater serves as a window into the subsurface world, revealing valuable information about geological formations and potential environmental impacts. Hydrogeologists analyze the major and minor ions present in groundwater, examining water quality parameters to assess the health of underground water resources. This aspect of hydrogeology is indispensable for understanding natural processes such as mineral dissolution, as well as identifying and mitigating anthropogenic influences like pollution and contamination.

2.2.6 Well Design and Construction: Gateways to the Subsurface World

Wells, meticulously designed and constructed, serve as gateways to the subterranean world explored by hydrogeologists. The science of well design involves considerations such as screen selection, casing materials, and grouting techniques. Well construction must strike a delicate balance between accessing groundwater resources for monitoring and extraction while preventing potential contamination. Hydrogeologists contribute their expertise to ensure the integrity of wells, essential for reliable data collection and sustainable groundwater management.

2.2.7 Groundwater Exploration Techniques: Technological Frontiers in Subsurface Mapping

Advancements in technology have propelled hydrogeologists to the forefront of subsurface exploration. Remote sensing, Geographic Information Systems (GIS), and geophysical methods like electrical resistivity and ground-penetrating radar enable the mapping of subsurface geology with unprecedented accuracy. These techniques aid in identifying potential groundwater resources, delineating aquifer boundaries, and assessing the feasibility of well sites, enhancing the efficiency of groundwater exploration.

2.2.8 Managed Aquifer Recharge: Safeguarding Groundwater Resources

As the demand for freshwater rises and climate variability poses challenges to water availability, managed aquifer recharge has emerged as a strategic solution. Hydrogeologists play a central role in designing and implementing managed aquifer recharge projects. By intentionally directing surface water into aquifers during periods of abundance, hydrogeologists contribute to the replenishment of groundwater resources, creating a buffer against droughts and over-extraction.

2.2.9 Groundwater and Environmental Challenges: Navigating Complexities for Sustainability

Hydrogeology confronts an array of environmental challenges, ranging from over-extraction of groundwater to saltwater intrusion, subsidence, and contamination. The expertise of hydrogeologists is instrumental in developing strategies to mitigate these challenges, ensuring the sustainable use and protection of groundwater resources. Balancing the needs of growing populations with environmental stewardship requires the nuanced understanding provided by hydrogeology.

2.2.10 Emerging Trends and Future Directions: A Dynamic Landscape

The field of hydrogeology is not static; it evolves in response to technological innovations, interdisciplinary collaborations, and the imperative for sustainable practices. Hydrogeologists find themselves at the intersection of traditional geoscience and cutting-edge technologies. From integrating remote sensing and data analytics to advancing modeling techniques, hydrogeology is at the forefront of addressing emerging challenges and unlocking new avenues for the sustainable management of Earth's water resources.

2.3 HISTORICAL EVOLUTION OF HYDROGEOLOGICAL STUDIES

The historical evolution of hydrogeological studies is a captivating journey that mirrors humanity's quest to understand and harness the hidden dynamics of groundwater. From ancient civilizations' reliance on wells and springs to the modern interdisciplinary science of hydrogeology, the exploration of subsurface water has evolved significantly, driven by the need for reliable water sources and a deeper comprehension of Earth's intricate hydrological processes. The historical evolution of hydrogeological studies reflects a continuum of human curiosity and ingenuity. From the practical wells of ancient civilizations to the sophisticated modeling techniques of the 21st century, each era has contributed to our understanding of the hidden pathways of groundwater. Hydrogeology, once a pragmatic pursuit for survival, has transformed into a complex and interdisciplinary science, addressing contemporary challenges and paving the way for a more sustainable relationship between humanity and Earth's subsurface waters. As we continue to unveil the mysteries beneath our feet, the historical journey of hydrogeological studies serves as a testament to the enduring human quest for knowledge and stewardship of one of our most vital resources.

2.3.1 Ancient Wells and Springs: The Birth of Subsurface Exploration

The roots of hydrogeological studies can be traced back to ancient civilizations that recognized the vital importance of groundwater. In Mesopotamia, the cradle of civilization, well drilling was a common practice as early as 4000 BCE. Ancient Egyptians, with their advanced understanding of hydraulic engineering, created intricate systems of wells and canals to tap into the Nile's groundwater. Similarly, the Indus Valley Civilization employed wells for both domestic use and irrigation. Springs, seen as sources of divine purity, were also revered and utilized for sustenance.

2.3.2 Greco-Roman Contributions: Early Philosophies and Engineering Marvels

The classical era marked a transition from practical utilitarianism to philosophical inquiry into the nature of groundwater. Greek philosophers such as Thales and Anaximander pondered the origins of springs and the cyclical nature of water. The Romans, renowned for their engineering prowess, developed an extensive aqueduct system that transported water from distant sources to urban centers. Their understanding of aquifers and groundwater flow laid the foundation for later hydrogeological advancements.

2.3.3 Medieval Advances: Wells, Aquifers, and Islamic Scholars

During the medieval period, hydrogeological knowledge continued to evolve. Islamic scholars made significant contributions to groundwater exploration, with Al-Razi (Rhazes) and Al-Farabi conducting experiments on water filtration and studying the behavior of wells. In Europe, wells became central to medieval communities, providing access to groundwater for drinking and agricultural purposes. Aquifers, although not yet explicitly understood, played a crucial role in sustaining these wells.

2.3.4 Renaissance and Enlightenment: Emergence of Hydrogeology as a Science

The Renaissance marked a revival of interest in natural sciences, and hydrogeology began to emerge as a distinct scientific discipline. Leonardo da Vinci's studies on groundwater and wells demonstrated an early understanding of aquifer properties. In the 17th and 18th centuries, European thinkers like Bernard Palissy and Edme Mariotte contributed to the scientific understanding of groundwater, laying the groundwork for more systematic exploration.

2.3.5 18th-19th Century: Rise of Hydrogeological Concepts and Technologies

The 18th and 19th centuries witnessed a surge in hydrogeological exploration, with scientists developing conceptual frameworks and tools for studying groundwater. Henry Darcy's groundbreaking work in the mid-19th century established the foundational principles of groundwater flow, encapsulated in what is now known as Darcy's Law. This era also saw the rise of artesian wells and the conceptualization of confined and unconfined aquifers, key distinctions that shaped hydrogeological theory.

2.3.6 20th Century: Technological Revolution and Interdisciplinary Integration

The 20th century witnessed a technological revolution that profoundly impacted hydrogeological studies. The development of electric logging tools, ground-penetrating radar, and sophisticated analytical techniques enabled hydrogeologists to delve deeper into the subsurface. The advent of numerical modeling in the latter half of the century allowed for a more comprehensive understanding of groundwater flow patterns and aquifer behavior.

Moreover, the 20th century marked a shift towards interdisciplinary collaboration. Hydrogeologists began working in tandem with experts in chemistry, physics, engineering, and environmental science to address complex groundwater challenges. This integration broadened the scope of hydrogeological studies, encompassing not only the physical properties of aquifers but also the chemical, biological, and environmental aspects of groundwater systems.

2.3.7 Contemporary Challenges: Hydrogeology in the 21st Century

In the 21st century, hydrogeological studies face new challenges driven by population growth, climate change, and anthropogenic impacts. Hydrogeologists are at the forefront of addressing issues such as over-extraction of groundwater, contamination from industrial and agricultural activities, and the sustainable management of water resources. Advanced technologies, including satellite remote sensing and sophisticated modeling software, empower hydrogeologists to tackle these challenges with unprecedented precision.

2.3.8 Future Prospects: Integrating Innovation and Sustainability

The future of hydrogeological studies holds exciting prospects as the field continues to integrate innovation and sustainability. Advancements in artificial intelligence, machine learning, and data analytics offer new avenues for modeling and understanding groundwater systems. Hydrogeologists are increasingly focused on developing strategies for climate-resilient water resource management and exploring innovative approaches like managed aquifer recharge to ensure the sustainable use of groundwater.

CHAPTER 3

THE HYDROLOGIC CYCLE

3.1 COMPONENTS OF THE HYDROLOGIC CYCLE

The water cycle, also known as the hydrologic cycle or the hydrological cycle, is a biogeochemical cycle that describes the continuous movement of water on, above and below the surface of the Earth. The mass of water on Earth remains fairly constant over time but the partitioning of the water into the major reservoirs of ice, fresh water, saline water (salt water) and atmospheric water is variable depending on a wide range of climatic variables. The water moves from one reservoir to another, such as from river to ocean, or from the ocean to the atmosphere, by the physical processes of evaporation, transpiration, condensation, precipitation, infiltration, surface runoff, and subsurface flow. In doing so, the water goes through different forms: liquid, solid (ice) and vapor. The ocean plays a key role in the water cycle as it is the source of 86% of global evaporation.

The water cycle involves the exchange of energy, which leads to temperature changes. When water evaporates, it takes up energy from its surroundings and cools the environment. When it condenses, it releases energy and warms the environment. These heat exchanges influence climate.

The evaporative phase of the cycle purifies water, causing salts and other solids picked up during the cycle to be left behind, and then the condensation phase in the atmosphere replenishes the land with freshwater. The flow of liquid water and ice transports minerals across the globe. It is also involved in reshaping the geological features of the Earth, through processes including erosion and sedimentation. The water cycle is also essential for the maintenance of most life and ecosystems on the planet.

3.1.1 Overall Process

The water cycle is powered from the energy emitted by the sun. This energy heats water in the ocean and seas. Water evaporates as water vapor into the air. Some ice and snow sublimates directly into water vapor. Evapotranspiration is water transpired from plants and evaporated from the soil. The water molecule H_2O has smaller molecular mass than the major components of the atmosphere, nitrogen (N_2) and oxygen (O_2) and hence is less dense. Due to the significant difference in density, buoyancy drives humid air higher. As altitude increases, air pressure decreases and the temperature drops. The lower temperature causes water vapor to condense into tiny liquid water droplets which are heavier than the air, and which fall unless supported by an updraft. A huge concentration of these droplets over a large area in the atmosphere becomes visible as cloud, while condensation near ground level is referred to as fog.

Atmospheric circulation moves water vapor around the globe; cloud particles collide, grow, and fall out of the upper atmospheric layers as precipitation. Some precipitation falls as snow, hail, or sleet, and can accumulate in ice caps and glaciers, which can store frozen water for thousands of years. Most water falls as rain back into the ocean or onto land, where the water flows over the ground as surface runoff. A portion of this runoff enters rivers, with streamflow moving water towards the oceans. Runoff and water emerging from the ground (groundwater) may be stored as freshwater in lakes. Not all runoff flows into rivers; much of it soaks into the ground as infiltration. Some water infiltrates deep into the ground and replenishes aquifers, which can store freshwater for long periods of time. Some infiltration stays close to the land surface and can seep back into surface-water bodies (and the ocean) as groundwater discharge or be taken up by plants and transferred back to the atmosphere as water vapor by transpiration. Some groundwater finds openings in the land surface and emerges as freshwater springs. In river valleys and floodplains, there is often continuous water exchange between surface water and ground water in the hyporheic zone. Over time, the water returns to the ocean, to continue the water cycle.

The ocean plays a key role in the water cycle. The ocean holds "97% of the total water on the planet; 78% of global precipitation occurs over the ocean, and it is the source of 86% of global evaporation".

Important physical processes within the water cycle include the following (in alphabetical order):

1. Advection: The movement of water through the atmosphere. Without advection, water that evaporated over the oceans could not precipitate over land. Atmospheric rivers that move large volumes of water vapor over long distances are an example of advection.
2. Condensation: The transformation of water vapor to liquid water droplets in the air, creating clouds and fog.
3. Evaporation: The transformation of water from liquid to gas phases as it moves from the ground or bodies of water into the overlying atmosphere. The source of energy for evaporation is primarily solar radiation. Evaporation often implicitly includes transpiration from plants, though together they are specifically referred to as evapotranspiration. Total annual evapotranspiration amounts to approximately 505,000 km

(121,000 cu mi) of water, 434,000 km (104,000 cu mi) of which evaporates from the oceans. 86% of global evaporation occurs over the ocean.

4. Infiltration: The flow of water from the ground surface into the ground. Once infiltrated, the water becomes soil moisture or groundwater. A recent global study using water stable isotopes, however, shows that not all soil moisture is equally available for groundwater recharge or for plant transpiration.
5. Percolation: Water flows vertically through the soil and rocks under the influence of gravity.
6. Precipitation: Condensed water vapor that falls to the Earth's surface. Most precipitation occurs as rain, but also includes snow, hail, fog drip, graupel, and sleet. Approximately 505,000 km (121,000 cu mi) of water falls as precipitation each year, 398,000 km (95,000 cu mi) of it over the oceans. The rain on land contains 107,000 km (26,000 cu mi) of water per year and a snowing only 1,000 km (240 cu mi). 78% of global precipitation occurs over the ocean.
7. Runoff: The variety of ways by which water moves across the land. This includes both surface runoff and channel runoff. As it flows, the water may seep into the ground, evaporate into the air, become stored in lakes or reservoirs, or be extracted for agricultural or other human uses.
8. Subsurface flow: The flow of water underground, in the vadose zone and aquifers. Subsurface water may return to the surface (e.g., as a spring or by being pumped) or eventually seep into the oceans. Water returns to the land surface at lower elevation than where it infiltrated, under the force of gravity or gravity induced pressures. Groundwater tends to move slowly and is replenished slowly, so it can remain in aquifers for thousands of years.
9. Transpiration: The release of water vapor from plants and soil into the air.

3.1.2 Residence Times

The residence time of a reservoir within the hydrologic cycle is the average time a water molecule will spend in that reservoir. It is a measure of the average age of the water in that reservoir.

Groundwater can spend over 10,000 years beneath Earth's surface before leaving. Particularly old groundwater is called fossil water. Water stored in the soil remains there very briefly, because it is spread thinly across the Earth, and is readily lost by evaporation, transpiration, stream flow, or groundwater recharge. After evaporating, the residence time in the atmosphere is about 9 days before condensing and falling to the Earth as precipitation.

The major ice sheets – Antarctica and Greenland – store ice for very long periods. Ice from Antarctica has been reliably dated to 800,000 years before present, though the average residence time is shorter.

In hydrology, residence times can be estimated in two ways. The more common method relies on the principle of conservation of mass (water balance) and assumes the amount of water in a given reservoir is roughly constant. With this method, residence times are estimated by dividing the volume of the reservoir by the rate by which water either enters or exits the reservoir. Conceptually, this is equivalent to timing how long it would take the reservoir to become filled from empty if no water were to leave (or how long it would take the reservoir to empty from full if no water were to enter).

An alternative method to estimate residence times, which is gaining in popularity for dating groundwater, is the use of isotopic techniques. This is done in the subfield of isotope hydrology.

3.1.3 Water in Storage

The water cycle describes the processes that drive the movement of water throughout the hydrosphere. However, much more water is "in storage" (or in "pools") for long periods of time than is actually moving through the cycle. The storehouses for the vast majority of all water on Earth are the oceans. It is estimated that of the 1,386,000,000 km of the world's water supply, about 1,338,000,000 km is stored in oceans, or about 97%. It is also estimated that the oceans supply about 90% of the evaporated water that goes into the water cycle. The Earth's ice caps, glaciers, and permanent snowpack stores another 24,064,000 km accounting for only 1.7% of the planet's total water volume. However, this quantity of water is 68.7% of all freshwater on the planet.

3.1.4 Water Cycle Intensification due to Climate Change

Since the middle of the 20th century, human-caused climate change has resulted in observable changes in the global water cycle. The IPCC Sixth Assessment Report in 2021 predicted that these changes will continue to grow significantly at the global and regional level. These findings are a continuation of scientific consensus expressed in the IPCC Fifth Assessment Report from 2007 and other special reports by the Intergovernmental Panel on Climate Change which had already stated that the water cycle will continue to intensify throughout the 21st century.

The effects of climate change on the water cycle are profound and have been described as an intensification or a strengthening of the water cycle (also called hydrologic cycle). This effect has been observed since at least 1980. One example is the intensification of heavy precipitation events. This has important negative effects on the availability of freshwater resources, as well as other water reservoirs such as oceans, ice sheets, atmosphere and land surface. The water cycle is essential to life on Earth and plays a large role in the global climate and the ocean circulation. The warming of our planet is expected to cause changes in the water cycle for various reasons. For example, warmer atmosphere can contain more water vapor which has effects on evaporation and rainfall.

The underlying cause of the intensifying water cycle is the increased amount of greenhouse gases, which lead to a warmer atmosphere through the greenhouse effect. Physics dictates that saturation vapor pressure increases by 7% when temperature rises by 1°C.

The strength of the water cycle and its changes over time are of considerable interest, especially as the climate changes. The essence of the overall hydrological cycle is the evaporation of moisture in one place and the precipitation in other places. In particular, evaporation exceeds precipitation over the oceans, which allows moisture to be transported by the atmosphere from the oceans onto land where precipitation exceeds evapotranspiration, and the runoff flows into streams and rivers and discharges into the ocean, completing the cycle. The water cycle is a key part of Earth's energy cycle through the evaporative cooling at the surface which provides latent heat to the atmosphere, as atmospheric systems play a primary role in moving heat upward.

3.1.5 Changes due to other Human Activities

Human activities, other than those that lead to global warming from greenhouse gas emissions, can also alter the water cycle. The IPCC Sixth Assessment Report stated that there is "abundant evidence that changes in land use and land cover alter the water cycle globally, regionally and locally, by changing precipitation, evaporation, flooding, groundwater, and the availability of freshwater for a variety of uses". Examples for such land use changes are converting fields to urban areas or clearing forests. Such changes can affect the ability of soils to soak up surface water. Deforestation has local as well as regional effects. For example it reduces soil moisture, evaporation and rainfall at the local level. Furthermore, deforestation causes regional temperature changes that can affect rainfall patterns.

Aquifer drawdown or overdrafting and the pumping of fossil water increase the total amount of water in the hydrosphere. This is because the water that was originally in the ground has now become available for evaporation as it is now in contact with the atmosphere.

3.1.6 Biogeochemical Cycling

While the water cycle is itself a biogeochemical cycle, flow of water over and beneath the Earth is a key component of the cycling of other biogeochemicals. Runoff is responsible for almost all of the transport of eroded sediment and phosphorus from land to waterbodies. The salinity of the oceans is derived from erosion and transport of dissolved salts from the land. Cultural eutrophication of lakes is primarily due to phosphorus, applied in excess to agricultural fields in fertilizers, and then transported overland and down rivers. Both runoff and groundwater flow play significant roles in transporting nitrogen from the land to waterbodies. The dead zone at the outlet of the Mississippi River is a consequence of nitrates from fertilizer being carried off agricultural fields and funnelled down the river system to the Gulf of Mexico. Runoff also plays a part in the carbon cycle, again through the transport of eroded rock and soil.

3.1.7 Slow Loss over Geologic Time

The hydrodynamic wind within the upper portion of a planet's atmosphere allows light chemical elements such as Hydrogen to move up to the exobase, the lower limit of the exosphere, where the gases can then reach escape velocity, entering outer space without impacting other particles of gas. This type of gas loss from a planet into space is known as planetary wind. Planets with hot lower atmospheres could result in humid upper atmospheres that accelerate the loss of hydrogen.

3.1.8 Historical Interpretations

Floating Land Mass

In ancient times, it was widely thought that the land mass floated on a body of water, and that most of the water in rivers has its origin under the earth. Examples of this belief can be found in the works of Homer (c.800 BCE).

Hebrew Bible

In the ancient Near East, Hebrew scholars observed that even though the rivers ran into the sea, the sea never became full. Some scholars conclude that the water cycle was described completely during this time in this passage: "The wind goeth toward the south, and turneth about unto the north; it whirleth about continually, and the wind returneth again according to its circuits. All the rivers run into the sea, yet the sea is not full; unto the place from whence the rivers come, thither they return again" (Ecclesiastes 1:6-7).

Scholars are not in agreement as to the date of Ecclesiastes, though most scholars point to a date during the time of King Solomon, son of David and Bathsheba, "three thousand years ago, there is some agreement that the time period is 962–922 BCE. Furthermore, it was also observed that when the clouds were full, they emptied rain on the earth (Ecclesiastes 11:3). In addition, during 793–740 BCE a Hebrew prophet, Amos, stated that water comes from the sea and is poured out on the earth (Amos 5:8).

In the Biblical Book of Job, dated between 7th and 2nd centuries BCE, there is a description of precipitation in the hydrologic cycle, "For he maketh small the drops of water: they pour down rain according to the vapour thereof; which the clouds do drop and distil upon man abundantly" (Job 36:27-28).

Understanding of Precipitation and Percolation

In the Adityahridayam (a devotional hymn to the Sun God) of Ramayana, a Hindu epic dated to the 4th century BCE, it is mentioned in the 22nd verse that the Sun heats up water and sends it down as rain. By roughly 500 BCE, Greek scholars were speculating that much of the water in rivers can be attributed to rain. The origin of rain was also known by then. These scholars maintained the belief, however, that water rising up through the earth contributed a great deal to rivers. Examples of this thinking included Anaximander (570 BCE) (who also speculated about the evolution of land animals from fish) and Xenophanes of Colophon (530 BCE). Chinese scholars such as Chi Ni Tzu (320 BCE) and Lu Shih Ch'un Ch'iu (239 BCE) had similar thoughts.

The idea that the water cycle is a closed cycle can be found in the works of Anaxagoras of Clazomenae (460 BCE) and Diogenes of Apollonia (460 BCE). Both Plato (390 BCE) and Aristotle (350 BCE) speculated about percolation as part of the water cycle. Aristotle correctly hypothesized that the sun played a role in the Earth's hydraulic cycle in his book Meteorology, writing "By it [the sun's] agency the finest and sweetest water is everyday carried up and is dissolved into vapor and rises to the upper regions, where it is condensed again by the cold and so returns to the earth.", and believed that clouds were composed of cooled and condensed water vapor.

Up to the time of the Renaissance, it was wrongly assumed that precipitation alone was insufficient to feed rivers, for a complete water cycle, and that underground water pushing upwards from the oceans were the main contributors to river water. Bartholomew of England held this view (1240 CE), as did Leonardo da Vinci (1500 CE) and Athanasius Kircher (1644 CE).

Discovery of the Correct Theory

The first published thinker to assert that rainfall alone was sufficient for the maintenance of rivers was Bernard Palissy (1580 CE), who is often credited as the discoverer of the modern theory of the water cycle. Palissy's theories were not tested scientifically until 1674, in a study commonly attributed to Pierre Perrault. Even then, these beliefs were not accepted in mainstream science until the early nineteenth century.

3.2 INTERACTIONS BETWEEN SURFACE WATER & GROUNDWATER

The intricate interplay between surface water and groundwater forms a dynamic interface crucial to understanding Earth's hydrosphere. This interaction, often referred to as the hyporheic zone or riparian zone, encompasses a complex network where surface water and groundwater meet, exchanging energy, nutrients, and contaminants. Interactions between surface water and groundwater form an intricate and vital connection within Earth's hydrosphere. From nutrient cycling and temperature regulation to contaminant filtration and biodiversity support, the hyporheic zone plays a multifaceted role in maintaining the health of aquatic ecosystems. As we navigate an era of environmental challenges and human impacts, recognizing the significance of this dynamic interface becomes paramount. Through scientific understanding, thoughtful management, and a commitment to sustainable practices, we can preserve the delicate balance between surface water and groundwater, ensuring the vitality of Earth's water systems for generations to come.

3.2.1 The Hyporheic Zone: Bridging Surface and Subsurface Realms

At the heart of interactions between surface water and groundwater lies the hyporheic zone, a transitional area that acts as a bridge between surface water bodies such as rivers, lakes, and streams, and the underlying subsurface aquifers. This zone, often characterized by porous sediments and gravel, serves as a critical interface where water from the surface and subsurface coexists and exchanges components essential for ecosystem health.

3.2.2 Flow Dynamics: A Dance of Water Movements

Surface water and groundwater engage in a continuous dance of exchange within the hyporheic zone. As rivers flow, water infiltrates the streambed, descending into the subsurface. Simultaneously, groundwater may discharge into the stream, creating a dynamic exchange that varies with seasons, precipitation patterns, and geological conditions. The intricate flow dynamics in this zone contribute to the maintenance of water balance and the transport of nutrients, sediments, and dissolved substances.

3.2.3 Nutrient Cycling: Sustaining Aquatic Ecosystems

The hyporheic zone serves as a hotspot for nutrient cycling, playing a pivotal role in sustaining aquatic ecosystems. As surface water infiltrates the streambed, it carries essential nutrients like nitrogen and phosphorus into the subsurface. Microbial activity within the sediments facilitates the transformation and cycling of these nutrients, influencing the productivity of downstream water bodies. This nutrient exchange is vital for supporting diverse aquatic life and maintaining the ecological balance of riparian ecosystems.

3.2.4 Temperature Regulation: A Thermal Buffer for Aquatic Habitats

One of the remarkable functions of the hyporheic zone is its role in regulating water temperature. During hot seasons, the cooler groundwater entering the stream can lower the overall temperature, providing a thermal buffer for aquatic habitats. Conversely, in colder seasons, the warmer groundwater can mitigate extreme temperature drops, creating a more stable environment for aquatic organisms. This thermal regulation is particularly crucial for species that are sensitive to temperature variations and contributes to the overall resilience of riparian ecosystems.

3.2.5 Contaminant Filtration and Transformation: Nature's Water Purification System

The hyporheic zone acts as a natural filter and transformation site for contaminants present in surface water. As water percolates through the streambed, sediments and microbial communities within the zone trap, break down, or transform various pollutants. This process, often referred to as natural attenuation, helps maintain water quality and reduces the impact of contaminants on downstream ecosystems. Understanding the contaminant filtration capacity of the hyporheic zone is essential for managing water resources and protecting vulnerable aquatic habitats.

3.2.6 Biotic Interactions: Supporting Biodiversity

The dynamic interactions between surface water and groundwater create a mosaic of habitats that support diverse biological communities. The hyporheic zone provides unique niches for various species of invertebrates, microorganisms, and fish. These organisms contribute to the breakdown of organic matter, nutrient cycling, and energy transfer within the ecosystem. The health and biodiversity of riparian habitats are intricately linked to the functional roles played by the hyporheic zone in fostering these biotic interactions.

3.2.7 Human Impacts: Altering the Natural Balance

Human activities can significantly alter the interactions between surface water and groundwater, often disrupting the delicate balance of the hyporheic zone. Urbanization, land-use changes, and excessive groundwater extraction can lead to modifications in flow patterns, temperature regimes, and nutrient dynamics. These alterations can have cascading effects on aquatic ecosystems, affecting the composition of biological communities, water quality, and the overall resilience of riparian environments.

3.2.8 Management and Restoration: Balancing Ecological Health and Human Needs

The effective management and restoration of interactions between surface water and groundwater require a holistic approach that considers both ecological health and human needs. Sustainable water resource management involves understanding the complex dynamics of the hyporheic zone and implementing strategies to mitigate human-induced impacts. Practices such as riparian buffer zones, groundwater recharge enhancement, and pollution control measures play pivotal roles in maintaining the functionality of the hyporheic zone and ensuring the long-term health of aquatic ecosystems.

3.2.9 Research Frontiers: Advancing Our Understanding

Ongoing research endeavors continue to advance our understanding of the interactions between surface water and groundwater. Hydrogeologists, ecologists, and environmental scientists collaborate to explore the intricacies of the hyporheic zone using cutting-edge technologies such as environmental monitoring networks, isotopic tracing, and numerical modeling. These tools enable researchers to unravel the complexities of water movement, nutrient cycling, and contaminant dynamics within this critical interface.

3.2.10 Climate Change Considerations: Adapting to a Dynamic Future

In the context of climate change, the interactions between surface water and groundwater become even more crucial. Altered precipitation patterns, temperature extremes, and shifts in hydrological cycles pose challenges to the resilience of riparian ecosystems. Understanding how these changes impact the hyporheic zone is essential for developing adaptive strategies that preserve the functionality of this critical interface and safeguard the health of aquatic environments.

3.3 GLOBAL AND REGIONAL PATTERNS IN THE HYDROLOGIC CYCLE

The hydrologic cycle, a rhythmic dance of water between the Earth's surface and the atmosphere, manifests itself in intricate global and regional patterns. This continuous movement of water, driven by solar energy, connects oceans, land, and the atmosphere, shaping the dynamics of climate, ecosystems, and human activities. Understanding the nuances of global and regional patterns in the hydrologic cycle is essential for managing water resources, addressing climate variability, and navigating the complexities of Earth's water systems. The hydrologic cycle, with its global and regional patterns, is a symphony of water that shapes the landscapes we inhabit. From the rhythmic dance of precipitation to the intricate movements of rivers and aquifers, understanding the nuances of this hydrological ballet is fundamental to our stewardship of water resources. As we confront the challenges of a changing climate and seek sustainable solutions for water management, harmonizing with nature's hydrologic symphony becomes not only a scientific imperative but a collective responsibility for the well-being of our planet and future generations.

3.3.1 Global Patterns: The Grand Choreography of Water on a Planetary Scale

At a global scale, the hydrologic cycle orchestrates a grand ballet of water movements, encompassing the vast expanses of oceans, continents, and the atmosphere. The cycle begins with solar radiation, which heats the Earth's surface, causing water to evaporate from oceans, lakes, and rivers. This invisible water vapor rises into the atmosphere, where it cools and condenses to form clouds. As these clouds gather moisture, precipitation occurs, returning water to the Earth's surface in the form of rain or snow.

The water that reaches the Earth's surface follows various paths. Some of it flows over the land as surface runoff, replenishing rivers and lakes. Another portion infiltrates the soil, contributing to soil moisture and groundwater recharge. Plants play a vital role in the cycle through transpiration, releasing water vapor from their leaves into the atmosphere. The cycle comes full circle as water returns to the oceans through runoff and river discharge.

These global patterns are not uniform across the planet. The equator, for example, experiences intense solar radiation, leading to high rates of evaporation and precipitation. The tropics witness heavy rainfall and lush vegetation, while subtropical regions may face arid conditions and deserts. Polar regions, with lower solar radiation, experience lower evaporation rates and contribute less moisture to the global hydrologic cycle.

3.3.2 Regional Patterns: Local Variations in the Water Ballet

While the global hydrologic cycle sets the stage, regional patterns introduce variations and localized dynamics. The geography, topography, and climate of a specific region influence how the hydrologic cycle manifests locally.

Mountainous regions, for instance, play a crucial role in regional hydrology. As air is lifted over mountains, it cools and releases moisture in the form of precipitation. This phenomenon, known as orographic precipitation, results in the wet windward side of the mountain and a rain shadow on the leeward side, where drier conditions prevail. The Himalayas, for example, significantly impact the hydrology of the Indian subcontinent.

Coastal areas exhibit distinct hydrological patterns influenced by proximity to oceans. Coastal regions often experience higher humidity, leading to increased evaporation and precipitation. The interaction between land and sea, known as land-sea breezes, can enhance rainfall in coastal zones. Additionally, coastal ecosystems are influenced by saltwater intrusion, where seawater infiltrates into freshwater aquifers, impacting both water quality and availability.

In arid and semi-arid regions, such as deserts, the hydrologic cycle takes on unique characteristics. Evaporation rates can exceed precipitation, leading to limited surface water availability. These regions rely heavily on groundwater as a crucial water source. Sustainable water management in arid regions requires a deep understanding of the local hydrologic patterns and careful consideration of human activities that impact water availability.

In polar regions, the hydrologic cycle exhibits distinctive features due to cold temperatures. Much of the precipitation in these areas falls as snow, contributing to ice accumulation and the formation of glaciers. The melting of glaciers and ice caps influences sea levels and freshwater availability in downstream areas. Understanding these regional variations is essential for predicting the impacts of climate change on polar hydrology.

3.3.3 Human Impacts: Disrupting the Natural Choreography

Human activities have introduced disruptions to the natural choreography of the hydrologic cycle, leading to alterations in global and regional patterns. Urbanization, deforestation, and changes in land use can affect surface runoff and infiltration, impacting local hydrology. The construction of dams and reservoirs alters river flow patterns and can influence groundwater recharge. Agricultural practices, such as irrigation and the drainage of wetlands, significantly impact the distribution of water in landscapes.

Climate change further exacerbates these disruptions, with shifts in temperature, precipitation patterns, and extreme weather events. Rising global temperatures can increase evaporation rates, altering the balance between precipitation and evaporation in different regions. Changes in the frequency and intensity of storms can lead to more extreme rainfall events or prolonged droughts, affecting both surface water and groundwater resources.

Understanding and mitigating the impacts of human activities on the hydrologic cycle are critical for ensuring water security, preserving ecosystems, and adapting to a changing climate. Sustainable water management practices must consider the intricate interplay between global and regional hydrological patterns.

3.3.4 Implications for Water Resource Management: Navigating the Water Mosaic

Global and regional patterns in the hydrologic cycle have profound implications for water resource management. Recognizing the interconnectedness of water systems on a planetary scale is essential for addressing water scarcity, ensuring sustainable agriculture, and safeguarding freshwater ecosystems.

In regions where water scarcity is a pressing issue, understanding local hydrologic patterns is crucial for developing effective water management strategies. This may involve implementing water conservation measures, improving irrigation efficiency, and exploring alternative water sources such as treated wastewater or desalination.

For areas prone to floods or droughts, adaptive water management strategies are essential. This may include the development of early warning systems, the construction of reservoirs for water storage, and the implementation of sustainable land-use practices to reduce the risk of extreme events.

Preserving and restoring natural ecosystems, such as wetlands and forests, can contribute to maintaining healthy hydrological cycles. These ecosystems play a vital role in regulating water flow, enhancing water quality, and providing crucial habitats for biodiversity.

International cooperation is increasingly important in managing shared water resources. Transboundary rivers and aquifers require collaborative efforts to address the needs of multiple regions and ensure equitable access to water.

As we navigate the complex mosaic of global and regional hydrological patterns, the integration of scientific knowledge, technological innovations, and sustainable policies becomes paramount.

A holistic approach that considers the interconnectedness of water systems and embraces adaptive management practices is key to navigating the challenges posed by a changing climate and growing human demands on water resources.

CHAPTER 4

PROPERTIES OF AQUIFERS

An aquifer is an underground layer of water-bearing material, consisting of permeable or fractured rock, or of unconsolidated materials (gravel, sand, or silt). Aquifers vary greatly in their characteristics. The study of water flow in aquifers and the characterization of aquifers is called hydrogeology. Related terms include aquitard, which is a bed of low permeability along an aquifer, and aquiclude (or aquifuge), which is a solid, impermeable area underlying or overlying an aquifer, the pressure of which could lead to the formation of a confined aquifer. The classification of aquifers is as follows: Saturated versus unsaturated; aquifers versus aquitards; confined versus unconfined; isotropic versus anisotropic; porous, karst, or fractured; transboundary aquifer.

Groundwater from aquifers can be sustainably harvested by humans through the use of qanats leading to a well. This groundwater is a major source of fresh water for many regions, however can present a number of challenges such as overdrafting (extracting groundwater beyond the equilibrium yield of the aquifer), groundwater-related subsidence of land, and the salinization or pollution of the groundwater.

4.1PROPERTIES

4.1.1 Depth

Aquifers occur from near-surface to deeper than 9,000 metres (30,000 ft). Those closer to the surface are not only more likely to be used for water supply and irrigation, but are also more likely to be replenished by local rainfall. Although aquifers are sometimes characterized as "underground rivers or lakes," they are actually porous rock saturated with water.

Many desert areas have limestone hills or mountains within them or close to them that can be exploited as groundwater resources. Part of the Atlas Mountains in North Africa, the Lebanon and Anti-Lebanon ranges between Syria and Lebanon, the Jebel Akhdar in Oman, parts of the Sierra Nevada and neighboring ranges in the United States' Southwest, have shallow aquifers that are exploited for their water. Overexploitation can lead to the exceeding of the practical sustained yield; i.e., more water is taken out than can be replenished.

Along the coastlines of certain countries, such as Libya and Israel, increased water usage associated with population growth has caused a lowering of the water table and the subsequent contamination of the groundwater with saltwater from the sea.

In 2013 large freshwater aquifers were discovered under continental shelves off Australia, China, North America and South Africa. They contain an estimated half a million cubic kilometers of "low salinity" water that could be economically processed into potable water. The reserves formed when ocean levels were lower and rainwater made its way into the ground in land areas that were not submerged until the ice age ended 20,000 years ago. The volume is estimated to be 100 times the amount of water extracted from other aquifers since 1900.

4.2 CLASSIFICATION

An aquitard is a zone within the Earth that restricts the flow of groundwater from one aquifer to another. An aquitard can sometimes, if completely impermeable, be called an aquiclude or aquifuge. Aquitards are composed of layers of either clay or non-porous rock with low hydraulic conductivity.

4.2.1 Saturated versus Unsaturated

Groundwater can be found at nearly every point in the Earth's shallow subsurface to some degree, although aquifers do not necessarily contain fresh water. The Earth's crust can be divided into two regions: the saturated zone or phreatic zone (e.g., aquifers, aquitards, etc.), where all available spaces are filled with water, and the unsaturated zone (also called the vadose zone), where there are still pockets of air that contain some water, but can be filled with more water.

Saturated means the pressure head of the water is greater than atmospheric pressure (it has a gauge pressure > 0). The definition of the water table is the surface where the pressure head is equal to atmospheric pressure (where gauge pressure = 0).

Unsaturated conditions occur above the water table where the pressure head is negative (absolute pressure can never be negative, but gauge pressure can) and the water that incompletely fills the pores of the aquifer material is under suction. The water content in the unsaturated zone is held in place by surface adhesive forces and it rises above the water table (the zero-gauge-pressure isobar) by capillary action to saturate a small zone above the phreatic surface (the capillary fringe) at less than atmospheric pressure. This is termed tension saturation and is not the same as saturation on a water-content basis. Water content in a capillary fringe decreases with increasing distance from the phreatic surface. The capillary head depends on soil pore size. In sandy soils with larger pores, the head will be less than in clay soils with very small pores. The normal capillary rise in a clayey soil is less than 1.8 m (6 ft) but can range between 0.3 and 10 m (1 and 33 ft).

The capillary rise of water in a small-diameter tube involves the same physical process. The water table is the level to which water will rise in a large-diameter pipe (e.g., a well) that goes down into the aquifer and is open to the atmosphere.

4.2.2 Aquifers versus Aquitards

Aquifers are typically saturated regions of the subsurface that produce an economically feasible quantity of water to a well or spring (e.g., sand and gravel or fractured bedrock often make good aquifer materials).

An aquitard is a zone within the Earth that restricts the flow of groundwater from one aquifer to another. A completely impermeable aquitard is called an aquiclude or aquifuge. Aquitards contain layers of either clay or non-porous rock with low hydraulic conductivity.

In mountainous areas (or near rivers in mountainous areas), the main aquifers are typically unconsolidated alluvium, composed of mostly horizontal layers of materials deposited by water processes (rivers and streams), which in cross-section (looking at a two-dimensional slice of the aquifer) appear to be layers of alternating coarse and fine materials. Coarse materials, because of the high energy needed to move them, tend to be found nearer the source (mountain fronts or rivers), whereas the fine-grained material will make it farther from the source (to the flatter parts of the basin or overbank areas—sometimes called the pressure area). Since there are less fine-grained deposits near the source, this is a place where aquifers are often unconfined (sometimes called the forebay area), or in hydraulic communication with the land surface.

4.2.3 Confined versus Unconfined

There are two end members in the spectrum of types of aquifers; confined and unconfined (with semi-confined being in between). Unconfined aquifers are sometimes also called water table or phreatic aquifers, because their upper boundary is the water table or phreatic surface. Typically (but not always) the shallowest aquifer at a given location is unconfined, meaning it does not have a confining layer (an aquitard or aquiclude) between it and the surface. The term "perched" refers to ground water accumulating above a low-permeability unit or strata, such as a clay layer. This term is generally used to refer to a small local area of ground water that occurs at an elevation higher than a regionally extensive aquifer. The difference between perched and unconfined aquifers is their size (perched is smaller). Confined aquifers are aquifers that are overlain by a confining layer, often made up of clay. The confining layer might offer some protection from surface contamination.

If the distinction between confined and unconfined is not clear geologically (i.e., if it is not known if a clear confining layer exists, or if the geology is more complex, e.g., a fractured bedrock aquifer), the value of storativity returned from an aquifer test can be used to determine it (although aquifer tests in unconfined aquifers should be interpreted differently than confined ones). Confined aquifers have very low storativity values (much less than 0.01, and as little as 10), which means that the aquifer is storing water using the mechanisms of aquifer matrix expansion and the compressibility of water, which typically are both quite small quantities. Unconfined aquifers have storativities (typically called specific yield) greater than 0.01 (1% of bulk volume); they release water from storage by the mechanism of actually draining the pores of the aquifer, releasing relatively large amounts of water (up to the drainable porosity of the aquifer material, or the minimum volumetric water content).

4.2.4 Isotropic versus Anisotropic

In isotropic aquifers or aquifer layers the hydraulic conductivity (K) is equal for flow in all directions, while in anisotropic conditions it differs, notably in horizontal (Kh) and vertical (Kv) sense.

Semi-confined aquifers with one or more aquitards work as an anisotropic system, even when the separate layers are isotropic, because the compound Kh and Kv values are different.

When calculating flow to drains or flow to wells in an aquifer, the anisotropy is to be taken into account lest the resulting design of the drainage system may be faulty.

4.2.5 Porous, Karst, or Fractured

To properly manage an aquifer its properties must be understood. Many properties must be known to predict how an aquifer will respond to rainfall, drought, pumping, and contamination. Where and how much water enters the groundwater from rainfall and snowmelt? How fast and what direction does the groundwater travel? How much water leaves the ground as springs? How much water can be sustainably pumped out? How quickly will a contamination incident reach a well or spring? Computer models can be used to test how accurately the understanding of the aquifer properties matches th actual aquifer performance. Environmental regulations require sites with potential sources of contamination to demonstrate that the hydrology has been characterized.

Porous

Porous aquifers typically occur in sand and sandstone. Porous aquifer properties depend on the depositional sedimentary environment and later natural cementation of the sand grains. The environment where a sand body was deposited controls the orientation of the sand grains, the horizontal and vertical variations, and the distribution of shale layers. Even thin shale layers are important barriers to groundwater flow. All these factors affect the porosity and permeability of sandy aquifers.

Sandy deposits formed in shallow marine environments and in windblown sand dune environments have moderate to high permeability while sandy deposits formed in river environments have low to moderate permeability. Rainfall and snowmelt enter the groundwater where the aquifer is near the surface. Groundwater flow directions can be determined from potentiometric surface maps of water levels in wells and springs. Aquifer tests and well tests can be used with Darcy's law flow equations to determine the ability of a porous aquifer to convey water.

Fig. 4.1: Water in Porous Aquifers Slowly Seeps through Pore Spaces between Sand Grains

Analyzing this type of information over an area gives an indication how much water can be pumped without overdrafting and how contamination will travel. In porous aquifers groundwater flows as slow seepage in pores between sand grains. A groundwater flow rate of 1 foot per day (0.3 m/d) is considered to be a high rate for porous aquifers.

Porosity is important, but, alone, it does not determine a rock's ability to act as an aquifer. Areas of the Deccan Traps (a basaltic lava) in west central India are good examples of rock formations with high porosity but low permeability, which makes them poor aquifers. Similarly, the micro-porous (Upper Cretaceous) Chalk Group of south east England, although having a reasonably high porosity, has a low grain-to-grain permeability, with its good water-yielding characteristics mostly due to micro-fracturing and fissuring.

Karst

Karst aquifers typically develop in limestone. Surface water containing natural carbonic acid moves down into small fissures in limestone. This carbonic acid gradually dissolves limestone thereby enlarging the fissures. The enlarged fissures allow a larger quantity of water to enter which leads to a progressive enlargement of openings. Abundant small openings store a large quantity of water. The larger openings form a conduit system that drains the aquifer to springs.

Characterization of karst aquifers requires field exploration to locate sinkholes, swallets, sinking streams, and springs in addition to studying geologic maps. Conventional hydrogeologic methods such as aquifer tests and potentiometric mapping are insufficient to characterize the complexity of karst aquifers. These conventional investigation methods need to be supplemented with dye traces, measurement of spring discharges, and analysis of water chemistry. U.S. Geological Survey dye tracing has determined that conventional groundwater models that assume a uniform distribution of porosity are not applicable for karst aquifers.

Linear alignment of surface features such as straight stream segments and sinkholes develop along fracture traces. Locating a well in a fracture trace or intersection of fracture traces increases the likelihood to encounter good water production. Voids in karst aquifers can be large enough to cause destructive collapse or subsidence of the ground surface that can initiate a catastrophic release of contaminants. Groundwater flow rate in karst aquifers is much more rapid than in porous aquifers. For example, in the Barton Springs Edwards aquifer, dye traces measured the karst groundwater flow rates from 0.5 to 7 miles per day (0.8 to 11.3 km/d). The rapid groundwater flow rates make karst aquifers much more sensitive to groundwater contamination than porous aquifers.

In the extreme case, groundwater may exist in underground rivers (e.g., caves underlying karst topography.

Fractured

If a rock unit of low porosity is highly fractured, it can also make a good aquifer (via fissure flow), provided the rock has a hydraulic conductivity sufficient to facilitate movement of water.

4.2.6 Transboundary Aquifer

When an aquifer transcends international boundaries, the term transboundary aquifer applies.

Transboundariness is a concept, a measure and an approach first introduced in 2017. The relevance of this approach is that the physical features of the aquifers become just additional variables among the broad spectrum of considerations of the transboundary nature of an aquifer:

1. social (population);
2. economic (groundwater productivity);

3. political (as transboundary);
4. available research or data;
5. water quality and quantity;
6. other issues governing the agenda (security, trade, immigration and so on).

The discussion changes from the traditional question of "is the aquifer transboundary?" to "how transboundary is the aquifer?". The socio-economic and political contexts effectively overwhelm the aquifer's physical features adding its corresponding geostrategic value (its transboundariness)

The criteria proposed by this approach attempt to encapsulate and measure all potential variables that play a role in defining the transboundary nature of an aquifer and its multidimensional boundaries.

4.3 HUMAN USE OF GROUNDWATER

Reliance on groundwater will only increase, mainly due to growing water demand by all sectors combined with increasing variation in rainfall patterns. Challenges for using groundwater include: overdrafting (extracting groundwater beyond the equilibrium yield of the aquifer), groundwater-related subsidence of land, groundwater becoming saline, groundwater pollution.

4.4 AQUIFER IN AFRICA

Aquifer depletion is a problem in some areas, especially in northern Africa, where one example is the Great Manmade River project of Libya. However, new methods of groundwater management such as artificial recharge and injection of surface waters during seasonal wet periods has extended the life of many freshwater aquifers, especially in the United States.

4.5 AQUIFER IN AUSTRALIA

The Great Artesian Basin situated in Australia is arguably the largest groundwater aquifer in the world (over 1.7 million km or 0.66 million sq mi). It plays a large part in water supplies for Queensland, and some remote parts of South Australia.

4.6 AQUIFER IN CANADA

Discontinuous sand bodies at the base of the McMurray Formation in the Athabasca Oil Sands region of northeastern Alberta, Canada, are commonly referred to as the Basal Water Sand (BWS) aquifers. Saturated with water, they are confined beneath impermeable bitumen-saturated sands that are exploited to recover bitumen for synthetic crude oil production. Where they are deep-lying and recharge occurs from underlying Devonian formations they are saline, and where they are shallow and recharged by surface water they are non-saline. The BWS typically pose problems for the recovery of bitumen, whether by open-pit mining or by in situ methods such as steam-assisted gravity drainage (SAGD), and in some areas they are targets for waste-water injection.

4.7 AQUIFER IN SOUTH AMERICA

The Guarani Aquifer, located beneath the surface of Argentina, Brazil, Paraguay, and Uruguay, is one of the world's largest aquifer systems and is an important source of fresh water. Named after the Guarani people, it covers 1,200,000 km (460,000 sq mi), with a volume of about 40,000 km (9,600 cu mi), a thickness of between 50 and 800 m (160 and 2,620 ft) and a maximum depth of about 1,800 m (5,900 ft).

4.8 AQUIFER IN UNITED STATES

The Ogallala Aquifer of the central United States is one of the world's great aquifers, but in places it is being rapidly depleted by growing municipal use, and continuing agricultural use. This huge aquifer, which underlies portions of eight states, contains primarily fossil water from the time of the last glaciation. Annual recharge, in the more arid parts of the aquifer, is estimated to total only about 10 percent of annual withdrawals. According to a 2013 report by the United States Geological Survey (USGS), the depletion between 2001 and 2008, inclusive, is about 32 percent of the cumulative depletion during the entire 20th century.

In the United States, the biggest users of water from aquifers include agricultural irrigation and oil and coal extraction. "Cumulative total groundwater depletion in the United States accelerated in the late 1940s and continued at an almost steady linear rate through the end of the century. In addition to widely recognized environmental consequences, groundwater depletion also adversely impacts the long-term sustainability of groundwater supplies to help meet the Nation's water needs."

An example of a significant and sustainable carbonate aquifer is the Edwards Aquifer in central Texas. This carbonate aquifer has historically been providing high quality water for nearly 2 million people, and even today, is full because of tremendous recharge from a number of area streams, rivers and lakes. The primary risk to this resource is human development over the recharge areas.

4.9 POROSITY AND PERMEABILITY

4.9.1 Porosity

Porosity or void fraction is a measure of the void (i.e., "empty") spaces in a material, and is a fraction of the volume of voids over the total volume, between 0 and 1, or as a percentage between 0% and 100%. Strictlyspeaking, some tests measure the "accessible void", the total amount of void space accessible from the surface.

There are many ways to test porosity in a substance or part, such as industrial CT scanning. The term porosity is used in multiple fields including pharmaceutics, ceramics, metallurgy, materials, manufacturing, petrophysics, hydrology, earth sciences, soil mechanics, rock mechanics, and engineering.

Void Fraction in Two-phase Flow

In gas-liquid two-phase flow, the void fraction is defined as the fraction of the flow-channel volume that is occupied by the gas phase or, alternatively, as the fraction of the cross-sectional area of the channel that is occupied by the gas phase.

Void fraction usually varies from location to location in the flow channel (depending on the two-phase flow pattern). It fluctuates with time and its value is usually time averaged.

In separated (i.e., non z homogeneous) flow, it is related to volumetric flow rates of the gas and the liquid phse, and to the ratio of thevelocity of the two phses(called slip ratio).

Porosity in Earth Sciences and Construction

Used in geoloy, hydrogeology, soil science, and building science, the porosity of a porous medium (such as rock or sediment) describes the fraction of void space in the material, where the void may contain, for example, air or water. It is defined by the ratio:

$$\phi = \frac{V_V}{V_T}$$

where V_V is the volume of void-space (such as fluids) and V_T is the total or bulk volume of material, including the solid and void components. Both the mathematical symbols ϕ and n are used to denote porosity.

Porosity is a fraction between 0 and 1, typically ranging from less than 0.005 for solid granite to more than 0.5 for peat and clay.

The porosity of a rock, or sedimentary layer, is an important consideration when attempting to evaluate the potential volume of water or hydrocarbons it may contain.

Sedimentary porosity is a complicated function of many factors, including but not limited to: rate of burial, depth of burial, the nature of the connate fluids, the nature of overlying sediments (which may impede fluid expulsion). One commonly used relationship between porosity and depth is given by the Athy (1930) equation:

$$\phi(z) = \phi_0 e^{-kz}$$

where ϕ_0 is the surface porosity, k is the compaction coefficient (m) and is depth (m).

A value for porosity can alternatively be calculated from the bulk density ρ_{bulk}, saturating fluid density ρ_{fluid} and particle density ρ_{particle} :

$$\phi = \frac{\rho_{\text{particle}} - \rho_{\text{bulk}}}{\rho_{\text{particle}} - \rho_{\text{fluid}}}$$

If the void space is filled with air, the following simpler form may be used:

$$\phi = 1 - \frac{\rho_{\text{bulk}}}{\rho_{\text{particle}}}$$

Normal particle density is assumed to be approximately 2.65 g/cm (silica), although a better estimation can be obtained by examinin the lithology of the particles.

Porosity and Hydraulic Conductivity

Porosity can be proportional to hydraulic conductivity; for two similar sandy aquifers, the one with a higher porosity will typically have a higher hydraulic conductivity (more open area for the flow of water), but there are many complications to this relationship. The principal complication is that there is not a direct proportionality between porosity and hydraulic conductivity but rather an inferred proportionality. There is a clear proportionality between pore throat radii and hydraulic conductivity. Also, there tends to be a proportionality between pore throat radii and pore volume. If the proportionality between pore throat radii and porosity exists then a proportionality between porosity and hydraulic conductivity may exist. However, as grain size or sorting decreases the proportionality between pore throat radii and porosity begins to fail and therefore so does the proportionality between porosity and hydraulic conductivity. For example: clays typically have very low hydraulic conductivity (due to their small pore throat radii) but also have very high porosities (due to the structured nature of clay minerals), which means clays can hold a large volume of water per volume of bulk material, but they do not release water rapidly and therefore have low hydraulic conductivity.

Sorting and Porosity

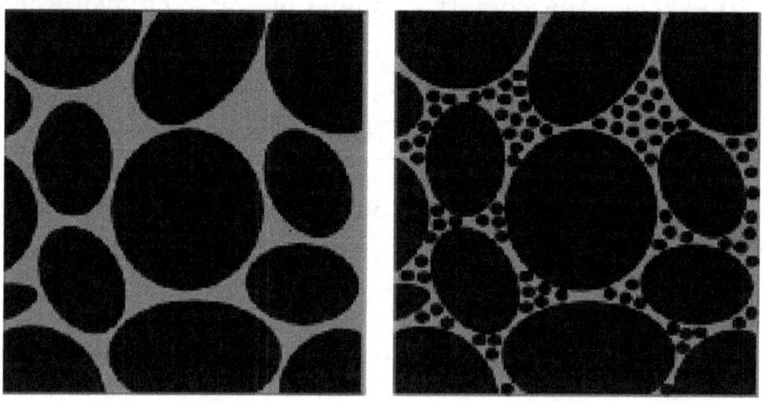

Fig. 4.2: Effects of Sorting on Alluvial Porosity. Black Represents Solids, Blue Represents Pore Space

Well sorted (grains of approximately all one size) materials have higher porosity than similarly sized poorly sorted materials (where smaller particles fill the gaps between larger particles). The graphic illustrates how some smaller grains can effectively fill the pores (where all water flow takes place), drastically reducing porosity and hydraulic conductivity, while only being a small fraction of the total volume of the material.

Porosity of Rocks

Consolidated rocks (e.g., sandstone, shale, granite or limestone) potentially have more complex "dual" porosities, as compared with alluvial sediment. This can be split into connected and unconnected porosity. Connected porosity is more easily measured through the volume of gas or liquid that can flow into the rock, whereas fluids cannot access unconnected pores.

Porosity is the ratio of pore volume to its total volume. Porosity is controlled by: rock type, pore distribution, cementation, diagenetic history and composition. Porosity is not controlled by grain size, as the volume of between-grain space is related only to the method of grain packing.

Rocks normally decrease in porosity with age and depth of burial. Tertiary age Gulf Coast sandstones are in general more porous than Cambrian age sandstones. There are exceptions to this rule, usually because of the depth of burial and thermal history.

Porosity of Soil

Porosity of surface soil typically decreases as particle size increases. This is due to soil aggregate formation in finer textured surface soils when subject to soil biological processes. Aggregation involves particulate adhesion and higher resistance to compaction. Typical bulk density of sandy soil is between 1.5 and 1.7 g/cm. This calculates to a porosity between 0.43 and 0.36. Typical bulk density of clay soil is between 1.1 and 1.3 g/cm. This calculates to a porosity between 0.58 and 0.51. This seems counterintuitive because clay soils are termed heavy, implying lower porosity. Heavy apparently refers to a gravitational moisture content effect in combination with terminology that harkens back to the relative force required to pull a tillage implement through the clayey soil at field moisture content as compared to sand.

Porosity of subsurface soil is lower than in surface soil due to compaction by gravity. Porosity of 0.20 is considered normal for unsorted gravel size material at depths below the biomantle. Porosity in finer material below the aggregating influence of pedogenesis can be expected to approximate this value.

Soil porosity is complex. Traditional models regard porosity as continuous. This fails to account for anomalous features and produces only approximate results.

Furthermore, it cannot help model the influence of environmental factors which affect pore geometry. A number of more complex models have been proposed, including fractals, bubble theory, cracking theory, Boolean grain process, packed sphere, and numerous other models. The characterisation of pore space in soil is an associated concept.

Types of Geologic Porosities

1. Primary porosity: The main or original porosity system in a rock or unconfined alluvial deposit.
2. Secondary porosity: A subsequent or separate porosity system in a rock, often enhancing overall porosity of a rock. This can be a result of chemical leaching of minerals or the generation of a fracture system. This can replace the primary porosity or coexist with it.
3. Fracture porosity: This is porosity associated with a fracture system or faulting. This can create secondary porosity in rocks that otherwise would not be reservoirs for hydrocarbons due to their primary porosity being destroyed (for example due to depth of burial) or of a rock type not normally considered a reservoir (for example igneous intrusions or metasediments).
4. Vuggy porosity: This is secondary porosity generated by dissolution of large features (such as macrofossils) in carbonate rocks leaving large holes, vugs, or even caves.
5. Effective porosity (also called open porosity): Refers to the fraction of the total volume in which fluid flow is effectively taking place and includes catenary and dead-end (as these pores cannot be flushed, but they can cause fluid movement by release of pressure like gas expansion) pores and excludes closed pores (or non-connected cavities). This is very important for groundwater and petroleum flow, as well as for solute transport.
6. Ineffective porosity (also called closed porosity): Refers to the fraction of the total volume in which fluids or gases are present but in which fluid flow can not effectively take place and includes the closed pores. Understanding the morphology of the porosity is thus very important for groundwater and petroleum flow.
7. Dual porosity: Refers to the conceptual idea that there are two overlapping reservoirs which interact. In fractured rock aquifers, the rock mass and fractures are often simulated as being two overlapping but distinct bodies. Delayed yield, and leaky aquifer flow solutions are both mathematically similar solutions to that obtained for dual porosity; in all three cases water comes from two mathematically different reservoirs (whether or not they are physically different).

8. Macroporosity: In solids (i.e., excluding aggregated materials such as soils), the term 'macroporosity' refers to pores greater than 50 nm in diameter. Flow through macropores is described by bulk diffusion.
9. Mesoporosity: In solids (i.e., excluding aggregated materials such as soils), the term 'mesoporosity' refers to pores greater than 2 nm and less than 50 nm in diameter. Flow through mesopores is described by Knudsen diffusion.
10. Microporosity: In solids (i.e., excluding aggregated materials such as soils), the term 'microporosity' refers to pores smaller than 2 nm in diameter. Movement in micropores is activated by diffusion.

Porosity of Fabric or Aerodynamic Porosity

The ratio of holes to solid that the wind "sees". Aerodynamic porosity is less than visual porosity, by an amount that depends on the constriction of holes.

Die Casting Porosity

Casting porosity is a consequence of one or more of the following: gasification of contaminants at molten-metal temperatures; shrinkage that takes place as molten metal solidifies; and unexpected or uncontrolled changes in temperature or humidity.

While porosity is inherent in die casting manufacturing, its presence may lead to component failure where pressure integrity is a critical characteristic. Porosity may take on several forms from interconnected micro-porosity, folds, and inclusions to macro porosity visible on the part surface. The end result of porosity is the creation of a leak path through the walls of a casting that prevents the part from holding pressure. Porosity may also lead to out-gassing during the painting process, leaching of plating acids and tool chatter in machining pressed metal components.

Measuring Porosity

Several methods can be employed to measure porosity:

1. Direct methods (determining the bulk volume of the porous sample, and then determining the volume of the skeletal material with no pores (ore volume = total volume – material volume).
2. Optical methods (e.g., determining the area of the material versus the area of the pores visible under the microscope). The "areal" and "volumetric" porosities are equal for porous media with random structure.
3. Computed tomography method (using industrial CT scanning to create a 3D rendering of external and internal geometry, including vods. Then implementing a defect analysis utilizing computer software)

4. Imbibition methods, i.e., immersion of the porous sample, under vacuum, in a fluid that preferentially wets the pores.
   a. Water saturation method (pore volume = total volume of water " volume of water left after soaking).
5. Water evaporation method (pore volume = (weight of saturated sample " weight of dried sample)/density of water)
6. Mercury intrusion porosimetry (several non-mercury intrusion techniques have been developed due to toxicological concerns, and the fact that mercury tends to form amalgams with several metals and alloys).
7. Gas expansion method. A sample of known bulk volume is enclosed in a container of known volume. It is connected to another container with a known volume which is evacuated (i.e., near vacuum pressure). When a valve connecting the two containers is opened, gas passes from the first container to the second until a uniform pressure distribution is attained. Using ideal gas law, the volume of the pores is calculated as

$$V_V = V_T - V_a - V_b \frac{P_2}{P_2 - P_1}$$

where

- V_V is the effective volume of the pores,
- V_T is the bulk volume of the sample,
- V_a is the volume of the container containing the sample,
- V_b is the volume of the evacuated container,
- P_1 is the initial pressure in the initial pressure in volume V_a and V_V, and
- P_2 is final pressure present in the entire system.
- The porosity follows straightforwardly by its proper definition

$$\phi = \frac{V_V}{V_T}$$

Note that this method assumes that gas communicates between the pores and the surrounding volume. In practice, this means that the pores must not be closed cavities.

1. Thermoporosimetry and cryoporometry: A small crystal of a liquid melts at a lower temperature than the bulk liquid, as given by the Gibbs-Thomson equation. Thus if a liquid is imbibed into a porous material, and frozen, the melting temperature will provide information on the pore-size distribution. The detection of the melting can be done by sensing the transient heat flows during phase-changes using differential scanning calorimetry – (DSC thermoporometry), measuring the quantity of mobile liquid using nuclear magnetic resonance – (NMR cryoporometry) or

measuring the amplitude of neutron scatterig from the imbibed crystalline or liquid phases – (ND cryoporometry).

4.9.2 Permeability

Permeability in fluid mechanics and the Earth sciences (commonly symbolized as k) is a measure of the ability of a porous material (often, a rock or an unconsolidated material) to allow fluids to pass through it.

Permeability

Permeability is a property of porous materials that is an indication of the ability for fluids (gas or liquid) to flow through them. Fluids can more easily flow through a material with high permeability than one with low permeability. The permeability of a medium is related to the porosity, but also to the shapes of the pores in the medium and their level of connectedness. Fluid flows can also be influenced in different lithological settings by brittle deformation of rocks in fault zones; the mechanisms by which this occurs are the subject of fault zone hydrogeology. Permeability is also affected by the pressure inside a material.

Units

The SI unit for permeability is the square metre (m). A practical unit for permeability is the darcy (d), or more commonly the millidarcy (md) . he name honors the French Engineer Henry Darcy who first described the flow of water through sand filters for potable water supply. Permeability values for most materials commonly range typically from a fraction to several thousand millidarcies. The unit of square centimetre (cm) is also sometimes used (1 cm = 10 m ≈ Hm 10 d).

Applications

The concept of permeability is of importance in determining the flow characteristics of hydrocarbons in oil and gas reservoirs, and of groundwater in aquifers. For a rock to be considered as an exploitable hydrocarbon reservoir without stimulation, its permeability must be greater than approximately 100 md (depending on the nature of thc hydrocarbon – gas reservoirs with lower permeabilities are still exploitable because of the lower viscosity of gas with respect to oil). Rocks with permeabilities significantly lower than 100 md can form efficient seals. Unconsolidated sands may have permeabilities of over 5000 md. The concept also has many practical applications outside of geology, for example in chemical engineering (e.g., filtration), as well as in Civil Engineering when determining whether the ground conditions of a site are suitable for construction.

Description

Permeability is part of the proportionality constant n Darcy's law which relates discharge (flow rate) and fluid physical prop η rties (e.g., viscosity), to a pressure gradient ρ applied to the porous media:

$$v = \frac{k}{\eta}\frac{\Delta P}{\Delta x} \quad \textbf{(for linear flow)}$$

Therefore:

$$k = v\frac{\eta \Delta x}{\Delta P}$$

where:

- v is the fluid velocity through the porous medium (i.e., the average flow velocity calculated as if the fluid was the only phase present in the porous medium) (m/s)
- k is the permeability of a medium (m)
- η is the dynamic viscosity of the fluid (Pa·s)
- Δx is the applied pressure difference (Pa)
- ΔP is the thickness of the bed of the porous medium (m)

In naturally occurring materials, the permebility values range over many orders of magnitude.

Relation to Hydraulic Conductivity

The global proportionality constant for the flow of water through a porous medium is called the hydraulic conductivity (K, unit: m/s). Permeability, or intrinsic permeability, (*k*, unit: m) is a part of this, and is a specific property characteristic of the solid skeleton and the microstructure of the porous medium itself, independently of the nature and properties of the fluid flowing through the pores of the medium.

This allows to take into account the effect of temperature on the viscosity of the fluid flowing though the porous medium and to address other fluids than pure water, *e.g.*, concentrated brines, petroleum, or organic solvents. Given the value of hydraulic conductivity for a studied system, the permeability can be calculated as follows:

$$k = K\frac{\eta}{\rho g}$$

where

- *k* is the permeability, m
- K is the hydraulic conductivity, m/s
- η is the dynamic viscosity of the fluid, Pa·s

- ρ is the density of the fluid, kg/m
- g is the acceleration due to gravity, m/s.

Anisotropic Permeability

Tissue such as brain, liver, muscle, etc can be treated as a heterogeneous porous medium. Describing the flow of biofluids (blood, cerebrospinal fluid, etc.) within such a medium requires a full 3-dimensional anisotropic treatment of the tissue. In this case the scalar hydraulic permeability is replaced with the hydraulic permeability tensor so that Darcy's Law reads

$$\mathbf{q} = -\frac{1}{\eta}\boldsymbol{\kappa} \cdot \nabla P$$

- q is the Darcy flux, or filtration velocity, which describes the bulk (not microscopic) velocity field of the fluid, $[\mathrm{Length}][\mathrm{Time}]^{-1}$
- η is the dynamic viscosity of the fluid, $[\mathrm{Mass}][\mathrm{L}]^{-1}[T]^{-1}$
- k is the hydraulic permeability tensor, $[\mathrm{L}]^{?}$
- ∇ is the gradient operator, $[\mathrm{L}]^{-1}$
- P is the pressure field in the fluid, $[\mathrm{M}][\mathrm{L}][\mathrm{T}]^{-1}$

Connecting this expression to the isotropic case, k = k 1 , where k is the scalar hydraulic permeability, and 1 is the identity tensor.

Determination

Permeability is typically determined in the lab by application of Darcy's law under steady state conditions or, more generally, by application of various solutions to the diffusion equation for unsteady flow conditions.

Permeability needs to be measured, either directly (using Darcy's law), or through estimation using empirically derived formulas. However, for some simple models of porous media, permeability can be calculated (e.g., random close packing of identical spheres).

Model based on Conduit Flow

Based on the Hagen–Poiseuille equation for viscous flow in a pipe, permeability can be expressed as:

$$k_I = C \cdot d^2$$

where:

- k_I is the intrinsic permeability [length]

- C is a dimensionless constant that is related to the configuration of the flow-paths
- d is the average, or effective pore diameter [length].

Absolute Permeability (Aka Intrinsic or Specific Permeability)

Absolute permeability denotes the permeability in a porous medium that is 100% saturated with a single-phase fluid. This may also be called the intrinsic permeability or specific permeability.

These terms refer to the quality that the permeability value in question is an intensive property of the medium and that it is a function of the material structure only (and not of the fluid). They explicitly distinguish the value from that of relative permeability.

Permeability to Gases

Sometimes permeability to gases can be somewhat different than those for liquids in the same media. One difference is attributable to "slippage" of gas at the interface with the solid when the gas mean free path is comparable to the pore size (about 0.01 to 0.1 μm at standard temperature and pressure). For example, measurement of permeability through sandstones and shales yielded values from 9.0 × 10m to 2.4 × 10m for water and between 1.7 × 10m to 2.6 × 10 m for nitrogen gas. Gas permeability of reservoir rock and source rock is important in petroleum engineering, when considering the optimal extraction of gas from unconventional sources such as shale gas, tight gas, or coalbed methane.

Permeability Tensor

To model permeability in anisotropic media, a permeability tensor is needed. Pressure can be applied in three directions, and for each direction, permeability can be measured (via Darcy's law in 3D) in three directions, thus leading to a 3 by 3 tensor. The tensor is realised using a 3 by 3 matrix being both symmetric and positive definite (SPD matrix):

1. The tensor is symmetric by the Onsager reciprocal relations
2. The tensor is positive definite because the energy being expended (the inner product of fluid flow and negative pressure gradient) is always positive

The permeability tensor is always diagonalizable (being both symmetric and positive definite). The eigenvectors will yield the principal directions of flow where flow is parallel to the pressure gradient, and the eigenvalues represent the principal permeabilities.

4.10 SEQUENCE STRATIGRAPHY

Sequence stratigraphy is a branch of geology, specifically a branch of stratigraphy, that attempts to discern and understand historic geology through time by subdividing and linking sedimentary deposits into unconformity bounded units on a variety of scales. The essence of the method is mapping of strata based on identification of surfaces which are assumed to represent time lines (e.g., subaerial unconformities, maximum flooding surfaces), thereby placing stratigraphy in chronostratigraphic framework allowing understanding of the evolution of the earth's surface in a particular region through time. Sequence stratigraphy is a useful alternative to a purely lithostratigraphic approach, which emphasizes solely based on the compositional similarity of the lithology of rock units rather than time significance. Unconformities are particularly important in understanding geologic history because they represent erosional surfaces where there is a clear gap in the record. Conversely within a sequence the geologic record should be relatively continuous and complete record that is genetically related.

Stratigraphers explain sequence boundaries and stratigraphic units primarily in terms of changes in relative sea level (the combination of global changes in eustatic sea level and regional subsidence caused by tectonic subsidence, thermal subsidence and load-induced subsidence as the weight of accumulated sediment and water cause isostatic subsidence as a sedimentary basin is filled). The net changes resulting from these vertical forces increases or reduces accommodation space for sediments to accumulate in a sedimentary basin. A secondary influence is the rate of sediment supply to the basin which determines the rate at which that space is filled.

4.10.1 Historical Development

The origin of sequence stratigraphy can be traced back to the work of L.L. Sloss on interregional unconformities of the North American craton. Sloss recognized six craton-wide sequences representing hundreds of millions of years of earth history. In the late 1960s Sloss had several students, notably Peter Vail, Robert Mitchum, and John Sangree, who completed dissertations studying the Pennsylvanian sedimentary rocks of the North American craton and became aware that global changes in sea level could have been responsible for the numerous widespread unconformities in those rocks. During their subsequent careers as research scientists at Exxon's research division Vail, Mitchum and others pioneered the practice of seismic stratigraphy, the stratigraphic interpretation of seismic reflection profiles to understand the layering and packaging of sedimentary rocks in the subsurface using acoustic imaging. The advent of seismic stratigraphy made it possible to identify sequences representing shorter period of time ranging in duration from tens of thousands to a few million years; and to compare the sequence stratigraphic history around the globe. This in turn led to sequence stratigraphy becoming systematized and understood to have widespread application to stratigraphic study of rock outcrops on the earth's surface as well. During the 1980s this ushered in a revolution in stratigraphy based on the delineation of regional physical surfaces that separate the sedimentary rock into packages representing discrete and sequential periods of time and predictable patterns of sediment depositional history.

4.10.2 Significant Surfaces

Sequence Boundaries

Sequence boundaries are deemed the most significant surfaces. Sequence boundaries are defined as unconformities or their correlative conformities. Sequence boundaries are formed due to the sea level fall. For example, multi-story fluvial sandstone packages often infill incised valleys formed by the sea level drop associated with sequence boundaries. The incised valleys of sequence boundaries correlate laterally with interfluves, palaeosols formed on the margins of incised valleys. The valley infills are not genetically related to underlying depositional systems as previous interpretations thought. There are four criteria distinguishing incised valley fills from other types of multi-story sandstone deposits: a widespread correlation with a regional, high relief erosional surface that is more widespread than the erosional bases of individual channels within the valley; facies associations reflect a basinward shift in facies when compared with underlying units; erosional base of the valley removes preceding systems tracts and marine bands producing a time gap, the removed units will be preserved beneath the interfluves; increasing channel fill and fine grained units upwards or changes in the character of the fluvial systems reflecting increasing accommodation space. Sandstone bodies associated with incised valleys can be good hydrocarbon reservoirs. There have been problems in the correlation and distribution of these bodies. Sequence stratigraphic principles and identification of significant surfaces have resolved some issues.

Parasequence Boundaries

Lesser importance is attached to parasequence boundaries, however, there is a suggestion that flooding surfaces representing parasequence boundaries may be more laterally extensive leaving more evidence than sequence boundaries because the coastal plain has a lower gradient than the inner continental shelf. Parasequence boundaries may be distinguished by differences in physical and chemical properties across the surface such as; formation water salinity, hydrocarbon properties, porosity, compressional velocities and mineralogy. Parasequence boundaries may not form a barrier to hydrocarbon accumulation but may inhibit vertical reservoir communication. After production begins the parasequences act as separate drainage units with the flooding surfaces, which are overlain by shales or carbonate-cemented horizons, forming a barrier to vertical reservoir communications. Sequence stratigraphic principles have optimized production potential once reservoir scale architecture is identified and separate drainage units identified.

4.10.3 Systems Tracts

The concept of systems tracts evolved to link the contemporaneous depositional systems. Systems tracts form subdivision in a sequence. Different kinds of systems tracts are assigned on the basis of stratal stacking pattern, position in a sequence, and in the sea level curve and types of bounding surfaces.

1. A lowstand systems tract (LST) forms when the rate of sedimentation outpaces the rate of sea level rise during the early stage of the sea level curve. It is bounded by a subaerial unconformity or its correlative conformity at the base and maximum regressive surface at the top.
2. A transgressive systems tract (TST) is bounded by maximum regressive surface at the base and maximum flooding surface at the top. This systems tracts forms when the rate of sedimentation is outpaced by the rate of sea level rise in the sea level curves.
3. A highstand systems tract (HST) occurs during the late stage of base level rise when the rate of sea level rise drops below the sedimentation rate. In this period of sea level highstand is formed. It is bounded by maximum flooding surface at the base and composite surface at the top.
4. Regressive systems tract forms in the marine part of the basin during the base level fall. Subaerial unconformities form in the landward side of the basin at the same time.

4.10.4 Parasequences and Stacking Patterns

A parasequence is a relatively conformable, genetically related succession of beds and bedsets bounded by marine flooding surfaces and their correlative surfaces. The flooding surfaces bounding parasequences are not of the same scale as the regional transgressive surface that is associated with a sequence boundary.

The parasequences are separated into stacking patterns:

1. Aggradational
2. Progradational
3. Retrogradational

Each stacking pattern will give different information on the behaviour of accommodation space, a major control of which is relative level. So a rapidly progradational pattern will be indicative of falling sea level, rapidly retrogradational is evidence for rapidly transgressing sea level and aggradational will be indicative of gently rising sea level.

4.10.5 Sea Level through Geologic Time

Sea level changes over geologic time. The modern age is depicted on the left side, labeled N for Neogene. The blue spikes near date zero represent the sea level changes associated with the most recent glacial period, which reached its maximum extent about 20,000 years Before Present (BP). During this glaciation event, the world's sea level was about 320 feet (98 meters) lower than today, due to the large amount of sea water that had evaporated and been deposited as snow and ice in Northern Hemisphere glaciers. When the world's sea level was at this "low stand", former sea bed sediments were subjected to subaerial weathering (erosion by rain, frost, rivers, etc.) and a new shoreline was established at the new level, sometimes miles basinward of the former shoreline if the sea floor was shallowly inclined.

Today, sea level is at a relative "high stand" within the Quaternary glacial cycles because of rapid end-Pleistocene and early-Holocene deglaciation. The ancient shoreline of the last glacial period is now under approximately 390 feet (120 meters) of water. Although there is debate among earth scientists whether we are currently experiencing a "high stand" it is generally accepted that the eustatic sea level is rising.

In the distant past, sea level has been significantly higher than today. During the Cretaceous, sea level was so high that a seaway extended across the center of North America from Texas to the Arctic Ocean.

These alternating high and low sea level stands repeat at several time scales. The smallest of these cycles is approximately 20,000 years, and corresponds to the rate of precession of the Earth's rotational axis and are commonly referred to as '5th order' cycles. The next larger cycle ('4th order') is about 40,000 years and approximately matches the rate at which the Earth's inclination to the Sun varies (again explained by Milankovitch). The next larger cycle ('3rd order') is about 110,000 years and corresponds to the rate at which the Earth's orbit oscillates from elliptical to circular. Lower order cycles are recognized, which seem to result from plate tectonic events like the opening of new ocean basins by splitting continental masses.

Hundreds of similar glacial cycles have occurred throughout the Earth's history. The earth scientists who study the positions of coastal sediment deposits through time ("sequence stratigraphers") have noted dozens of similar basinward shifts of shorelines associated with a later recovery. The largest of these sedimentary cycles can in some cases be correlated around the world with great confidence.

The three controls on stratigraphic architecture and sedimentary cycle development are:

1. Eustatic sea level changes
2. Subsidence rate of the basin
3. Sediment supply.

Eustatic sea level is the sea level with reference to a fixed point, the centre of the Earth. Relative sea level is measured with reference to the base level, above which erosion can occur and below which deposition can occur. Both eustatic sea level changes and subsidence rates tend to be longer cycles. Sediment supply is largely thought to be controlled by local climatic conditions and can vary rapidly. These variations in local sediment supply affect the local and relative sea level which causes local sedimentary cycles.

Smaller and localised sedimentary cycles are not related to worldwide (eustatic) sea level changes but more to the supply of sediment to the adjacent basins where these sediments are being supplied. For example, when the basinward (oceanward) shift with progradation of shorelines was occurring in the Book Cliffs area of Utah the shorelines were receding or transgressing northwards in Wyoming. These sedimentary cycles are representative of the amount of supply of sediment to the basin. In a transgression, less sediment is being supplied than the rate of increase in the depth of water, and thus the shoreline migrates landward. In a regression, if the water depth is decreasing, the shoreline migrates seaward (basinward) and the previous shoreline is eroded. A regression of the shoreline also occurs if more sediment is being supplied than the shoreline can erode, causing the shoreline to migrate seaward. The latter is called progradation. The cycle of strata deposited during repeated transgressions and regressions creates a depositional sequence.

4.10.6 Economic Significance

Sequence stratigraphy is and essential tool in the application of geology to the exploration for oil and gas, as a part of the field of Petroleum geology. Much of the development of this scientific discipline has occurred within or been funded by energy corporations and their geological research labs.

Sequence boundaries have economic significance because these changes in sea level cause large lateral shifts in the depositional patterns of seafloor sediments. These lateral shifts in deposition create alternating layers of good reservoir quality rock (porous and permeable sands) and poorer-quality mudstones (capable of providing a reservoir "seal" to prevent the leakage of any accumulated hydrocarbons that may have migrated into the sandstones). Hydrocarbon prospectors look for places in the world where porous and permeable sands are overlain by low permeability rocks, and where conditions are right for hydrocarbons to be generated and migrate into these "traps".

4.11 AQUIFER STORAGE AND RECOVERY

Aquifer storage and recovery (ASR) is the direct injection of surface water supplies such as potable water, reclaimed water (i.e., rainwater), or river water into an aquifer for later recovery and use. The injection and extraction is often done by means of a well. In areas where the rainwater cannot percolate the soil or where it is not capable of percolating it fast enough (i.e., urban areas) and where the rainwater is thus diverted to rivers, rainwater ASR could help to keep the rainwater within an area. ASR is used for municipal, industrial and agricultural purposes.

4.11.1 ASR Use in the United States

Colorado

The first ASR well with a downhole control valve was installed in Highlands Ranch, Colorado, in 1992 for Centennial Water and Sanitation District. Since then, over 40 ASR wells have been installed for many different municipalities. These wells range in depths from 1,000 ft (300 m) to 3,000 ft (910 m) below ground surface, with injection rates commonly between 100 US gal (380 L; 83 imp gal) and 500 US gal (1,900 L; 420 imp gal) per minute (gpm) per well.

Florida

The use of ASR in Florida has been examined to determine potential benefits for the Everglades and other Floridian water systems under the Comprehensive Everglades Restoration Plan (CERP). An estimate of 333 ASR wells would be implemented as part of CERP and used to store, treat and supply excess surface water to the Everglades and other systems of water during dry periods.

Controversy

Doubts remain about the benefit of introducing ASR in such large capacity to Florida due to a predominantly karst geography. Current or potential problems include:

1. Poor recovery due to mixing of the injected fresh water with the existing brackish to saline water in the aquifer;
2. pre-existing quality of water introduced to ASR; and
3. potential risk of resulting water quality due to mixing of injected freshwater and existing aquifer.

Oregon

The first agriculture ASR wells were put into service in Oregon in the autumn of 2006 and have injected well over 3,000 acre-feet (3,700,000 m) of water during the winter and spring flood flow times using artificial recharge (AR) of flood water as their water source. This shallow recharged water is then recovered as potable water and injected into the deep basalt aquifer.

During the injection process, electrical energy can be generated by the head pressure of the water flowing back into the aquifer. This stored water is recovered during late summer and early autumn for irrigation needs.

Both of these well types use a down-hole control valve. ASR can also be used to re-inject water used by HVAC systems to maintain the ground water levels and store the thermal differences from summertime cooling for winter time heating. Industry can also capture cold winter waters and store it for summertime use and avoid the need for cooling water in the summer for industrial processes. This may also free up short supplies of summer time water for other beneficial uses. This reinjection process may also avoid the cost of surface disposal and avoid the increased thermal load to the rivers and streams during the summer air conditioning season.

Texas

The Texan cities of El Paso, Kerrville and San Antonio use ASR, providing water resources to these water-vulnerable communities. A University of Florida report ranked daily per-capita water availability for 225 large urban areas across the U.S. The study weighed fresh water available to cities from naturally occurring and constructed sources such as reservoirs, aquifers and imports. Of the cities reviewed, San Antonio ranked last, or most vulnerable, and El Paso ranked as 10th-worst, though other Texas cities made the list. San Antonio stores drinking water in its Carrizo ASR facility, which contains more than 91,000 acre-feet of water and has a maximum capacity of 120,000 acre-feet.

A 2010 Texas Water Development Board (TWDB) survey of Texas water utilities found four primary objections to ASR in other parts of Texas: legal and physical limitations, the quality of the recovered water, cost-effectiveness and the potential for other pumpers to capture the utility's stored water.

4.11.2 ASR Use in Australia

In South Australia, the City of Salisbury in northern Adelaide has since 1994 played a pioneering role in establishing the viability of the capture and treatment of urban stormwater runoff within artificial wetlands, and injecting the treated water into Tertiary aquifers in winter, for later use by industry, and for irrigation of city parks and school playing fields in summer. In 2009 the capacity of the Salisbury ASR schemes was around 5 gigalitres per annum (expected to rise to 14 GL by 2014) and this success has led to other local government areas across Adelaide undertaking similar ASR projects.

4.11.3 ASR Use in Europe

In Spain, the SubSol ASR project is active and in the Netherlands, there is the COASTAR project.

4.12 AQUIFER TEST

In hydrogeology, an aquifer test (or a pumping test) is conducted to evaluate an aquifer by "stimulating" the aquifer through constant pumping, and observing the aquifer's "response" (drawdown) in observation wells. Aquifer testing is a common tool that hydrogeologists use to characterize a system of aquifers, aquitards and flow system boundaries.

A slug test is a variation on the typical aquifer test where an instantaneous change (increase or decrease) is made, and the effects are observed in the same well. This is often used in geotechnical engineering settings to get a quick estimate (minutes instead of days) of the aquifer properties immediately around the well.

Aquifer tests are typically interpreted by using an analytical model of aquifer flow (the most fundamental being the Theis solution) to match the data observed in the real world, then assuming that the parameters from the idealized model apply to the real-world aquifer. In more complex cases, a numerical model may be used to analyze the results of an aquifer test.

Aquifer testing differs from well testing in that the behaviour of the well is primarily of concern in the latter, while the characteristics of the aquifer are quantified in the former. Aquifer testing also often utilizes one or more monitoring wells, or piezometers ("point" observation wells).

A monitoring well is simply a well which is not being pumped (but is used to monitor the hydraulic head in the aquifer). Typically monitoring and pumping wells are screened across the same aquifers.

4.12.1 General Characteristics

Most commonly an aquifer test is conducted by pumping water from one well at a steady rate and for at least one day, while carefully measuring the water levels in the monitoring wells. When water is pumped from the pumping well the pressure in the aquifer that feeds that well declines. This decline in pressure will show up as drawdown (change in hydraulic head) in an observation well. Drawdown decreases with radial distance from the pumping well and drawdown increases with the length of time that the pumping continues.

The aquifer characteristics which are evaluated by most aquifer tests are:

1. Hydraulic conductivity: The rate of flow of water through a unit cross sectional area of an aquifer, at a unit hydraulic gradient. In US units the rate of flow is in gallons per day per square foot of cross sectional area; in

SI units hydraulic conductivity is usually quoted in m per day per m. Units are frequently shortened to metres per day or equivalent.

2. Specific storage or storativity: A measure of the amount of water a confined aquifer will give up for a certain change in head;
3. Transmissivity: The rate at which water is transmitted through whole thickness and unit width of an aquifer under a unit hydraulic gradient. It is equal to the hydraulic conductivity times the thickness of an aquifer;

Additional aquifer characteristics which are sometimes evaluated, depending on the type of aquifer, include:

1. ***Specific yield or drainable porosity:*** A measure of the amount of water an unconfined aquifer will give up when completely drained;
2. ***Leakage coefficient:*** Some aquifers are bounded by aquitards which slowly give up water to the aquifer, providing additional water to reduce drawdown;
3. The presence of aquifer boundaries (recharge or no-flow) and their distance from the pumped well and piezometers.

4.12.2 Analysis Methods

An appropriate model or solution to the groundwater flow equation must be chosen to fit to the observed data. There are many different choices of models, depending on what factors are deemed important including:

1. Leakyaquitards,
2. Unconfined flow (delayed yield),
3. Partial penetration of the pmping and monitoring wells,
4. Finite wellbore radius — which can lead to wellbore storage,
5. Dual porosity (typically in fractured rock),
6. Anisotropic aquifers,
7. Heterogeneous aquifers,
8. Finite aquifers (the effects of physical boundaries are seen in the test), and
9. combinations of the above situations.

Nearly all aquifer test solution methods are based on the Theis solution; it is built upon the most simplifying assumptions.

Other methods relax one or more of the assumptions the Theis solution is built on, and therefore they get a more flexible (and more complex) result.

Transient Theis Solution

The Theis equation was created by Charles Vernon Theis (working for the US Geological Survey) in 1935, from heat transfer literature (with the mathematical help of C.I. Lubin), for two-dimensional radial flow to a point sink in an infinite, homogeneous aquifer. It is simply

$$S = \frac{Q}{4\pi T} W(u)$$

$$u = \frac{r^2 S}{4Tt}$$

where s is the drawdown (change in hydraulic head at a point since the beginning of the test in units of distance), u is a dimensionless parameter, Q is the discharge (pumping) rate of the well (volume per unit time), T and S are the transmissivity and storativity of the aquifer around the well (distance squared per unit time and dimensionless, respectively), r is the distance from the pumping well to the point where the drawdown was observed, t is the time since pumping began, and W(u) is the "Well function" (called the incomplete gamma function, , in non-hydrogeology literature).

The well function is given by the infinite series

$$W(u) = \Gamma(0,u) = -\gamma - \ln(u) + u - \frac{u^2}{2\times 2!} + \frac{u^3}{3\times 3!} - \frac{u^4}{4\times 4!} + \cdots$$

where γ is the Euler constant (=0.577216...). Typically this equation is used to find the average *T* and *S* values near a pumping well, from drawdown data collected during an aquifer test. This is a simple form of inverse modeling, since the result (*s*) is measured in the well, *r*, *t*, and *Q* are observed, and values of *T* and *S* which best reproduce the measured data are put into the equation until a best fit between the observed data and the analytic solution is found.

The solution is based on the following assumptions:

1. The flow in the aquifer is adequately described by Darcy's law (i.e., Re<10).
2. Homogeneous, isotropic, confined aquifer,
3. Well is fully penetrating (open to the entire thickness (b) of aquifer),
4. The well has zero radius (it is approximated as a vertical line) — therefore no water can be stored in the well,
5. The well has a constant pumping rate Q,
6. The head loss over the well screen is negligible,
7. Aquifer is infinite in radial extent,
8. Horizontal (not sloping), flat, impermeable (non-leaky) top and bottom boundaries of aquifer,

9. Groundwater flow is horizontal
10. No other wells or long term changes in regional water levels (all changes in potentiometric surface are the result of the pumping well alone)

Even though these assumptions are rarely all met, depending on the degree to which they are violated (e.g., if the boundaries of the aquifer are well beyond the part of the aquifer which will be tested by the pumping test) the solution may still be useful.

Steady-state Thiem Solution

Steady-state radial flow to a pumping well is commonly called the Thiem solution, it comes about from application of Darcy's law to cylindrical shell control volumes (i.e., a cylinder with a larger radius which has a smaller radius cylinder cut out of it) about the pumping well; it is commonly written as:

$$h_0 - h = \frac{Q}{2\pi T}\ln\left(\frac{R}{r}\right)$$

In this expression h_0 is the background hydraulic head, h_0-h is the drawdown at the radial distance r from the pumping well, Q is the discharge rate of the pumping well (at the origin), T is the transmissivity, and R is the radius of influence, or the distance at which the head is still h_0. These conditions (steady-state flow to a pumping well with no nearby boundaries) never truly occur in nature, but it can often be used as an approximation to actual conditions; the solution is derived by assuming there is a circular constant head boundary (e.g., a lake or river in full contact with the aquifer) surrounding the pumping well at a distance R.

4.12.3 Sources of Error

Of critical importance in both aquifer and well testing is the accurate recording of data. Not only must water levels and the time of the measurement be carefully recorded, but the pumping rates must be periodically checked and recorded. An unrecorded change in pumping rate of as little as 2% can be misleading when the data are analysed.

4.13 ARTESIAN WELL

An artesian well is a well that brings groundwater to the surface without pumping because it is under pressure within a body of rock and/or sediment known as an aquifer. When trapped water in an aquifer is surrounded by layers of impermeable rock or clay, which apply positive pressure to the water, it is known as an artesian aquifer. If a well were to be sunk into an artesian aquifer, water in the well-pipe would rise to a height corresponding to the point where hydrostatic equilibrium is reached.

A well drilled into such an aquifer is called an artesian well. If water reaches the ground surface under the natural pressure of the aquifer, the well is termed a flowing artesian well.

Fossil water aquifers can also be artesian if they are under sufficient pressure from the surrounding rocks, similar to how many newly tapped oil wells are pressurized.

Not all aquifers are artesian (i.e., water table aquifers occur where the groundwater level at the top of the aquifer is at equilibrium with atmospheric pressure). The recharging of aquifers happens when the water table at its recharge zone is at a higher elevation than the head of the well.

4.14 SOLE SOURCE AQUIFER

A Sole Source Aquifer (SSA) is an aquifer that has been designated by the United States Environmental Protection Agency (EPA) as the sole or principal source of drinking water for an area. By definition, SSA is an aquifer that supplies at least 50% of the drinking water consumed in the area overlying the aquifer. There may be other factors to designate SSA. For example, in New York City, Kings and Queens Counties are designated as SSA not because the aquifer of these areas are the sole or principal sources of drinking water for these counties; they are designated as SSA because the geographic boundaries of Kings and Queens Counties are within the recharge zone for the aquifers underlying the southeastern portion of Queens County. Groundwater beneath Manhattan and the Bronx is not used for drinking or non-potable purposes.

These areas may have no alternative drinking water source(s) that could physically, legally and economically supply all those who depend on the aquifer for drinking water and therefore, if contamination occurs, using an alternative source would be extremely expensive. SSA designation works as a tool to protect drinking water supplies in these areas.

4.15 SURFICIAL AQUIFER

Surficial aquifers are shallow aquifers typically less than 50 feet (15 m) thick, but larger surficial aquifers of about 60 feet (18 m) have been mapped. They mostly consist of unconsolidated sand enclosed by layers of limestone, sandstone or clay and the water is commonly extracted for urban use. The aquifers are replenished by streams and from precipitation and can vary in volume considerably as the water table fluctuates. Being shallow, they are susceptible to contamination by fuel spills, industrial discharge, landfills, and saltwater. Parts of southeastern United States are dependent on surficial aquifers for their water supplies.

4.15.1 Composition

Surficial aquifers system consists mostly of beds of unconsolidated sand, cavity-riddled limestone and shells, sandstone, sand, and clay sand with minor clay or silt from the Pliocene to Holocene periods.

In most cases the flow system is undivided, though in places, clay beds are sufficiently thick and continuous to divide the system into two or three aquifers. Complex interbedding of fine and coarse-textured rocks is typical of the system. These rocks range from late Miocene to Holocene periods.

In Georgia and South Carolina, unnamed, sandy, marine terrace deposits of Pleistocene age and sand of Holocene age comprise the system. These sandy beds commonly contain clay and silt.

In Florida, these aquifers are shallow beds of sea shells and sand that lie less than 100 feet (30 m) underground. They are separated from the Floridan Aquifer by a confining bed of soil. Some have been contaminated by saltwater, yet they provide most of the public freshwater supply southwest of Lake Okeechobee and along the Atlantic coast north of Palm Beach.

In surficial aquifers, the groundwater continuously moves along the hydraulic gradient from areas of recharge to streams and other places of discharge. Surficial aquifers are recharged locally as the water table fluctuates in response to drought or rainfall. Therefore, the temperature and flow from water-table springs varies.

4.15.2 Important Surficial Aquifers

The Biscayne Aquifer is a surficial aquifer located in southeast Florida. It covers over 3,000 square miles (7,800 km), and is the most intensely used water source in Florida, supplying water to Dade County, Broward County, Palm Beach County and Monroe County. The aquifer lies close to the surface and is extremely vulnerable to pollutants that leach through the shallow limestone bedrock. In some areas, it has been contaminated by fuel spills, industrial discharge, landfills, and saltwater.

The Sand and Gravel Aquifer stretches 2,400 square miles (6,200 km) across the panhandle of Florida and is replenished with rainfall. Over the years water levels have dropped due to water-well use by pumping, and it has been contaminated by industrial waste and saltwater intrusion.

The Chokoloskee Aquifer is surficial aquifer that covers 3,000 square miles (7,800 km) in southwest Florida. It is recharged by rainfall. It is believed that artificial drainage canals have lowered water levels and increased saltwater intrusion.

CHAPTER 5

GROUNDWATER IN THE HYDROLOGICAL CYCLE

5.1 GROUNDWATER

Groundwater is the water present beneath Earth's surface in rock and soil pore spaces and in the fractures of rock formations. About 30 percent of all readily available freshwater in the world is groundwater. A unit of rock or an unconsolidated deposit is called an aquifer when it can yield a usable quantity of water. The depth at which soil pore spaces or fractures and voids in rock become completely saturated with water is called the water table. Groundwater is recharged from the surface; it may discharge from the surface naturally at springs and seeps, and can form oases or wetlands. Groundwater is also often withdrawn for agricultural, municipal, and industrial use by constructing and operating extraction wells. The study of the distribution and movement of groundwater is hydrogeology, also called groundwater hydrology.

Typically, groundwater is thought of as water flowing through shallow aquifers, but, in the technical sense, it can also contain soil moisture, permafrost (frozen soil), immobile water in very low permeability bedrock, and deep geothermal or oil formation water. Groundwater is hypothesized to provide lubrication that can possibly influence the movement of faults. It is likely that much of Earth's subsurface contains some water, which may be mixed with other fluids in some instances.

Groundwater is often cheaper, more convenient and less vulnerable to pollution than surface water. Therefore, it is commonly used for public water supplies. For example, groundwater provides the largest source of usable water storage in the United States, and California annually withdraws the largest amount of groundwater of all the states. Underground reservoirs contain far more water than the capacity of all surface reservoirs and lakes in the US, including the Great Lakes. Many municipal water supplies are derived solely from groundwater. Over 2 billion people rely on it as their primary water source worldwide.

Human use of groundwater causes environmental problems. For example, polluted groundwater is less visible and more difficult to clean up than pollution in rivers and lakes. Groundwater pollution most often results from improper disposal of wastes on land. Major sources include industrial and household chemicals and garbage landfills, excessive fertilizers and pesticides used in agriculture, industrial waste lagoons, tailings and process wastewater from mines, industrial fracking, oil field brine pits, leaking underground oil storage tanks and pipelines, sewage sludge and septic systems. Additionally, groundwater is susceptible to saltwater intrusion in coastal areas and can cause land subsidence when extracted unsustainably, leading to sinking cities (like Bangkok) and loss in elevation (such as the multiple meters lost in the Central Valley of California). These issues are made more complicated by sea level rise and other effects of climate change, particularly those on the water cycle. Earth's axial tilt has shifted 31 inches because of human groundwater pumping.

5.1.1 Definition

Groundwater is fresh water located in the subsurface pore space of soil and rocks. It is also water that is flowing within aquifers below the water table. Sometimes it is useful to make a distinction between groundwater that is closely associated with surface water, and deep groundwater in an aquifer (called "fossil water" if it infiltrated into the ground millennia ago).

5.1.2 Role in the Water Cycle

Groundwater can be thought of in the same terms as surface water: inputs, outputs and storage. The natural input to groundwater is seepage from surface water. The natural outputs from groundwater are springs and seepage to the oceans. Due to its slow rate of turnover, groundwater storage is generally much larger (in volume) compared to inputs than it is for surface water. This difference makes it easy for humans to use groundwater unsustainable for a long time without severe consequences. Nevertheless, over the long term the average rate of seepage above a groundwater source is the upper bound for average consumption of water from that source.

Groundwater is naturally replenished by surface water from precipitation, streams, and rivers when this recharge reaches the water table.

Groundwater can be a long-term 'reservoir' of the natural water cycle (with residence times from days to millennia), as opposed to short-term water reservoirs like the atmosphere and fresh surface water (which have residence times from minutes to years). Deep groundwater (which is quite distant from the surface recharge) can take a very long time to complete its natural cycle.

The Great Artesian Basin in central and eastern Australia is one of the largest confined aquifer systems in the world, extending for almost 2 million km. By analysing the trace elements in water sourced from deep underground, hydrogeologists have been abe to determine that water extracted from these aquifers can be more than 1 million years old.

By comparing the age of groundwater obtained from different parts of the Great Artesian Basin, hydrogeologists have found it increases in age across the basin. Where water recharges the aquifers along the Eastern Divide, ages are young. As groundwater flows westward across the continent, it increases in age, with the oldest groundwater occurring in the western parts. This means that in order to have travelled almost 1000 km from the source of recharge in 1 million years, the groundwater flowing through the Great Artesian Basin travels at an average rate of about 1 metre per year.

Groundwater Recharge

Groundwater recharge or deep drainage or deep percolation is a hydrologic process, where water moves downward from surface water to groundwater. Recharge is the primary method through which water enters an aquifer. This process usually occurs in the vadose zone below plant roots and is often expressed as a flux to the water table surface. Groundwater recharge also encompasses water moving away from the water table farther into the saturated zone. Recharge occurs both naturally (through the water cycle) and through anthropogenic processes (i.e., "artificial groundwater recharge"), where rainwater and or reclaimed water is routed to the subsurface.

The most common methods to estimate recharge rates are: chloride mass balance (CMB); soil physics methods; environmental and isotopic tracers; groundwater-level fluctuation methods; water balance (WB) methods (including groundwater models (GMs)); and the estimation of baseflow (BF) to rivers.

5.1.3 Characteristics

Fig. 5.1: The Entire Surface Water Fflow of the Alapaha River near Jennings, Florida

Temperature

The high specific heat capacity of water and the insulating effect of soil and rock can mitigate the effects of climate and maintain groundwater at a relatively steady temperature. In some places where groundwater temperatures are maintained by this effect at about 10°C (50°F), groundwater can be used for controlling the temperature inside structures at the surface. For example, during hot weather relatively cool groundwater can be pumped through radiators in a home and then returned to the ground in another well. During cold seasons, because it is relatively warm, the water can be used in the same way as a source of heat for heat pumps that is much more efficient than using air.

Availability

Groundwater makes up about thirty percent of the world's fresh water supply, which is about 0.76% of the entire world's water, including oceans and permanent ice. About 99% of the world's liquid fresh water is groundwater. Global groundwater storage is roughly equal to the total amount of freshwater stored in the snow and ice pack, including the north and south poles. This makes it an important resource that can act as a natural storage that can buffer against shortages of surface water, as in during times of drought.

The volume of groundwater in an aquifer can be estimated by measuring water levels in local wells and by examining geologic records from well-drilling to determine the extent, depth and thickness of water-bearing sediments and rocks. Before an investment is made in production wells, test wells may be drilled to measure the depths at which water is encountered and collect samples of soils, rock and water for laboratory analyses. Pumping tests can be performed in test wells to determine flow characteristics of the aquifer.

The characteristics of aquifers vary with the geology and structure of the substrate and topography in which they occur. In general, the more productive aquifers occur in sedimentary geologic formations.

By comparison, weathered and fractured crystalline rocks yield smaller quantities of groundwater in many environments. Unconsolidated to poorly cemented alluvial materials that have accumulated as valley-filling sediments in major river valleys and geologically subsiding structural basins are included among the most productive sources of groundwater.

Fluid flows can be altered in different lithological settings by brittle deformation of rocks in fault zones; the mechanisms by which this occurs are the subject of fault zone hydrogeology.

5.1.4 Uses by Humans

Reliance on groundwater will only increase, mainly due to growing water demand by all sectors combined with increasing variation in rainfall patterns.

Quantities

Groundwater is the most accessed source of freshwater around the world, including as drinking water, irrigation, and manufacturing. Groundwater accounts for about half of the world's drinking water, 40% of its irrigation water, and a third of water for industrial purposes.

Another estimate stated that globally groundwater accounts for about one third of all water withdrawals, and surface water for the other two thirds. Groundwater provides drinking water to at least 50% of the global population. About 2.5 billion people depend solely on groundwater resources to satisfy their basic daily water needs. A similar estimate was published in 2021 which stated that "groundwater is estimated to supply between a quarter and a third of the world's annual freshwater withdrawals to meet agricultural, industrial and domestic demands."

Global freshwater withdrawal was probably around 600 km^3 per year in 1900 and increased to 3,880 km^3 per year in 2017. The rate of increase was especially high (around 3% per year) during the period 1950–1980, partly due to a higher population growth rate, and partly to rapidly increasing groundwater development, particularly for irrigation. The rate of increase is (as per 2022) approximately 1% per year, in tune with the current population growth rate.

Global groundwater depletion has been calculated to be between 100 and 300 km per year. This depletion is mainly caused by "expansion of irrigated agriculture in drylands". The Asia-Pacific region is the largest groundwater abstractor in the world, containing seven out of the ten countries that extract most groundwater (Bangladesh, China, India, Indonesia, Iran, Pakistan and Turkey). These countries alone account for roughly 60% of the world's total groundwater withdrawal.

Drinking Water Quality Aspects

Groundwater may or may not be a safe water source. In fact, there is considerable uncertainty with groundwater in different hydrogeologic contexts: the widespread presence of contaminants such as arsenic, fluoride and salinity can reduce the suitability of groundwater as a drinking water source. Arsenic and fluoride have been considered as priority contaminants at a global level, although priority chemicals will vary by country.

There is a lot of heterogeneity of hydrogeologic properties. For this reason, salinity of groundwater is often highly variable over space. This contributes to highly variable groundwater security risks even within a specific region. Salinity in groundwater makes the water unpalatable and unusable and often occurs in coastal areas, for example in Bangladesh and East and West Africa.

Water Supply for Municipal and Industrial Uses

Municipal and industrial water supplies are provided through large wells. Multiple wells for one water supply source are termed "wellfields", which may withdraw water from confined or unconfined aquifers. Using groundwater from deep, confined aquifers provides more protection from surface water contamination. Some wells, termed "collector wells", are specifically designed to induce infiltration of surface (usually river) water.

Aquifers that provide sustainable fresh groundwater to urban areas and for agricultural irrigation are typically close to the ground surface (within a couple of hundred metres) and have some recharge by fresh water. This recharge is typically from rivers or meteoric water (precipitation) that percolates into the aquifer through overlying unsaturated materials.

Irrigation

In general, the irrigation of 20% of farming land (with various types of water sources) accounts for the production of 40% of food production. Irrigation techniques across the globe includes canals redirecting surface water, groundwater pumping, and diverting water from dams. Aquifers are critically important in agriculture. Deep aquifers in arid areas have long been water sources for irrigation. A majority of extracted groundwater, 70%, is used for agricultural purposes.

In India, 65% of the irrigation is from groundwater and about 90% of extracted groundwater is used for irrigation.

Occasionally, sedimentary or "fossil" aquifers are used to provide irrigation and drinking water to urban areas. In Libya, for example, Muammar Gaddafi's Great Manmade River project has pumped large amounts of groundwater from aquifers beneath the Sahara to populous areas near the coast. Though this has saved Libya money over the alternative, desalination, the aquifers are likely to run dry in 60 to 100 years. Aquifer depletion has been cited as one of the causes of the food price rises of 2011.

In Developing Countries

Groundwater provides critical freshwater supply, particularly in dry regions where surface water availability is limited. Globally, more than one-third of the water used originates from underground. In the mid-latitude arid and semi-arid regions lacking sufficient surface water supply from rivers and reservoirs, groundwater is critical for sustaining global ecology and meeting societal needs of drinking water and food production. The demand for groundwater is rapidly increasing with population growth, while climate change is imposing additional stress on water resources and raising the probability of severe drought occurrence.

The anthropogenic effects on groundwater resources are mainly due to groundwater pumping and the indirect effects of irrigation and land use changes.

Groundwater plays a central role in sustaining water supplies and livelihoods in sub-Saharan Africa. In some cases, groundwater is an additional water source that was not used previously.

Reliance on groundwater is increasing in Sub-Saharan Africa as development programs work towards improving water access and strengthening resilience to climate change. In lower-income areas, groundwater supplies are typically installed without water quality treatment infrastructure or services. This practice is underpinned by an assumption that untreated groundwater is typically suitable for drinking due to the relative microbiological safety of groundwater compared to surface water; however, chemistry risks are largely disregarded. Chemical contaminants occur widely in groundwaters that are used for drinking but are not regularly monitored. Example priority parameters are fluoride, arsenic, nitrate, or salinity.

5.1.5 Challenges

First, flood mitigation schemes, intended to protect infrastructure built on floodplains, have had the unintended consequence of reducing aquifer recharge associated with natural flooding. Second, prolonged depletion of groundwater in extensive aquifers can result in land subsidence, with associated infrastructure damage – as well as, third, saline intrusion. Fourth, draining acid sulphate soils, often found in low-lying coastal plains, can result in acidification and pollution of formerly freshwater and estuarine streams.

Overdraft

Groundwater is a highly useful and often abundant resource. Most land areas on Earth have some form of aquifer underlying them, sometimes at significant depths. In some cases, these aquifers are rapidly being depleted by the human population. Such over-use, over-abstraction or overdraft can cause major problems to human users and to the environment. The most evident problem (as far as human groundwater use is concerned) is a lowering of the water table beyond the reach of existing wells. As a consequence, wells must be drilled deeper to reach the groundwater; in some places (e.g., California, Texas, and India) the water table has dropped hundreds of feet because of extensive well pumping. The GRACE satellites have collected data that demonstrates 21 of Earth's 37 major aquifers are undergoing depletion. In the Punjab region of India, for example, groundwater levels have dropped 10 meters since 1979, and the rate of depletion is accelerating. A lowered water table may, in turn, cause other problems such as groundwater-related subsidence and saltwater intrusion.

Another cause for concern is that groundwater drawdown from over-allocated aquifers has the potential to cause severe damage to both terrestrial and aquatic ecosystems – in some cases very conspicuously but in others quite imperceptibly because of the extended period over which the damage occurs. The importance of groundwater to ecosystems is often overlooked, even by freshwater biologists and ecologists. Groundwaters sustain rivers, wetlands, and lakes, as well as subterranean ecosystems within karst or alluvial aquifers.

Not all ecosystems need groundwater, of course. Some terrestrial ecosystems – for example, those of the open deserts and similar arid environments – exist on irregular rainfall and the moisture it delivers to the soil, supplemented by moisture in the air. While there are other terrestrial ecosystems in more hospitable environments where groundwater plays no central role, groundwater is in fact fundamental to many of the world's major ecosystems. Water flows between groundwaters and surface waters. Most rivers, lakes, and wetlands are fed by, and (at other places or times) feed groundwater, to varying degrees. Groundwater feeds soil moisture through percolation, and many terrestrial vegetation communities depend directly on either groundwater or the percolated soil moisture above the aquifer for at least part of each year. Hyporheic zones (the mixing zone of streamwater and groundwater) and riparian zones are examples of ecotones largely or totally dependent on groundwater.

A 2021 study found that of ~39 million investigated groundwater wells 6-20% are at high risk of running dry if local groundwater levels decline by a few meters, or – as with many areas and possibly more than half of major aquifers – continue to decline.

Fresh-water aquifers, especially those with limited recharge by snow or rain, also known as meteoric water, can be over-exploited and depending on the local hydrogeology, may draw in non-potable water or saltwater intrusion from hydraulically connected aquifers or surface water bodies. This can be a serious problem, especially in coastal areas and other areas where aquifer pumping is excessive.

Subsidence

Subsidence occurs when too much water is pumped out from underground, deflating the space below the above-surface, and thus causing the ground to collapse. The result can look like craters on plots of land. This occurs because, in its natural equilibrium state, the hydraulic pressure of groundwater in the pore spaces of the aquifer and the aquitard supports some of the weight of the overlying sediments. When groundwater is removed from aquifers by excessive pumping, pore pressures in the aquifer drop and compression of the aquifer may occur. This compression may be partially recoverable if pressures rebound, but much of it is not. When the aquifer gets compressed, it may cause land subsidence, a drop in the ground surface.

In unconsolidated aquifers, groundwater is produced from pore spaces between particles of gravel, sand, and silt. If the aquifer is confined by low-permeability layers, the reduced water pressure in the sand and gravel causes slow drainage of water from the adjoining confining layers. If these confining layers are composed of compressible silt or clay, the loss of water to the aquifer reduces the water pressure in the confining layer, causing it to compress from the weight of overlying geologic materials. In severe cases, this compression can be observed on the ground surface as subsidence. Unfortunately, much of the subsidence from groundwater extraction is permanent (elastic rebound is small). Thus, the subsidence is not only permanent, but the compressed aquifer has a permanently reduced capacity to hold water.

The city of New Orleans, Louisiana is actually below sea level today, and its subsidence is partly caused by removal of groundwater from the various aquifer/aquitard systems beneath it. In the first half of the 20th century, the San Joaquin Valley experienced significant subsidence, in some places up to 8.5 metres (28 feet) due to groundwater removal. Cities on river deltas, including Venice in Italy, and Bangkok in Thailand, have experienced surface subsidence; Mexico City, built on a former lake bed, has experienced rates of subsidence of up to 40 cm (1'3") per year.

For coastal cities, subsidence can increase the risk of other environmental issues, such as sea level rise. For example, Bangkok is expected to have 5.138 million people exposed to coastal flooding by 2070 because of these combining factors.

Groundwater becoming Saline due to Evaporation

If the surface water source is also subject to substantial evaporation, a groundwater source may become saline. This situation can occur naturally under endorheic bodies of water, or artificially under irrigated farmland. In coastal areas, human use of a groundwater source may cause the direction of seepage to ocean to reverse which can also cause soil salinization.

As water moves through the landscape, it collects soluble salts, mainly sodium chloride. Where such water enters the atmosphere through evapotranspiration, these salts are left behind. In irrigation districts, poor drainage of soils and surface aquifers can result in water tables' coming to the surface in low-lying areas. Major land degradation problems of soil salinity and waterlogging result, combined with increasing levels of salt in surface waters. As a consequence, major damage has occurred to local economies and environments.

Aquifers in surface irrigated areas in semi-arid zones with reuse of the unavoidable irrigation water losses percolating down into the underground by supplemental irrigation from wells run the risk of salination.

Surface irrigation water normally contains salts in the order of 0.5 g/L or more and the annual irrigation requirement is in the order of 10,000 m/ha or more so the annual import of salt is in the order of 5,000 kg/ha or more.

Under the influence of continuous evaporation, the salt concentration of the aquifer water may increase continually and eventually cause an environmental problem.

For salinity control in such a case, annually an amount of drainage water is to be discharged from the aquifer by means of a subsurface drainage system and disposed of through a safe outlet. The drainage system may be horizontal (i.e., using pipes, tile drains or ditches) or vertical (drainage by wells). To estimate the drainage requirement, the use of a groundwater model with an agro-hydro-salinity component may be instrumental, e.g., SahysMod.

Seawater Intrusion

Aquifers near the coast have a lens of freshwater near the surface and denser seawater under freshwater. Seawater penetrates the aquifer diffusing in from the ocean and is denser than freshwater. For porous (i.e., sandy) aquifers near the coast, the thickness of freshwater atop saltwater is about 12 metres (40 ft) for every 0.3 m (1 ft) of freshwater head above sea level. This relationship is called the Ghyben-Herzberg equation. If too much groundwater is pumped near the coast, salt-water may intrude into freshwater aquifers causing contamination of potable freshwater supplies. Many coastal aquifers, such as the Biscayne Aquifer near Miami and the New Jersey Coastal Plain aquifer, have problems with saltwater intrusion as a result of overpumping and sea level rise.

Seawater intrusion is the flow or presence of seawater into coastal aquifers; it is a case of saltwater intrusion. It is a natural phenomenon but can also be caused or worsened by anthropogenic factors, such as sea level rise due to climate change. In the case of homogeneous aquifers, seawater intrusion forms a saline wedge below a transition zone to fresh groundwater, flowing seaward on the top. These changes can have other effects on the land above the groundwater. For example, coastal groundwater in California would rise in many aquifers, increasing risks of flooding and runoff challenges.

Sea level rise causes the mixing of sea water into the coastal groundwater, rendering it unusable once it amounts to more than 2-3% of the reservoir. Along an estimated 15% of the US coastline, the majority of local groundwater levels are already below the sea level.

Pollution

Groundwater pollution (also called groundwater contamination) occurs when pollutants are released to the ground and make their way into groundwater. This type of water pollution can also occur naturally due to the presence of a minor and unwanted constituent, contaminant, or impurity in the groundwater, in which case it is more likely referred to as contamination rather than pollution. Groundwater pollution can occur from on-site sanitation systems, landfill leachate, effluent from wastewater treatment plants, leaking sewers, petrol filling stations, hydraulic fracturing (fracking) or from over application of fertilizers in agriculture. Pollution (or contamination) can also occur from naturally occurring contaminants, such as arsenic or fluoride. Using polluted groundwater causes hazards to public health through poisoning or the spread of disease (water-borne diseases).

The pollutant often produces a contaminant plume within an aquifer. Movement of water and dispersion within the aquifer spreads the pollutant over a wider area. Its advancing boundary, often called a plume edge, can intersect with groundwater wells and surface water, such as seeps and springs, making the water supplies unsafe for humans and wildlife. The movement of the plume, called a plume front, may be analyzed through a hydrological transport model or groundwater model. Analysis of groundwater pollution may focus on soil characteristics and site geology, hydrogeology, hydrology, and the nature of the contaminants. Different mechanisms have influence on the transport of pollutants, e.g., diffusion, adsorption, precipitation, decay, in the groundwater.

Climate Change

The impacts of climate change on groundwater may be greatest through its indirect effects on irrigation water demand via increased evapotranspiration. There is an observed declined in groundwater storage in many parts of the world. This is due to more groundwater being used for irrigation activities in agriculture, particularly in drylands. Some of this increase in irrigation can be due to water scarcity issues made worse by effects of climate change on the water cycle. Direct redistribution of water by human activities amounting to ~24,000 km per year is about double the global groundwater recharge each year.

Climate change causes changes to the water cycle which in turn affect groundwater in several ways: There can be a decline in groundwater storage, and reduction in groundwater recharge and water quality deterioration due to extreme weather events. In the tropics intense precipitation and flooding events appear to lead to more groundwater recharge.

However, the exact impacts of climate change on groundwater are still under investigation. This is because scientific data derived from groundwater monitoring is still missing, such as changes in space and time, abstraction data and "numerical representations of groundwater recharge processes".

Effects of climate change could have different impacts on groundwater storage: The expected more intense (but fewer) major rainfall events could lead to increased groundwater recharge in many environments. But more intense drought periods could result in soil drying-out and compaction which would reduce infiltration to groundwater.

For the higher altitudes regions, the reduced duration and amount of snow may lead to reduced recharge of groundwater in spring. The impacts of receding alpine glaciers on groundwater systems are not well understood.

Global sea level rise due to climate change has induced seawater intrusion into coastal aquifers around the world, particularly in low-lying areas and small islands. However, groundwater abstraction is usually the main reason for seawater intrusion, rather than sea level rise. Seawater intrusion threatens coastal ecosystems and livelihood resilience. Bangladesh is a vulnerable country for this issue, and mangrove forest of Sundarbans is a vulnerable ecosystem.

Groundwater pollution may also increase indirectly due to climate change: More frequent and intense storms can pollute groundwater by mobilizing contaminants, for example fertilizers, wastewater or human excreta from pit latrines. Droughts reduce river dilution capacities and groundwater levels, increasing the risk of groundwater contamination. Aquifer systems that are vulnerable to climate change include the following examples (the first four are largely independent of human withdrawals, unlike examples 5 to 8 where the intensity of human groundwater withdrawals plays a key role in amplifying vulnerability to climate change):

1. low-relief coastal and deltaic aquifer systems,
2. aquifer systems in continental northern latitudes or alpine and polar regions,
3. aquifers in rapidly expanding low-income cities and large displaced and informal communities,
4. shallow alluvial aquifers underlying seasonal rivers in drylands,
5. intensively pumped aquifer systems for groundwater-fed irrigation in drylands,
6. intensively pumped aquifers for dryland cities,
7. intensively pumped coastal aquifers, and
8. low-storage/low-recharge aquifer systems in drylands.

Climate Change Adaptation

Using more groundwater, particularly in Sub-Saharan Africa, is seen as a method for climate change adaptation in the case that climate change causes more intense or frequent droughts. Groundwater-based adaptations to climate change exploit distributed groundwater storage and the capacity of aquifer systems to store seasonal or episodic water surpluses. They incur substantially lower evaporative losses than conventional infrastructure, such as surface dams. For example in tropical Africa, pumping water from groundwater storage can help to improve the climate resilience of water and food supplies.

Climate Change Mitigation

The development of geothermal energy, a sustainable energy source, plays an important role in reducing CO_2 emissions and thus mitigating climate change. Groundwater is an agent in the storage, movement, and extraction of geothermal energy. In pioneering nations, such as the Netherlands and Sweden, the ground/groundwater is increasingly seen as just one component (a seasonal source, sink or thermal 'buffer') in district heating and cooling networks. Deep aquifers can also be used for carbon capture and sequestration, the process of storing carbon to curb accumulation of carbon dioxide in the atmosphere.

5.1.6 Groundwater Governance

Groundwater governance processes enable groundwater management, planning and policy implementation. It takes place at multiple scales and geographic levels, including regional and transboundary scales. Groundwater management is action-oriented, focusing on practical implementation activities and day-to-day operations. Because groundwater is often perceived as a private resource (that is, closely connected to land ownership, and in some jurisdictions treated as privately owned), regulation and top-down governance and management are difficult. Governments need to fully assume their role as resource custodians in view of the common-good aspects of groundwater.

Domestic laws and regulations regulate access to groundwater as well as human activities that impact the quality of groundwater. Legal frameworks also need to include protection of discharge and recharge zones and of the area surrounding water supply wells, as well as sustainable yield norms and abstraction controls, and conjunctive use regulations. In some jurisdictions, groundwater is regulated in conjunction with surface water, including rivers.

5.2 GROUNDWATER OCCURRENCE IN THE UPPER CONTINENTAL CRUST

Groundwater, a vital component of the Earth's hydrological cycle, plays a crucial role in sustaining life and ecosystems. The upper continental crust serves as a reservoir for this essential resource, influencing its occurrence and distribution. Understanding the geological and hydrogeological factors that control groundwater in the upper continental crust is crucial for managing water resources sustainably. The occurrence of groundwater in the upper continental crust is a complex interplay of geological processes, hydrogeological characteristics, and human interactions. Understanding the dynamics of aquifer formation, groundwater movement, and the impacts of human activities is crucial for sustainable water resource management. As the global demand for freshwater continues to rise, safeguarding and replenishing groundwater resources in the upper continental crust becomes imperative for ensuring the well-being of ecosystems and human societies. Continued research, monitoring, and collaborative efforts are essential to address the challenges posed by changing climates, population growth, and anthropogenic pressures on groundwater systems.

5.2.1 Geological Processes and Formation of Aquifers

The upper continental crust, comprising the Earth's outermost layer, is a heterogeneous mix of rocks, sediments, and minerals. Geological processes such as weathering, erosion, and sedimentation contribute to the formation of aquifers—permeable geological formations capable of storing and transmitting groundwater. Porosity, a key property of aquifers, is influenced by factors such as rock type, mineral composition, and the degree of cementation. Sandstone and limestone, for example, often exhibit high porosity, providing ideal conditions for groundwater storage.

Fractures and faults also play a crucial role in the development of aquifers. The presence of fractures allows for increased permeability, facilitating the movement of water through the subsurface. Fault zones, where rocks have experienced significant deformation, can act as conduits for groundwater flow. Understanding the geological structure of the upper continental crust is essential for identifying potential aquifers and predicting groundwater occurrence.

5.2.2 Hydrogeological Characteristics

Hydrogeology, the study of the distribution and movement of groundwater, further elucidates the complexities of groundwater occurrence in the upper continental crust. Aquifers can be classified into different types based on their hydrogeological characteristics, such as confined and unconfined aquifers. Confined aquifers are characterized by an impermeable layer above and below the saturated zone, creating pressure within the aquifer. Unconfined aquifers, on the other hand, lack an impermeable layer above the saturated zone, allowing water to rise and fall based on natural variations in recharge and discharge. Recharge processes, including precipitation and infiltration, replenish groundwater in aquifers. The rate of recharge is influenced by factors such as climate, topography, and land use.

In arid regions, where precipitation is scarce, recharge rates may be lower, making groundwater resources more vulnerable to depletion.

5.2.3 Groundwater Movement and Flow Dynamics

Groundwater movement within the upper continental crust is governed by the principles of Darcy's law, which describes the relationship between hydraulic conductivity, gradient, and flow rate. Hydraulic conductivity, a measure of a material's ability to transmit water, varies among different geological formations. Coarse-grained materials like sand and gravel generally exhibit higher hydraulic conductivity, allowing for faster groundwater flow.

The structure of aquifers, including the presence of interconnected pore spaces, fractures, and conduits, influences the flow dynamics of groundwater. Understanding these factors is crucial for modeling groundwater flow patterns, predicting contaminant transport, and managing water resources effectively.

5.2.4 Human Interactions and Impacts

Human activities, such as urbanization, industrialization, and agriculture, significantly impact groundwater occurrence in the upper continental crust. Excessive groundwater extraction for agricultural irrigation, municipal water supply, and industrial processes can lead to aquifer depletion and land subsidence. Over-extraction may also induce saltwater intrusion in coastal aquifers, jeopardizing freshwater resources.

Pollution from anthropogenic sources, including industrial discharges and agricultural runoff, poses a threat to groundwater quality. Contaminants can infiltrate the subsurface and affect the quality of groundwater resources, necessitating effective management strategies to protect this vital water source.

5.3 SUBMARINE GROUNDWATER DISCHARGE

Submarine groundwater discharge (SGD) is a hydrological process which commonly occurs in coastal areas. It is described as submarine inflow of fresh-, and brackish groundwater from land into the sea. Submarine Groundwater Discharge is controlled by several forcing mechanisms, which cause a hydraulic gradient between land and sea. Considering the different regional settings the discharge occurs either as:

1. a focused flow along fractures in karst and rocky areas,
2. a dispersed flow in soft sediments, or
3. a recirculation of seawater within marine sediments.

Submarine Groundwater Discharge plays an important role in coastal biogeochemical processes and hydrological cycles such as the formation of offshore plankton blooms, hydrological cycles, and the release of nutrients, trace elements and gases. It affects coastal ecosystems and has been used as a freshwater resource by some local communities for millennia.

5.3.1 Methods

The first study about submarine groundwater discharge was done by Sonrel (1868), who speculated on the risk of submarine springs for sailors. However, until the mid-1990s, SGD remained rather unrecognized by the scientific community because it was hard to detect and measure the freshwater discharge.

The first elaborated method to study SGD was done by Moore (1996), who used radium-226 as a tracer for groundwater. Since then several methods and instruments have been developed to attempt to detect and quantify discharge rates.

Radium-226

The first study which detected and quantified submarine groundwater discharge on a regional basis was done by Moore (1996) in the South Atlantic Bight off South Carolina. He measured enhanced radium-226 concentrations within the water column near shore and up to about 100 kilometres (62 mi) from the shoreline.

Radium-226 is a decay product of thorium-230, which is produced within sediments and supplied by rivers.

However, these sources could not explain the high concentrations present in the study area. Moore (1996) hypothesized that submarine groundwater, enriched in radium-226, was responsible for the high concentrations. This hypothesis has been tested numerous times at sites around the world and confirmed at each site.

Seepage Meter

Lee (1977) designed a seepage meter, which consists of a chamber which is connected to a sampling port and a plastic bag. The chamber is inserted into the sediment and water discharging through the sediments is caught within the plastic bag. The change in volume of water which is caught in the plastic bag over time represents the freshwater flux.

Pore Water Profiles

According to Schlüter et al. (2004) chloride pore water profiles can be used to investigate submarine groundwater discharge. Chloride can be used as a conservative tracer, as it is enriched in seawater and depleted in groundwater. Three different shapes of chloride pore water profiles reflect three different transport modes within marine sediments. A chloride profile showing constant concentrations with depth indicates that no submarine groundwater is present. A chloride profile with a linear decline indicates a diffusive mixing between groundwater and seawater and a concave shaped chloride profile represents an advective admixture of submarine groundwater from below. Stable isotope ratios in the water molecule may also be used to trace and quantify the sources of a submarine groundwater discharge.

5.4 GLOBAL GROUNDWATER MATERIAL AND ELEMENTAL FLUXES

Groundwater, a hidden but critical component of the Earth's hydrological cycle, serves as a reservoir for various materials and elements essential for life and environmental processes. The study of global groundwater material and elemental fluxes involves unraveling the intricate dynamics of subsurface water movement, chemical interactions, and the transport of elements through aquifers. Global groundwater material and elemental fluxes represent a captivating realm of study that unveils the hidden dynamics of the Earth's subsurface. The movement and transport of materials and elements within aquifers influence water quality, ecosystem health, and human well-being. As we navigate an era of increasing water stress and environmental challenges, a deeper understanding of these fluxes is imperative for sustainable water resource management. By addressing the complexities associated with global groundwater material and elemental fluxes, we can contribute to safeguarding this invaluable resource for current and future generations.

5.4.1 Material Fluxes in Global Groundwater

Material fluxes in global groundwater encompass the movement and transport of various substances, including dissolved ions, organic matter, and gases. These materials interact with geological formations, influencing groundwater chemistry and participating in complex biogeochemical processes. Understanding material fluxes is essential for comprehending the global cycling of nutrients, contaminants, and essential elements within the subsurface environment.

One of the primary material fluxes in groundwater is the movement of dissolved ions. Groundwater often contains dissolved minerals such as calcium, magnesium, sodium, and bicarbonate. The dissolution of these minerals from geological formations contributes to the mineralization of groundwater, influencing its chemical composition. Additionally, the leaching of nutrients, such as nitrogen and phosphorus, from the soil into groundwater plays a vital role in sustaining terrestrial and aquatic ecosystems.

5.4.2 Elemental Fluxes in Global Groundwater

Elemental fluxes in groundwater involve the transport of individual chemical elements through aquifers, influencing the composition of subsurface waters. Key elements, including carbon, sulfur, nitrogen, and trace metals, undergo complex transformations as they interact with rocks, sediments, and microbial communities in the subsurface.

Carbon cycling in global groundwater is a significant elemental flux. Dissolved organic carbon (DOC) and inorganic carbon (DIC) are essential components of groundwater chemistry. Microbial activity in the subsurface, including processes such as respiration and organic matter decomposition, contributes to the cycling of carbon in aquifers. Understanding carbon fluxes in groundwater is crucial for comprehending the carbon balance of the Earth's subsurface and its implications for global carbon cycling.

Similarly, the flux of nitrogen in global groundwater plays a vital role in nutrient cycling. Nitrogen compounds, such as nitrate and ammonium, can be transported through aquifers, affecting both groundwater quality and ecosystem health. Excessive nitrogen in groundwater can lead to water contamination and pose risks to human health, making the study of nitrogen fluxes essential for effective water resource management.

Sulfur cycling in groundwater involves the transport of sulfate and other sulfur compounds through the subsurface. Microbial processes, including sulfate reduction and sulfur oxidation, influence the speciation and mobility of sulfur in aquifers. Understanding sulfur fluxes is critical for assessing the impact of groundwater on sulfur cycling in the Earth's crust and its implications for ecosystem dynamics.

Trace metals, including iron, manganese, and arsenic, also undergo elemental fluxes in global groundwater. The mobilization and transport of these elements are often influenced by redox conditions, mineral interactions, and anthropogenic activities. Elevated concentrations of certain trace metals in groundwater can pose risks to human health and ecosystems, underscoring the importance of studying their elemental fluxes in the subsurface.

5.4.3 Challenges and Implications

Despite the critical importance of understanding global groundwater material and elemental fluxes, several challenges exist in studying the subsurface environment. The complexity of subsurface processes, the heterogeneity of geological formations, and the limited accessibility to groundwater reservoirs pose challenges to accurate measurement and modeling of material and elemental fluxes.

Furthermore, anthropogenic activities, such as groundwater extraction, land use changes, and contamination from industrial discharges, can significantly alter material and elemental fluxes in aquifers. Over-extraction of groundwater, for instance, may induce changes in flow patterns and lead to the intrusion of contaminants, affecting both water quality and the sustainability of groundwater resources.

The implications of material and elemental fluxes in global groundwater are far-reaching. Changes in groundwater chemistry can influence the quality of drinking water supplies, impact the health of ecosystems, and pose challenges to agriculture. Understanding these fluxes is crucial for implementing effective water management strategies, mitigating contamination risks, and ensuring the sustainable use of groundwater resources.

5.4.4 Future Directions and Research Needs

Advancements in technology, including improved analytical techniques and modeling approaches, hold promise for enhancing our understanding of global groundwater material and elemental fluxes. High-resolution monitoring, isotopic tracing, and remote sensing technologies can provide valuable data for studying subsurface processes with greater accuracy.

Integrating interdisciplinary approaches that combine hydrogeology, geochemistry, microbiology, and modeling is essential for gaining a comprehensive understanding of groundwater dynamics. Collaborative efforts between researchers, policymakers, and water resource managers are crucial for addressing the complex challenges associated with material and elemental fluxes in global groundwater.

Research priorities should include investigating the impact of climate change on groundwater dynamics, understanding the role of microbial communities in subsurface processes, and assessing the long-term consequences of anthropogenic activities on groundwater quality. Additionally, there is a need for developing sustainable groundwater management practices that consider the complex interplay of material and elemental fluxes within the subsurface environment.

5.5 HUMAN INFLUENCE ON THE WATER CYCLE

The water cycle, a fundamental Earth process, plays a crucial role in sustaining life and ecosystems. Over the centuries, human activities have significantly altered the natural dynamics of the water cycle, leading to complex challenges that impact water availability, quality, and distribution. Human influence on the water cycle is a complex and multifaceted challenge that requires a comprehensive and coordinated response. Balancing the needs of growing populations, sustaining ecosystems, and adapting to a changing climate necessitates innovative approaches to water management. By understanding the interconnected nature of human activities and the water cycle, societies can develop sustainable practices that ensure the availability and quality of water resources for future generations. Through responsible stewardship and collective efforts, we can navigate the challenges posed by human influence on the water cycle and foster a more resilient and sustainable relationship with our planet's most precious resource.

5.5.1 Altering Landscapes through Urbanization and Agriculture

Urbanization and agricultural expansion have transformed landscapes, affecting the water cycle at various stages. The replacement of natural vegetation with impervious surfaces in urban areas disrupts infiltration processes, leading to increased surface runoff. This alteration contributes to heightened flood risks, diminished groundwater recharge, and altered streamflow patterns.

In agriculture, widespread irrigation practices significantly impact the water cycle. The withdrawal of large volumes of water for irrigation purposes can lead to groundwater depletion and changes in local hydrological patterns. Additionally, the application of fertilizers and pesticides in agriculture introduces contaminants into water bodies, affecting water quality and ecosystem health.

5.5.2 Deforestation and Land Use Changes

Deforestation, driven by logging, agriculture, and urban development, disrupts the water cycle by altering the balance between evaporation and transpiration. Trees play a crucial role in absorbing and releasing water through the process of transpiration, influencing regional precipitation patterns. The removal of forests can lead to reduced rainfall, increased soil erosion, and changes in local climates.

Land use changes, such as the conversion of natural ecosystems to urban or industrial areas, also contribute to alterations in the water cycle. Impervious surfaces prevent water from infiltrating the soil, leading to increased runoff and a decrease in groundwater recharge. These changes have cascading effects on the availability and quality of water resources.

5.5.3 Water Extraction and Dam Construction

Human activities, particularly water extraction for agriculture, industry, and municipal use, have a direct impact on the water cycle. Excessive groundwater pumping can lead to the depletion of aquifers, causing land subsidence and reduced baseflow in rivers. Unregulated extraction practices can exacerbate water scarcity in regions already facing water stress.

The construction of dams and reservoirs is another significant human-induced alteration to the water cycle. While dams serve essential functions such as water storage, flood control, and energy generation, they also disrupt natural river flow. Changes in sediment transport, altered downstream ecosystems, and impacts on aquatic habitats are some of the consequences associated with dam construction.

5.5.4 Climate Change and Its Impacts

Human-induced climate change is altering the global water cycle, contributing to extreme weather events, shifts in precipitation patterns, and changes in temperature. Rising global temperatures intensify evaporation rates, affecting the timing and intensity of rainfall. This has implications for drought frequency, water availability, and the sustainability of ecosystems.

Melting glaciers and changes in snowpack dynamics, attributed to climate change, impact the timing and magnitude of river flows. Glacial retreat reduces the availability of freshwater resources for downstream communities, particularly in regions dependent on glacier-fed rivers. These changes underscore the interconnectedness of human-induced climate change and alterations in the water cycle.

5.5.5 Pollution and Contamination

Human activities introduce pollutants into water bodies, compromising water quality and posing risks to both human health and ecosystems. Industrial discharges, agricultural runoff, and inadequate waste management contribute to the pollution of rivers, lakes, and groundwater. Contaminants such as heavy metals, nutrients, and organic pollutants disrupt aquatic ecosystems and affect the suitability of water for various uses.

The presence of contaminants in water sources necessitates advanced water treatment technologies, increasing the energy and resource requirements for supplying safe drinking water. Addressing pollution requires comprehensive approaches, including regulatory measures, sustainable agricultural practices, and public awareness campaigns to reduce the impact of human activities on water quality.

5.5.6 Over-extraction and Depletion of Aquifers

The increasing demand for freshwater resources has led to over-extraction of groundwater from aquifers worldwide. Excessive pumping for agriculture, industry, and domestic use can result in aquifer depletion, leading to land subsidence and reduced groundwater levels. In many regions, the rate of groundwater extraction exceeds the natural recharge rate, creating long-term sustainability challenges.

The depletion of aquifers has far-reaching consequences for ecosystems that depend on groundwater, such as wetlands and riparian habitats. It also jeopardizes the resilience of communities that rely on groundwater as a primary source of drinking water. Sustainable groundwater management practices, including aquifer recharge and the implementation of water conservation measures, are essential to mitigate the impacts of over-extraction.

5.5.7 Integrated Water Management Solutions

Addressing the challenges posed by human influence on the water cycle requires integrated and sustainable water management solutions. Policymakers, scientists, and communities must collaborate to develop strategies that balance the competing demands for water resources while preserving ecological integrity. Some key approaches include:

1. Water Conservation and Efficiency: Implementing water conservation measures and promoting efficient water use can reduce the overall demand for freshwater resources. Technologies such as drip irrigation, rainwater harvesting, and water-efficient appliances contribute to sustainable water management.
2. Restoration of Natural Ecosystems: Protecting and restoring natural ecosystems, including forests, wetlands, and watersheds, can enhance the resilience of the water cycle. Natural vegetation plays a critical role in regulating water flow, preventing soil erosion, and maintaining biodiversity.
3. Adaptive Infrastructure: Developing adaptive infrastructure that considers the dynamic nature of the water cycle is essential. This includes designing dams and reservoirs with ecological considerations, implementing groundwater recharge projects, and incorporating climate resilience into water management practices.
4. Policy and Regulation: Enacting and enforcing policies that promote sustainable water use, prevent pollution, and safeguard ecosystems is crucial. Integrated water resource management frameworks that consider the interconnectedness of surface water and groundwater are essential for effective governance.
5. Climate Mitigation and Adaptation: Addressing the root causes of climate change through mitigation efforts is essential to minimize its impact on the water cycle. Additionally, communities must develop adaptive strategies to cope with the changing climate, including altered precipitation patterns and more frequent extreme weather events.

5.6 GLOBAL GROUNDWATER RESOURCES

Groundwater, a hidden but vital component of the Earth's water cycle, plays a crucial role in sustaining ecosystems, providing drinking water, and supporting agriculture worldwide. As a vast and often overlooked resource, global groundwater reserves are critical for meeting the increasing demand for freshwater. Global groundwater resources, a linchpin of water security, face complex challenges driven by human activities, climate change, and inadequate governance. The imperative to address these challenges and ensure the sustainability of groundwater resources requires a multi-faceted approach. Sustainable management practices, international collaboration, and integrated water resource management are essential components of a resilient strategy to safeguard this invaluable resource for current and future generations. Through concerted efforts, responsible governance, and the adoption of innovative solutions, we can strike a balance between meeting the water needs of a growing global population and preserving the health and longevity of our planet's groundwater resources.

5.6.1 Importance of Global Groundwater Resources

1. Primary Source of Drinking Water: Groundwater serves as a primary source of drinking water for a significant portion of the global population. In many regions, especially arid and semi-arid areas, groundwater is a reliable and consistent supply, offering resilience against climatic variability and surface water shortages. Access to safe and sustainable groundwater is essential for ensuring global water security.
2. Support for Agriculture: Agriculture is a major consumer of freshwater, and globally, groundwater plays a crucial role in irrigation. Groundwater irrigation sustains crop production, enhances food security, and supports the livelihoods of millions of people. In regions where surface water availability is limited, such as parts of India and the Middle East, groundwater is a lifeline for agricultural activities.
3. Ecosystem Sustenance: Groundwater sustains diverse ecosystems, contributing to the health of wetlands, rivers, and lakes. It provides baseflow to rivers, maintaining streamflow during dry periods and supporting aquatic habitats. Groundwater discharge into coastal areas also influences estuarine ecosystems, playing a vital role in maintaining the balance of saltwater and freshwater.
4. Industrial and Commercial Uses: Many industries and commercial enterprises rely on groundwater for various processes, including manufacturing, mining, and energy production. Groundwater's reliability and consistent quality make it an attractive resource for these sectors. However, excessive groundwater extraction for industrial purposes can pose challenges to sustainability.

5.6.2 Challenges Facing Global Groundwater Resources

1. Over-Extraction and Depletion: One of the primary challenges facing global groundwater resources is over-extraction, wherein the rate of pumping exceeds the natural recharge rate. This unsustainable practice leads to groundwater depletion, lowered water tables, and increased risks of land subsidence. Over-extraction is particularly prevalent in regions facing water stress and rapid population growth.
2. Contamination and Pollution: Groundwater is susceptible to contamination from various sources, including industrial discharges, agricultural runoff, and inadequate waste disposal. Contaminants such as nitrates, pesticides, and heavy metals can infiltrate the subsurface, compromising the quality of groundwater. Contamination poses risks to human health, ecosystems, and the overall sustainability of groundwater resources.

3. Climate Change Impacts: Climate change introduces additional stressors to global groundwater resources. Altered precipitation patterns, rising temperatures, and changes in evaporation rates impact groundwater recharge and discharge. Changes in the frequency and intensity of extreme weather events, such as droughts and floods, further complicate the management of groundwater resources.
4. Subsidence and Land Degradation: Over-extraction of groundwater can lead to land subsidence, a phenomenon where the land surface sinks. This subsidence can result in infrastructure damage, increased flood risks, and changes in the topography of the affected areas. Additionally, the loss of groundwater can contribute to land degradation, affecting soil fertility and ecosystem health.
5. Lack of Monitoring and Data: In many regions, there is a lack of comprehensive monitoring and data collection systems for groundwater. Incomplete understanding of aquifer characteristics, groundwater flow dynamics, and water quality hinders effective management. Investing in robust monitoring networks is crucial for informed decision-making and sustainable groundwater management.

5.6.3 Sustainable Management Practices

1. Aquifer Recharge: Implementing artificial recharge techniques, such as spreading basins, injection wells, and recharge ponds, can enhance groundwater levels. By allowing excess surface water to percolate into the ground, aquifer recharge replenishes groundwater reserves and mitigates the impacts of over-extraction.
2. Water Use Efficiency and Conservation: Promoting water use efficiency and conservation practices is essential for sustainable groundwater management. This includes adopting precision irrigation methods, implementing water-saving technologies, and raising awareness about the importance of responsible water use in agriculture, industry, and households.
3. Integrated Water Resource Management: Integrating surface water and groundwater management is crucial for a holistic approach to water resource management. Recognizing the interconnectedness of these systems ensures that actions taken in one realm do not inadvertently harm the other. Integrated approaches consider the cumulative impact on the entire water cycle.
4. Protecting and Restoring Watersheds: Protecting natural ecosystems, such as forests and wetlands, contributes to groundwater sustainability. Vegetation plays a role in regulating water flow, preventing soil erosion, and enhancing

groundwater recharge. Restoring degraded watersheds helps maintain ecological balance and supports the health of groundwater resources.

5. Groundwater Pollution Prevention: Implementing measures to prevent groundwater pollution is critical. This includes proper disposal of industrial waste, implementing agricultural best practices to reduce runoff, and establishing regulatory frameworks to monitor and control contaminant sources. Public awareness campaigns can also contribute to pollution prevention.
6. Climate-Resilient Strategies: Developing climate-resilient strategies for groundwater management involves anticipating and adapting to the impacts of climate change. This may include adjusting extraction rates based on changing recharge patterns, implementing adaptive infrastructure, and incorporating climate considerations into long-term groundwater planning.

5.6.4 International Collaboration and Governance

Addressing the challenges and ensuring the sustainability of global groundwater resources requires international collaboration and effective governance. Key considerations include:

1. Transboundary Aquifers: Many aquifers extend across international borders, requiring collaborative management approaches. Establishing agreements and frameworks for sustainable use and equitable sharing of transboundary aquifers is crucial. The United Nations' International Shared Aquifer Resource Management (ISARM) initiative is an example of efforts to promote cooperation in this regard.
2. Data Sharing and Monitoring: International cooperation should involve the sharing of data and information related to groundwater resources. Collaborative monitoring programs, joint research initiatives, and the exchange of best practices contribute to a better understanding of global groundwater dynamics.
3. Policy Harmonization: Harmonizing policies and regulations related to groundwater management at the international level helps create a consistent framework. Aligning goals and strategies facilitates effective cooperation and ensures that efforts to protect and sustain groundwater are cohesive across borders.
4. Capacity Building: Building the capacity of nations to manage their groundwater resources is essential. This involves providing technical expertise, sharing knowledge and best practices, and supporting the development of local and national policies that prioritize sustainable groundwater management.

5.7 GLOBAL GROUNDWATER ABSTRACTION

Global groundwater abstraction, the extraction of water from underground aquifers, plays a pivotal role in meeting the growing demand for freshwater resources. As an essential component of the Earth's water cycle, groundwater serves as a primary source of drinking water, supports agriculture, and sustains ecosystems. However, the increasing rates of groundwater extraction worldwide have raised concerns about sustainability, environmental impacts, and the potential for over-exploitation.

Global groundwater abstraction is at a crossroads, facing both immense challenges and opportunities for sustainable management. As the demand for freshwater continues to rise, it is imperative to address the environmental, social, and economic implications of groundwater extraction. By embracing sustainable solutions, fostering international collaboration, and implementing robust governance frameworks, the world can strike a balance between meeting current water needs and ensuring the long-term health of global groundwater resources.

Responsible and equitable groundwater abstraction practices are essential for securing water availability for future generations and fostering a resilient and sustainable water future.

5.7.1 The Significance of Groundwater Abstraction

1. Primary Source of Drinking Water: Groundwater abstraction is a critical source of drinking water for millions of people globally. In both urban and rural areas, wells and boreholes tapping into aquifers provide a reliable and often high-quality water supply. In regions where surface water is scarce or unreliable, groundwater abstraction becomes indispensable for ensuring access to safe drinking water.
2. Agricultural Irrigation: Agriculture is a major consumer of freshwater, and groundwater abstraction plays a crucial role in sustaining irrigation practices. In many parts of the world, especially arid and semi-arid regions, groundwater supports crop production, enhances food security, and contributes to the livelihoods of farming communities. The ability to access groundwater for irrigation provides resilience against erratic rainfall patterns.
3. Industrial and Commercial Uses: Industries and commercial enterprises depend on groundwater for various processes, including manufacturing, mining, and energy production. The reliability and consistent quality of groundwater make it an attractive resource for these sectors. However, the increased demand from industrial activities also contributes to the overall stress on global groundwater resources.

4. Ecosystem Sustenance: Groundwater abstraction affects the flow regimes of rivers and sustains ecosystems by contributing to baseflow. This baseflow is crucial for maintaining streamflow during dry periods, supporting aquatic habitats, and preserving the ecological health of rivers and lakes. Groundwater discharge into coastal areas influences estuarine ecosystems, playing a vital role in maintaining ecological balance.

5.7.2 Challenges Associated with Global Groundwater Abstraction

1. Over-Extraction and Depletion: Over-extraction of groundwater occurs when the rate of withdrawal exceeds the natural recharge rate of aquifers. This leads to groundwater depletion, a phenomenon where the water table drops, reducing the amount of water available for extraction. Over-extraction is a pervasive issue globally, driven by the increasing demand for water in agriculture, industry, and urban areas.
2. Land Subsidence: Over-extraction of groundwater can result in land subsidence, a gradual sinking of the Earth's surface. As aquifers are depleted, the voids created by the withdrawn water are often filled with air, causing the land above to settle. Land subsidence poses risks to infrastructure, including buildings, roads, and pipelines, and can result in increased flood vulnerability.
3. Saltwater Intrusion: In coastal areas, over-extraction of groundwater can lead to saltwater intrusion, where saline water infiltrates freshwater aquifers. This phenomenon is exacerbated by excessive pumping, reducing the freshwater pressure and allowing saltwater to migrate inland. Saltwater intrusion compromises the quality of groundwater and can render aquifers unusable for drinking water and agriculture.
4. Ecosystem Impact: The alteration of natural groundwater flow patterns can impact ecosystems dependent on groundwater discharge. Reductions in baseflow to rivers and streams can harm aquatic habitats, affecting fish migration, nutrient cycling, and overall biodiversity. Sensitive ecosystems such as wetlands are particularly vulnerable to changes in groundwater levels.
5. Contamination Risks: The intensive extraction of groundwater can increase the risk of contaminant intrusion into aquifers. Industrial discharges, agricultural runoff, and improper waste disposal practices contribute to groundwater pollution. Over-extraction may exacerbate these contamination risks by creating pathways for pollutants to enter the subsurface.
6. Social and Economic Disparities: Groundwater abstraction can lead to social and economic disparities, especially in regions with limited water resources. Large-scale groundwater extraction by industries or agricultural enterprises may adversely impact local communities, causing conflicts over

water access. Small-scale farmers and marginalized communities may face challenges in accessing and competing for groundwater resources.

5.7.3 Sustainable Solutions for Groundwater Abstraction

1. Managed Aquifer Recharge (MAR): Managed Aquifer Recharge involves intentionally directing water to infiltration areas to replenish aquifers. This can be achieved through techniques such as recharge basins, injection wells, and spreading ponds. MAR enhances groundwater levels, mitigates the impacts of over-extraction, and contributes to sustainable water management.
2. Water Use Efficiency and Conservation: Promoting water use efficiency and conservation practices is crucial for sustainable groundwater abstraction. Implementing technologies such as drip irrigation, adopting precision agriculture, and improving water-use practices in urban areas can reduce overall water demand, lessening the strain on groundwater resources.
3. Regulation and Governance: Effective regulation and governance are essential for managing groundwater abstraction sustainably. Establishing and enforcing limits on extraction rates, implementing groundwater monitoring systems, and developing comprehensive water management plans help prevent over-extraction and ensure responsible use of groundwater resources.
4. Community Engagement and Education: Engaging local communities in water management decisions and providing education about sustainable water practices foster a sense of ownership and responsibility. Community-led initiatives can contribute to the protection and conservation of groundwater resources, ensuring that solutions align with local needs.
5. Integration of Surface Water and Groundwater Management: Integrating the management of surface water and groundwater is essential for a holistic approach to water resource management. Recognizing the interconnections between these two systems helps avoid unintended consequences and ensures a balanced and sustainable use of both surface and subsurface water.
6. Research and Data Collection: Investing in research and data collection efforts is crucial for understanding aquifer dynamics, groundwater recharge mechanisms, and the impacts of extraction. Improved knowledge about local hydrogeological conditions allows for more informed decision-making and the development of targeted strategies for sustainable groundwater abstraction.
7. Climate-Resilient Strategies: Developing climate-resilient strategies acknowledges the impact of climate change on groundwater resources. This includes adapting to changing precipitation patterns, considering

the potential for increased variability, and incorporating climate considerations into long-term water management planning.

5.7.4 International Collaboration and Policy Frameworks

1. Global Collaboration on Transboundary Aquifers: Many aquifers extend across international borders, necessitating collaborative efforts to manage them sustainably. The international community, through organizations such as the United Nations, can facilitate dialogue, cooperation, and the development of frameworks for responsible transboundary aquifer management.
2. Development of Comprehensive Policies: Governments and policymakers need to develop and implement comprehensive policies that address the challenges associated with groundwater abstraction. These policies should include regulations on extraction rates, pollution prevention measures, and incentives for sustainable practices. Integrating groundwater considerations into broader water management frameworks is crucial.
3. Incentives for Sustainable Practices: Governments can provide incentives for sustainable groundwater abstraction practices. This may include subsidies for water-efficient technologies, tax incentives for industries adopting responsible water use, and financial support for communities implementing groundwater recharge projects.
4. Capacity Building: Building the capacity of communities, water management authorities, and institutions involved in groundwater management is vital. This involves providing training, technical support, and resources to enhance understanding, monitoring capabilities, and effective governance of groundwater resources.

5.8 GLOBAL GROUNDWATER DEPLETION AND SEA LEVEL RISE

5.8.1 Global Groundwater Depletion

Groundwater, a crucial component of the Earth's freshwater resources, is under increasing stress due to global groundwater depletion. As populations grow, agricultural demands increase, and industrial activities expand, the unsustainable extraction of groundwater has become a widespread issue, posing significant challenges to water security and ecosystem health. Global groundwater depletion poses significant challenges to water security, ecosystems, and communities. Addressing this issue requires a comprehensive and collaborative approach that combines sustainable practices, effective governance, and community engagement. By adopting managed aquifer recharge, promoting water use efficiency, enforcing regulations, and integrating surface water and groundwater management, societies can work towards a more sustainable and responsible use of this essential resource.

The consequences of groundwater depletion are far-reaching, impacting everything from agricultural productivity to ecosystem health, and it is crucial that we act collectively to ensure the resilience and longevity of global groundwater resources for current and future generations.

Causes of Global Groundwater Depletion

1. Agricultural Demand: Agricultural irrigation is a primary driver of global groundwater depletion. In many regions, farmers heavily rely on groundwater to irrigate crops, especially in arid and semi-arid areas where surface water availability is limited. The demand for water to sustain agriculture often exceeds natural recharge rates, leading to excessive groundwater extraction.
2. Population Growth and Urbanization: Rapid population growth and urbanization contribute to increased water demand for municipal use. As cities expand and populations rise, the demand for groundwater for drinking water, industrial processes, and municipal services escalates. Urban areas often rely on wells and boreholes, depleting local aquifers and impacting the surrounding ecosystems.
3. Industrial Activities: Industries extract groundwater for various processes, including manufacturing, energy production, and mining. The reliability and consistent quality of groundwater make it an attractive resource for industrial operations. However, the intensive extraction for industrial purposes can lead to groundwater depletion and environmental consequences.
4. Lack of Regulation and Governance: In many regions, inadequate regulation and governance contribute to over-extraction of groundwater. The absence of effective management practices, limited monitoring, and a lack of enforcement mechanisms allow for unsustainable pumping rates. The absence of clear policies on groundwater use exacerbates the problem.
5. Climate Change: Climate change introduces additional challenges to groundwater resources. Altered precipitation patterns, increased temperatures, and changes in evaporation rates impact groundwater recharge rates. Extended periods of drought may further intensify the demand for groundwater, as surface water sources become less reliable.

Consequences of Global Groundwater Depletion

1. Land Subsidence: One of the most visible consequences of groundwater depletion is land subsidence. As aquifers are over-pumped, the voids left by the extracted water are not immediately filled. This causes the land above to sink, leading to subsidence. Subsidence poses risks to infrastructure, increases flood vulnerability, and can result in long-term changes to the landscape.

2. Saltwater Intrusion: Coastal areas are particularly susceptible to saltwater intrusion when groundwater is over-extracted. As freshwater is withdrawn from coastal aquifers, the reduced pressure allows saltwater to infiltrate, contaminating the freshwater resources. Saltwater intrusion can render aquifers unsuitable for drinking water and agriculture.
3. Ecological Impacts: Groundwater depletion affects ecosystems that depend on groundwater discharge. Reduced baseflow to rivers and streams during dry periods can harm aquatic habitats, impacting fish migration, nutrient cycling, and overall biodiversity. Wetlands, which rely on groundwater, may shrink or disappear, leading to the loss of valuable ecosystems.
4. Threats to Agriculture: Groundwater depletion poses a direct threat to agriculture, as many regions heavily rely on groundwater for irrigation. Reduced availability of groundwater can lead to declining water tables, increased pumping costs for farmers, and, in severe cases, the need to transition to less water-intensive crops. This has implications for food security and rural livelihoods.
5. Impact on Drinking Water Supplies: Over extraction of groundwater can lead to diminished drinking water supplies. As aquifers are depleted, wells and boreholes may run dry, especially in regions where groundwater is a primary source of drinking water. This jeopardizes the water security of communities and necessitates the development of alternative water sources.
6. Economic Consequences: The economic consequences of groundwater depletion are widespread. Agriculture, which relies heavily on groundwater, may face decreased productivity, increased costs, and economic losses for farmers. Industries dependent on groundwater may experience disruptions, and communities relying on groundwater for drinking water may face higher expenses for alternative water sources.

5.8.2 Sea Level Rise

Between 1901 and 2018, the average global sea level rose by 15–25 cm (6–10 in), or an average of 1–2 mm per year. This rate accelerated to 4.62 mm/yr for the decade 2013–2022. Climate change due to human activities is the main cause. Between 1993 and 2018, thermal expansion of water accounted for 42% of sea level rise. Melting temperate glaciers accounted for 21%, with Greenland accounting for 15% and Antarctica 8%. Sea level rise lags changes in the Earth's temperature. So sea level rise will continue to accelerate between now and 2050 in response to warming that is already happening. What happens after that depends on human greenhouse gas emissions. Sea level rise may slow down between 2050 and 2100 if there are deep cuts in emissions. It could then reach a little over 30 cm (1 ft) from now by 2100. With high emissions it may accelerate. It could rise by 1 m (3+1/2 ft) or even 2 m

(6+1/2 ft) by then. In the long run, sea level rise would amount to 2–3 m (7–10 ft) over the next 2000 years if warming amounts to 1.5°C (2.7 °F). It would be 19–22 metres (62–72 ft) if warming peaks at 5°C (9.0°F).

Rising seas ultimately impact every coastal and island population on Earth. This can be through flooding, higher storm surges, king tides, and tsunamis. These have many follow-on effects. They lead to loss of coastal ecosystems like mangroves. Crop production falls because of salinization of irrigation water and damage to ports disrupts sea trade. The sea level rise projected by 2050 will expose places currently inhabited by tens of millions of people to annual flooding. Without a sharp reduction in greenhouse gas emissions, this may increase to hundreds of millions in the latter decades of the century. Areas not directly exposed to rising sea levels could be affected by large scale migrations and economic disruption.

At the same time, local factors like tidal range or land subsidence, as well as the varying resilience and adaptive capacity of individual ecosystems, sectors, and countries will greatly affect the severity of impacts. For instance, sea level rise in the United States (particularly along the US East Coast) is already higher than the global average, and is expected to be 2 to 3 times greater than the global average by the end of the century. Yet, of the 20 countries with the greatest exposure to sea level rise, 12 are in Asia. Bangladesh, China, India, Indonesia, Japan, the Philippines, Thailand and Vietnam collectively account for 70% of the global population exposed to sea level rise and land subsidence. Finally, the greatest near-term impact on human populations will occur in the low-lying Caribbean and Pacific islands—many of those would be rendered uninhabitable by sea level rise later this century.

Societies can adapt to sea level rise in three ways: by managed retreat, by accommodating coastal change, or by protecting against sea level rise through hard-construction practices like seawalls or soft approaches such as dune rehabilitation and beach nourishment. Sometimes these adaptation strategies go hand in hand; at other times choices must be made among different strategies. A managed retreat strategy is difficult if an area's population is quickly increasing: this is a particularly acute problem for Africa, where the population of low-lying coastal areas is projected to increase by around 100 million people within the next 40 years. Poorer nations may also struggle to implement the same approaches to adapt to sea level rise as richer states, and sea level rise at some locations may be compounded by other environmental issues, such as subsidence in so-called sinking cities. Coastal ecosystems typically adapt to rising sea levels by moving inland; but may not always be able to do so, due to natural or artificial barriers.

Observations

Between 1901 and 2018, the global mean sea level rose by about 20 cm (or 8 inches). More precise data gathered from satellite radar measurements found a rise of 7.5 cm (3 in) from 1993 to 2017 (average of 2.9 mm/yr). This accelerated to 4.62 mm/yr for 2013–2022.

Regional Variations

Sea level rise is not uniform around the globe. Some land masses are moving up or down as a consequence of subsidence (land sinking or settling) or post-glacial rebound (land rising as melting ice reduces weight). Therefore, local relative sea level rise may be higher or lower than the global average. Changing ice masses also affect the distribution of sea water around the globe through gravity.

When a glacier or ice sheet melts, it loses mass. This reduces its gravitational pull. In some places near current and former glaciers and ice sheets, this has caused water levels to drop. At the same time water levels will increase more than average further away from the ice sheet. Thus ice loss in Greenland affects regional sea level differently than the equivalent loss in Antarctica. On the other hand, the Atlantic is warming at a faster pace than the Pacific. This has consequences for Europe and the U.S. East Coast. The East Coast sea level is rising at 3–4 times the global average. Scientists have linked extreme regional sea level rise on the US Northeast Coast to the downturn of the Atlantic meridional overturning circulation (AMOC).

Many ports, urban conglomerations, and agricultural regions stand on river deltas. Here land subsidence contributes to much higher relative sea level rise. Unsustainable extraction of groundwater and oil and gas is one cause. Levees and other flood management practices are another. They prevent sediments from accumulating. These would otherwise compensate for the natural settling of deltaic soils. Estimates for total human-caused subsidence in the Rhine-Meuse-Scheldt delta (Netherlands) are 3–4 m (10–13 ft), over 3 m (10 ft) in urban areas of the Mississippi River Delta (New Orleans), and over 9 m (30 ft) in the Sacramento–San Joaquin River Delta. On the other hand, relative sea level around the Hudson Bay in Canada and the northern Baltic is falling due to post-glacial isostatic rebound.

Projections

There are two complementary ways to model sea level rise (SLR) and project the future. The first uses process-based modeling. This combines all relevant and well-understood physical processes in a global physical model. This approach calculates the contributions of ice sheets with an ice-sheet model and computes rising sea temperature and expansion with a general circulation model. The processes are not fully understood. But this approach can predict non-linearities and long delays in the response, which studies of the recent past will miss.

The other approach employs semi-empirical techniques. These use historical geological data to determine likely sea level responses to a warming world, and some basic physical modeling. These semi-empirical sea level models rely on statistical techniques. They use relationships between observed past contributions to global mean sea level and temperature. Scientists developed this type of modeling because most physical models in previous Intergovernmental Panel on Climate Change (IPCC) literature assessments had underestimated the amount of sea level rise compared to 20th century observations.

Projections for the 21st Century

Intergovernmental Panel on Climate Change is the largest and most influential scientific organization on climate change, and since 1990, it provides several plausible scenarios of 21st century sea level rise in each of its major reports. The differences between scenarios are mainly due to uncertainty about future greenhouse gas emissions. These depend on future economic developments, and also future political action which is hard to predict. Each scenario provides an estimate for sea level rise as a range with a lower and upper limit to reflect the unknowns. The scenarios in the 2013-2014 Fifth Assessment Report (AR5) were called Representative Concentration Pathways, or RCPs and the scenarios in the IPCC Sixth Assessment Report (AR6) are known as Shared Socioeconomic Pathways, or SSPs. A large difference between the two was the addition of SSP1-1.9 to AR6, which represents meeting the best Paris climate agreement goal of 1.5 °C (2.7 °F). In that case, the likely range of sea level rise by 2100 is 28–55 cm (11–21+1/2 in).

The lowest scenario in AR5, RCP2.6, would see greenhouse gas emissions low enough to meet the goal of limiting warming by 2100 to 2 °C. It shows sea level rise in 2100 of about 44 cm with a range of 28–61 cm. The "moderate" scenario, where CO_2 emissions take a decade or two to peak and its atmospheric concentration does not plateau until 2070s is called RCP 4.5. Its likely range of sea level rise is 36–71 cm (14–28 in). The highest scenario in RCP8.5 pathway sea level would rise between 52 and 98 cm (20+1/2 and 38+1/2 in). AR6 had equivalents for both scenarios, but it estimated larger sea level rise under both. In AR6, the SSP1-2.6 pathway results in a range of 32–62 cm (12+1/2–24+1/2 in) by 2100. The "moderate" SSP2-4.5 results in a 44–76 cm (17+1/2–30 in) range by 2100 and SSP5-8.5 led to 65–101 cm (25+1/2–40 in).

Further, AR5 was criticized by multiple researchers for excluding detailed estimates the impact of "low-confidence" processes like marine ice sheet and marine ice cliff instability, which can substantially accelerate ice loss to potentially add "tens of centimeters" to sea level rise within this century. AR6 includes a version of SSP5-8.5 where these processes take place, and in that case, sea level rise of over 2 m (6+1/2 ft) by 2100 could not be ruled out. The general increase of projections in AR6 was caused by the observed ice-sheet erosion in Greenland and Antarctica matching the upper-end range of the AR5 projections by 2020, and the finding that AR5 projections were likely too slow next to an extrapolation of observed sea level rise trends, while the subsequent reports had improved in this regard.

Notably, some scientists believe that ice sheet processes may accelerate sea level rise even at temperatures below the highest possible scenario, though not as much. For instance, a 2017 study from the University of Melbourne researchers suggested that these processes increase RCP2.6 sea level rise by about one quarter, RCP4.5 sea level rise by one half and practically double RCP8.5 sea level rise. A 2016 study led by Jim Hansen hypothesized that vulnerable ice sheet section collapse can lead to near-term exponential sea level rise acceleration, with a doubling time of 10, 20 or 40 years. Such acceleration would lead to multi-meter sea level rise in 50, 100 or 200 years, respectively, but it remains a minority view amongst the scientific community.

For comparison, a major scientific survey of 106 experts in 2020 found that even when accounting for instability processes they had estimated a median sea level rise of 45 cm (17+1/2 in) by 2100 for RCP2.6, with a 5%-95% range of 21–82 cm (8+1/2–32+1/2 in). For RCP8.5, the experts estimated a median of 93 cm (36+1/2 in) by 2100 and a 5%-95% range of 45–165 cm (17+1/2–65 in). Similarly, NOAA in 2022 had suggested that there is a 50% probability of 0.5 m (19+1/2 in) sea level rise by 2100 under 2°C (3.6°F), which increases to >80% to >99% under 3–5 °C (5.4–9.0 °F). Year 2019 elicitation of 22 ice sheet experts suggested a median SLR of 30 cm (12 in) by 2050 and 70 cm (27+1/2 in) by 2100 in the low emission scenario and the median of 34 cm (13+1/2 in) by 2050 and 110 cm (43+1/2 in) by 2100 in a high emission scenario. They also estimated a small chance of sea levels exceeding 1 meter by 2100 even in the low emission scenario and of going beyond 2 metres in the high emission scenario, with the latter causing the displacement of 187 million people.

Post-2100 Sea Level Rise

Even if the temperature stabilizes, significant sea-level rise (SLR) will continue for centuries. This is what models consistent with paleo records of sea level rise. After 500 years, sea level rise from thermal expansion alone may have reached only half of its eventual level. Models suggest this may lie within ranges of 0.5–2 m (1+1/2–6+1/2 ft). Additionally, tipping points of Greenland and Antarctica ice sheets are likely to play a larger role over such timescales. Ice loss from Antarctica is likely to dominate very long-term SLR, especially if the warming exceeds 2 °C (3.6 °F). Continued carbon dioxide emissions from fossil fuel sources could cause additional tens of metres of sea level rise, over the next millennia. The available fossil fuel on

Earth is enough to melt the entire Antarctic ice sheet, causing about 58 m (190 ft) of sea level rise.

In the next 2,000 years, sea level is predicted to rise by 2–3 m (6+1/2–10 ft) if the temperature increase peaks at its current 1.5 °C (2.7 °F), It would rise by 2–6 m (6+1/2–19+1/2 ft) if it peaks at 2 °C (3.6 °F) and by 19–22 m (62+1/2–72 ft) if it peaks at 5 °C (9.0 °F). If the temperature rise stops at 2 °C (3.6 °F) or at 5 °C (9.0 °F), the sea level would still continue to rise for about 10,000 years. In the first case it will reach 8–13 m (26–42+1/2 ft) above pre-industrial level, and in the second 28–37 m (92–121+1/2 ft). With better models and observational records, several studies have attempted to project SLR for the centuries immediately after 2100. This remains largely speculative. An April 2019 expert elicitation asked 22 experts about total sea level rise projections for the years 2200 and 2300 under its high, 5 °C warming scenario. It ended up with 90% confidence intervals of +10 cm (4 in) to 740 cm (24+1/2 ft) and +9 cm (3+1/2 in) to 970 cm (32 ft), respectively. Negative values represent the extremely low probability of very large increases in the ice sheet surface mass balance due to climate change-induced increase in precipitation. An elicitation of 106 experts led by Stefan Rahmstorf also included 2300 for RCP2.6 and RCP8.5. The former had the median of 118 cm (46+1/2 in), and a 5%-95% range of 24–311 cm (9+1/2–122+1/2 in). The latter had the median of 329 cm (129+1/2 in), and a 5%-95% range of 88–783 cm (34+1/2–308+1/2 in).

By 2021, AR6 was also able to provide estimates for sea level rise in 2150 alongside the 2100 estimates for the first time. This showed that keeping warming at 1.5 °C under the SSP1-1.9 scenario would result in sea level rise in the 17-83% range of 37–86 cm (14+1/2–34 in). In the SSP1-2.6 pathway the range would be 46–99 cm (18–39 in), for SSP2-4.5 a 66–133 cm (26–52+1/2 in) range by 2100 and for SSP5-8.5 a rise of 98–188 cm (38+1/2–74 in). It stated that a "low-confidence" projection of over 2 m (6+1/2 ft) by 2100, would accelerate further to potentially 5 m (16+1/2 ft) by 2150. AR6 also provided lower-confidence estimates for year 2300 sea level rise under SSP1-2.6 and SSP5-8.5. The former had a range between 0.5 m (1+1/2 ft) and 3.2 m (10+1/2 ft), while the latter ranged from just under 2 m (6+1/2 ft) to just under 7 m (23 ft). The low-confidence projections of SSP5-8.5 project sea level rise exceeding 15 m (49 ft) by then.

A 2018 paper estimated that sea level rise in 2300 would increase by a median of 20 cm (8 in) for every five years CO_2 emissions increase before peaking. It shows a 5% likelihood of a 1 m (3+1/2 ft) increase due to the same. The same estimate found that if the temperature stabilized below 2 °C (3.6 °F), 2300 sea level rise would still exceed 1.5 m (5 ft). Early net zero and slowly falling temperatures could limit it to 70–120 cm (27+1/2–47 in).

Measurements

Variations in the amount of water in the oceans, changes in its volume, or varying land elevation compared to the sea surface can drive sea level changes. Over a consistent time period, assessments can attribute contributions to sea level rise and provide early indications of change in trajectory. This helps to inform adaptation plans. The different techniques used to measure changes in sea level do not measure exactly the same level. Tide gauges can only measure relative sea level. Satellites can also measure absolute sea level changes. To get precise measurements for sea level, researchers studying the ice and oceans factor in ongoing deformations of the solid Earth. They look in particular at landmasses still rising from past ice masses retreating, and the Earth's gravity and rotation.

Satellites

Since the launch of TOPEX/Poseidon in 1992, an overlapping series of altimetric satellites has been continuously recording the sea level and its changes. These satellites can measure the hills and valleys in the sea caused by currents and detect trends in their height. To measure the distance to the sea surface, the satellites send a microwave pulse towards Earth and record the time it takes to return after reflecting off the ocean's surface. Microwave radiometers correct the additional delay caused by water vapor in the atmosphere. Combining these data with the location of the spacecraft determines the sea-surface height to within a few centimetres. These satellite measurements have estimated rates of sea level rise for 1993–2017 at 3.0 ± 0.4 millimetres (1D 8 ± 1/64 in) per year.

Satellites are useful for measuring regional variations in sea level. An example is the substantial rise between 1993 and 2012 in the western tropical Pacific. This sharp rise has been linked to increasing trade winds. These occur when the Pacific Decadal Oscillation (PDO) and the El Niño–Southern Oscillation (ENSO) change from one state to the other. The PDO is a basin-wide climate pattern consisting of two phases, each commonly lasting 10 to 30 years. The ENSO has a shorter period of 2 to 7 years.

Tide Gauges

The global network of tide gauges is the other important source of sea-level observations. Compared to the satellite record, this record has major spatial gaps but covers a much longer period. Coverage of tide gauges started mainly in the Northern Hemisphere. Data for the Southern Hemisphere remained scarce up to the 1970s. The longest running sea-level measurements, NAP or Amsterdam Ordnance Datum were established in 1675, in Amsterdam. Record collection is also extensive in Australia. They including measurements by an amateur meteorologist beginning in 1837. They also include measurements taken from a sea-level benchmark struck on a small cliff on the Isle of the Dead near the Port Arthur convict settlement in 1841.

Together with satellite data for the period after 1992, this network established that global mean sea level rose 19.5 cm (7.7 in) between 1870 and 2004 at an average rate of about 1.44 mm/yr. (For the 20th century the average is 1.7 mm/yr.) By 2018, data collected by Australia's Commonwealth Scientific and Industrial Research Organisation (CSIRO) had shown that the global mean sea level was rising by 3.2 mm (1/8 in) per year. This was double the average 20th century rate. The 2023 World Meteorological Organization report found further acceleration to 4.62 mm/yr over the 2013–2022 period. These observations help to check and verify predictions from climate change simulations.

Regional differences are also visible in the tide gauge data. Some are caused by local sea level differences. Others are due to vertical land movements. In Europe, only some land areas are rising while the others are sinking. Since 1970, most tidal stations have measured higher seas. However sea levels along the northern Baltic Sea have dropped due to post-glacial rebound.

Past Sea Level Rise

An understanding of past sea level is an important guide to where current changes in sea level will end up. In the recent geological past, thermal expansion from increased temperatures and changes in land ice are the dominant reasons of sea level rise. The last time that the Earth was 2°C (3.6°F) warmer than pre-industrial temperatures was 120,000 years ago. This was when warming due to Milankovitch cycles (changes in the amount of sunlight due to slow changes in the Earth's orbit) caused the Eemian interglacial. Sea levels during that warmer interglacial were at least 5 m (16 ft) higher than now. The Eemian warming was sustained over a period of thousands of years. The size of the rise in sea level implies a large contribution from the Antarctic and Greenland ice sheets. Levels of atmospheric carbon dioxide of around 400 parts per million had increased temperature by over 2–3 °C (3.6–5.4 °F) around three million years ago. This temperature increase eventually melted one third of Antarctica's ice sheet, causing sea levels to rise 20 meters above the preindustrial levels.

Since the Last Glacial Maximum, about 20,000 years ago, sea level has risen by more than 125 metres (410 ft). Rates vary from less than 1 mm/year during the pre-industrial era to 40+ mm/year when major ice sheets over Canada and Eurasia melted. Meltwater pulses are periods of fast sea level rise caused by the rapid disintegration of these ice sheets. The rate of sea level rise started to slow down about 8,200 years before today. Sea level was almost constant for the last 2,500 years. The recent trend of rising sea level started at the end of the 19th or beginning of the 20th century.

Causes

The three main reasons warming causes global sea level to rise are the expansion of oceans due to heating, water inflow from melting ice sheets and water inflow from glaciers. Glacier retreat and ocean expansion have dominated sea level rise since the start of the 20th century. Some of the losses from glaciers are offset when precipitation falls as snow, accumulates and over time forms glacial ice. If precipitation, surface processes and ice loss at the edge balance each other, sea level remains the same. Because of this precipitation began as water vapor evaporated from the ocean surface, effects of climate change on the water cycle can even increase ice build-up. However, this effect is not enough to fully offset ice losses, and sea level rise continues to accelerate.

The contributions of the two large ice sheets, in Greenland and Antarctica, are likely to increase in the 21st century. They store most of the land ice (~99.5%) and have a sea-level equivalent (SLE) of 7.4 m (24 ft 3 in) for Greenland and 58.3 m (191 ft 3 in) for Antarctica. Thus, melting of all the ice on Earth would result in about 70 m (229 ft 8 in) of sea level rise, although this would require at least 10,000 years and up to of global warming.

Ocean Heating

The oceans store more than 90% of the extra heat added to Earth's climate system by climate change and act as a buffer against its effects. This means that the same amount of heat that would increase the average world ocean temperature by 0.01 °C (0.018 °F) would increase atmospheric temperature by approximately 10 °C (18 °F). So a small change in the mean temperature of the ocean represents a very large change in the total heat content of the climate system. Winds and currents move heat into deeper parts of the ocean. Some of it reaches depths of more than 2,000 m (6,600 ft).

When the ocean gains heat, the water expands and sea level rises. Warmer water and water under great pressure (due to depth) expand more than cooler water and water under less pressure. Consequently, cold Arctic Ocean water will expand less than warm tropical water. Different climate models present slightly different patterns of ocean heating. So their projections do not agree fully on how much ocean heating contributes to sea level rise.

Antarctic Ice Loss

The large volume of ice on the Antarctic continent stores around 60% of the world's fresh water. Excluding groundwater this is 90%. Antarctica is experiencing ice loss from coastal glaciers in the West Antarctica and some glaciers of East Antarctica. However it is gaining mass from the increased snow build-up inland, particularly in the East. This leads to contradicting trends. There are different satellite methods for measuring ice mass and change. Combining them helps to reconcile the differences.

However, there can still be variations between the studies. In 2018, a systematic review estimated average annual ice loss of 43 billion tons (Gt) across the entire continent between 1992 and 2002. This tripled to an annual average of 220 Gt from 2012 to 2017. However, a 2021 analysis of data from four different research satellite systems (Envisat, European Remote-Sensing Satellite, GRACE and GRACE-FO and ICESat) indicated annual mass loss of only about 12 Gt from 2012 to 2016. This was due to greater ice gain in East Antarctica than estimated earlier.

In the future, it is known that West Antarctica at least will continue to lose mass, and the likely future losses of sea ice and ice shelves, which block warmer currents from direct contact with the ice sheet, can accelerate declines even in the East. Altogether, Antarctica is the source of the largest uncertainty for future sea level projections. By 2019, several studies attempted to estimate 2300 sea level rise caused by ice loss in Antarctica alone. They suggest a median rise of 16 cm (6+1/2 in) and maximum rise of 37 cm (14+1/2 in) under the low-emission scenario. The highest emission scenario results in a median rise of 1.46 m (5 ft) metres, with a minimum of 60 cm (2 ft) and a maximum of 2.89 m (9+1/2 ft)).

East Antarctica

The world's largest potential source of sea level rise is the East Antarctic Ice Sheet (EAIS). It is 2.2 km thick on average and holds enough ice to raise global sea levels by 53.3 m (174 ft 10 in) Its great thickness and high elevation make it more stable than the other ice sheets. As of the early 2020s, most studies show that it is still gaining mass. Some analyses have suggested it began to lose mass in the 2000s. However they over-extrapolated some observed losses on to the poorly observed areas. A more complete observational record shows continued mass gain.

In spite of the net mass gain, some East Antarctica glaciers have lost ice in recent decades due to ocean warming and declining structural support from the local sea ice, such as Denman Glacier, and Totten Glacier. Totten Glacier is particularly important because it stabilizes the Aurora Subglacial Basin. Subglacial basins like Aurora and Wilkes Basin are major ice reservoirs together holding as much ice as all of West Antarctica. They are more vulnerable than the rest of East Antarctica. Their collective tipping point probably lies at around 3 °C (5.4 °F) of global warming. It may be as high as 6 °C (11 °F) or as low as 2 °C (3.6 °F). Once this tipping point is crossed, the collapse of these subglacial basins could take place over as little as 500 or as much as 10,000 years. The median timeline is 2000 years. Depending on how many subglacial basins are vulnerable, this causes sea level rise of between 1.4 m (4 ft 7 in) and 6.4 m (21 ft 0 in).

On the other hand, the whole EAIS would not definitely collapse until global warming reaches 7.5 °C (13.5 °F), with a range between 5 °C (9.0 °F) and 10 °C (18 °F).

It would take at least 10,000 years to disappear. Some scientists have estimated that warming would have to reach at least 6 °C (11 °F) to melt two thirds of its volume.

West Antarctica

East Antarctica contains the largest potential source of sea level rise. However the West Antarctica ice sheet (WAIS) is substantially more vulnerable. Temperatures on West Antarctica have increased significantly, unlike East Antarctica and the Antarctic Peninsula. The trend is between 0.08 °C (0.14 °F) and 0.96 °C (1.73 °F) per decade between 1976 and 2012. Satellite observations recorded a substantial increase in WAIS melting from 1992 to 2017. This resulted in 7.6 ± 3.9 mm (19/64 ± 5/32 in) of Antarctica sea level rise. Outflow glaciers in the Amundsen Sea Embayment played a disproportionate role.

The 2021 IPCC Sixth Assessment Report estimated that the median increase in sea level rise from the West Antarctic ice sheet melt by 2100 is ~11 cm (5 in). There is no difference between scenarios, because the increased warming would intensify the water cycle and increase snowfall accumulation over the ice sheet at about the same rate as it would increase ice loss. However, most of the bedrock underlying the WAIS lies well below sea level, and it has to be buttressed by the Thwaites and Pine Island glaciers. If these glaciers collapse, the entire ice sheet would as well. They would take at least several centuries to disappear. But scientists consider this almost inevitable, as their bedrock topography deepens inland and becomes more vulnerable to meltwater.

The contribution of these glaciers to global sea levels has already accelerated since the beginning of the 21st century. The Thwaites Glacier now accounts for 4% of global sea level rise. It could start to lose even more ice if the Thwaites Ice Shelf fails, potentially in mid-2020s. This is due to the marine ice sheet instability hypothesis. This suggests that warm water enters between the seafloor and the base of the ice sheet once it is no longer heavy enough to displace the flow. This causes accelerated melting and collapse. Marine ice cliff instability, when ice cliffs with heights greater than 100 m (330 ft) collapse under their own weight once they are no longer buttressed by ice shelves, may also occur. However it has never been observed. More detailed modelling has ruled it out.

Another hard-to-model process is hydrofracturing. Here meltwater collects atop the ice sheet, pools into fractures and forces them open. Another is changes in the ocean circulation at a smaller scale. A combination of these processes could cause the WAIS to contribute as much as 41 cm (16 in) by 2100 under the low-emission scenario and 57 cm (22 in) under the highest-emission one.

The melting of all the ice in West Antarctica would increase the total sea level rise to 4.3 m (14 ft 1 in). However, mountain ice caps not in contact with water are less vulnerable than the majority of the ice sheet, which is located below the sea level. Its collapse would cause ~3.3 m (10 ft 10 in) of sea level rise. Scientists consider this collapse practically inevitable. It seems to have already occurred during the Eemian period 125,000 years ago. At that time temperatures were similar to the early 21st century. This disappearance would take an estimated 2000 years. The absolute minimum for the loss of West Antarctica ice is 500 years, and the potential maximum is 13,000 years.

The only way to stop ice loss from West Antarctica once triggered is by lowering the global temperature to 1 °C (1.8 °F) below the preindustrial level. This would be 2 °C (3.6 °F) below the temperature of 2020. Other researchers suggested that a climate engineering intervention to stabilize the ice sheet's glaciers may delay its loss by centuries and give more time to adapt. However this is an uncertain proposal, and would end up as one of the most expensive projects ever attempted.

Isostatic Rebound

2021 research indicates that isostatic rebound after the loss of the main portion of the West Antarctic ice sheet would ultimately add another 1.02 m (3 ft 4 in) to global sea levels. This effect would start to increase sea levels before 2100. However it would take 1000 years for it to cause 83 cm (2 ft 9 in) of sea level rise. At this point, West Antarctica itself would be 610 m (2,001 ft 4 in) higher than now. Estimates of isostatic rebound after the loss of East Antarctica's subglacial basins suggest increases of between 8 cm (3.1 in) and 57 cm (1 ft 10 in)

Greenland Ice Sheet Loss

Most ice on Greenland is in the Greenland ice sheet which is 3 km (10,000 ft) at its thickest. The rest of Greenland ice forms isolated glaciers and ice caps. The average annual ice loss in Greenland more than doubled in the early 21st century compared to the 20th century. Its contribution to sea level rise correspondingly increased from 0.07 mm per year between 1992 and 1997 to 0.68 mm per year between 2012 and 2017. Total ice loss from the Greenland ice sheet between 1992 and 2018 amounted to 3,902 gigatons (Gt) of ice. This is equivalent to a SLR contribution of 10.8 mm. The contribution for the 2012–2016 period was equivalent to 37% of sea level rise from land ice sources (excluding thermal expansion). This observed rate of ice sheet melting is at the higher end of predictions from past IPCC assessment reports.

In 2021, AR6 estimated that the Greenland ice sheet would continue to melt and add to sea level rise even under the SSP1-2.6 emission scenario. This largely fulfils the Paris Agreement goals. Under this scenario Greenland ice sheet melt adds around 6 cm (2+1/2 in) to global sea level rise by the end of the century. It shows a plausible maximum of 15 cm (6 in). It also shows a very small chance of the ice sheet reducing sea levels by around 2 cm (1 in) by gaining mass through surface mass balance feedback. The scenario associated with the highest global warming, SSP5-8.5, would see Greenland add at least 5 cm (2 in) to sea level rise. It shows a likely median of 13 cm (5 in) cm and a plausible maximum of 23 cm (9 in).

Greenland's peripheral glaciers and ice caps crossed an irreversible tipping point around 1997. So they are already causing unstoppable sea level rise. However the temperature changes in future, the warming of 2000–2019 had already damaged the ice sheet enough for it to eventually lose ~3.3% of its volume. This is leading to 27 cm (10+1/2 in) of future sea level rise. At a certain level of global warming, the Greenland ice sheet will almost completely melt. Ice cores show this happened at least once during the last million years, even though temperatures have not been higher than 2.5 °C (4.5 °F) above preindustrial levels.

2012 research suggested that the tipping point of the ice sheet was between 0.8 °C (1.4 °F) and 3.2 °C (5.8 °F). 2023 modelling has narrowed the tipping threshold to a 1.7 °C (3.1 °F)-2.3 °C (4.1 °F) range. If temperatures reach or exceed that level, reducing the global temperature to 1.5 °C (2.7 °F) above pre-industrial levels or lower would prevent the loss of the entire ice sheet. One way to do this in theory would be large-scale carbon dioxide removal. But it would also cause greater losses and sea level rise from Greenland than if the threshold was not breached in the first place. Otherwise, the ice sheet would take between 10,000 and 15,000 years to disintegrate entirely once the tipping point had been crossed. The most likely estimate is 10,000 years. If climate change continues along its worst trajectory and temperatures continue to rise fast for a long time, it would only take 1,000 years.

Mountain Glacier Loss

There are roughly 200,000 glaciers on Earth, which are spread out across all continents. Less than 1% of glacier ice is in mountain glaciers, compared to 99% in Greenland and Antarctica. However, this small size also makes mountain glaciers more vulnerable to melting than the larger ice sheets. This means they have had a disproportionate contribution to historical sea level rise and are set to contribute a smaller, but still significant fraction of sea level rise in the 21st century. Observational and modelling studies of mass loss from glaciers and ice caps show they contribute 0.2-0.4 mm per year to sea level rise, averaged over the 20th century. The contribution for the 2012–2016 period was nearly as large as that of Greenland. It was 0.63 mm of sea level rise per year, equivalent to 34% of sea level rise from land ice sources. Glaciers contributed around 40% to sea level rise during the 20th century, with estimates for the 21st century of around 30%.

In 2023, a Science paper estimated that at 1.5 °C (2.7 °F), one quarter of mountain glacier mass would be lost by 2100 and nearly half would be lost at 4 °C (7.2 °F), contributing ~9 cm (3+1/2 in) and ~15 cm (6 in) to sea level rise, respectively. Glacier mass is disproportionately concentrated in the most resilient glaciers. So in practice this would remove 49-83% of glacier formations. It further estimated that the current likely trajectory of 2.7 °C (4.9 °F) would result in the SLR contribution of ~11 cm (4+1/2 in) by 2100. Mountain glaciers are even more vulnerable over the longer term. In 2022, another Science paper estimated that almost no mountain glaciers could survive once warming crosses 2 °C (3.6 °F). Their complete loss is largely inevitable around 3 °C (5.4 °F). There is even a possibility of complete loss after 2100 at just 1.5 °C (2.7 °F). This could happen as early as 50 years after the tipping point is crossed, although 200 years is the most likely value, and the maximum is around 1000 years.

Sea Ice Loss

Sea ice loss contributes very slightly to global sea level rise. If the melt water from ice floating in the sea was exactly the same as sea water then, according to Archimedes' principle, no rise would occur. However melted sea ice contains less dissolved salt than sea water and is therefore less dense, with a slightly greater volume per unit of mass. If all floating ice shelves and icebergs were to melt sea level would only rise by about 4 cm (1+1/2 in).

Changes to Land Water Storage

Human activity impacts how much water is stored on land. Dams retain large quantities of water, which is stored on land rather than flowing into the sea, though the total quantity stored will vary from time to time. On the other hand, humans extract water from lakes, wetlands and underground reservoirs for food production. This often causes subsidence. Furthermore, the hydrological cycle is influenced by climate change and deforestation. This can increase or reduce contributions to sea level rise. In the 20th century, these processes roughly balanced, but dam building has slowed down and is expected to stay low for the 21st century. Water redistribution caused by irrigation from 1993 to 2010 caused a drift of Earth's rotational pole by 78.48 centimetres (30.90 in). This caused groundwater depletion equivalent to a global sea level rise of 6.24 millimetres (0.246 in).

Impacts

The impacts of sea level rise include higher and more frequent high-tide and storm-surge flooding, increased coastal erosion, inhibition of primary production processes, more extensive coastal inundation, along with changes in surface water quality and groundwater. These can lead to a greater loss of property and coastal habitats, loss of life during floods and loss of cultural resources. Agriculture and aquaculture can also be impacted. There can also be loss of tourism, recreation, and transport related functions. Coastal flooding impacts are exacerbated by land use changes such as urbanisation or deforestation of low-lying coastal zones. Regions that are already vulnerable to the rising sea level also struggle with coastal flooding washing away land and altering the landscape.

Because the projected extent of sea level rise by 2050 will be only slightly affected by any changes in emissions, there is confidence that 2050 levels of SLR combined with the 2010 population distribution (i.e., absent the effects of population growth and human migration) would result in ~150 million people under the water line during high tide and ~300 million in places which are flooded every year—an increase of 40 and 50 million people relative to 2010 values for the same. By 2100, there would be another 40 million people under the water line during high tide if sea level rise remains low, and 80 million for a high estimate of the median sea level rise. If ice sheet processes under the highest emission scenario result in sea level rise of well over one metre (3+1/4 ft) by 2100, with a chance of levels over two metres (6+1/2 ft), then as many as 520 million additional people would end up under the water line during high tide and 640 million in places which are flooded every year, when compared to the 2010 population distribution.

Over the longer term, coastal areas are particularly vulnerable to rising sea levels, changes in the frequency and intensity of storms, increased precipitation, and rising ocean temperatures. Ten percent of the world's population live in coastal areas that are less than 10 metres (33 ft) above sea level. Furthermore, two-thirds of the world's cities with over five million people are located in these low-lying coastal areas. In total, approximately 600 million people live directly on the coast around the world. Cities such as Miami, Rio de Janeiro, Osaka and Shanghai will be especially vulnerable later in the century under the warming of 3 °C (5.4 °F), which is close to the current trajectory. Altogether, LiDAR-based research had established in 2021 that 267 million people worldwide lived on land less than 2 m (6+1/2 ft) above sea level and that with a 1 m (3+1/2 ft) sea level rise and zero population growth, that number could increase to 410 million people.

Even populations who live further inland may be impacted by a potential disruption of sea trade, and by migrations. In 2023, United Nations secretary general António Guterres warned that sea level rises risk causing human migrations on a "biblical scale". Sea level rise will inevitably affect ports, but the current research into this subject is limited. Not enough is known about the investments required to protect the ports currently in use, and for how they may be protected before it becomes more reasonable to build new port facilities elsewhere. Moreover, some coastal regions are rich agricultural lands, whose loss to the sea can result in food shortages elsewhere. This is a particularly acute issue for river deltas such as Nile Delta in Egypt and Red River and Mekong Deltas in Vietnam, which are disproportionately affected by saltwater intrusion into the soil and irrigation water.

Ecosystems

When seawater reaches inland, coastal plants, birds, and freshwater/estuarine fish are threatened with habitat loss due to flooding and soil/water salinization.

So-called ghost forests emerge when coastal forest areas become inundated with saltwater to the point no trees can survive. Starting around 2050, some nesting sites in Florida, Cuba, Ecuador and the island of Sint Eustatius for leatherback, loggerhead, hawksbill, green and olive ridley turtles are expected to be flooded, and the proportion would only increase over time. And in 2016, Bramble Cay islet in the Great Barrier Reef was inundated, flooding the habitat of a rodent named Bramble Cay melomys. In 2019, it was officially declared extinct.

While some ecosystems can move land inward with the high-water mark, many are prevented from migrating due to natural or artificial barriers. This coastal narrowing, sometimes called 'coastal squeeze' when considering human-made barriers, could result in the loss of habitats such as mudflats and tidal marshes. Mangrove ecosystems on the mudflats of tropical coasts nurture high biodiversity, yet they are particularly vulnerable due to mangrove plants' reliance on breathing roots or pneumatophores, which will be submerged if the rate is too rapid for them to migrate upward, resulting in the loss of an ecosystem. Both mangroves and tidal marshes protect against storm surges, waves and tsunamis, so their loss makes the effects of sea level rise worse. Human activities, such as dam building, may restrict sediment supplies to wetlands, and thereby prevent natural adaptation processes. The loss of some tidal marshes is unavoidable as a consequence.

Likewise, corals, important for bird and fish life, need to grow vertically to remain close to the sea surface in order to get enough energy from sunlight. The corals have so far been able to keep up the vertical growth with the rising seas, but might not be able to do so in the future.

Regional Impacts

In Africa

In Africa, risk from sea level rise is amplified by the future population growth. It is believed that 54.2 million people lived in the highly exposed low elevation coastal zones (LECZ) around 2000, but this number will effectively double to around 110 million people by 2030, and by 2060 it will be around 185 to 230 million people, depending on the extent of population growth. While the average regional sea level rise by 2060 will be around 21 cm (with climate change scenarios making little difference at that point), local geography and population trends interact to increase the exposure to hazards like 100-year floods in a complex way.

Table 5.1: Populations within 100-year Floodplains

Country	2000	2030	2060	Growth 2000-2060
Egypt	7.4	13.8	20.7	0.28
Nigeria	0.1	0.3	0.9	0.84
Senegal	0.4	1.1	2.7	0.76
Benin	0.1	0.6	1.6	1.12
Tanzania	0.2	0.9	4.3	2.3
Somalia	0.2	0.6	2.7	1.7
Cote d'Ivoire	0.1	0.3	0.7	0.65
Mozambique	0.7	1.4	2.5	0.36

In the near term, some of the largest displacement is projected to occur in the East Africa region, where at least 750,000 people are likely to be displaced from the coasts between 2020 and 2050. It was also estimated that by 2050, 12 major African cities (Abidjan, Alexandria, Algiers, Cape Town, Casablanca, Dakar, Dar es Salaam, Durban, Lagos, Lomé, Luanda and Maputo) would collectively sustain cumulative damages of US$65 billion for the "moderate" climate change scenario RCP4.5 and US$86.5 billion for the high-emission scenario RCP8.5: the version of the high-emission scenario with additional impacts from high ice sheet instability would involve up to US$137.5 billion in damages. Additional accounting for the "low-probability, high-damage events" may increase aggregate risks to US$187 billion for the "moderate" RCP4.5, US$206 billion for RCP8.5 and US$397 billion under the high-end instability scenario.

In all of these estimates, the Egyptian city of Alexandria alone amounts for around half of this figure: hundreds of thousands of people in its low-lying areas may already have to be relocated in the coming decade. Across sub-Saharan Africa as a whole, damages from sea level rise could reach 2-4% of GDP by 2050, although this is strongly affected by the extent of future economic growth and adaptation.

In the longer term, Egypt, Mozambique and Tanzania are also projected to have the largest number of people affected by annual flooding amongst all African countries if global warming reaches 4 °C by the end of the century (a level associated with the RCP8.5 scenario).

Under RCP8.5, 10 important cultural sites (Casbah of Algiers, Carthage Archaeological site, Kerkouane, Leptis Magna Archaeological site, Medina of Sousse, Medina of Tunis, Sabratha Archaeological site, Robben Island, Island of Saint-Louis and Tipasa) would be at risk of flooding and erosion by the end of the century, along with a total of 15 Ramsar sites and other natural heritage sites (Bao Bolong Wetland Reserve, Delta du Saloum National Park, Diawling National Park, Golfe de Boughrara, Kalissaye, Lagune de Ghar el Melh et Delta de la Mejerda, Marromeu Game Reserve, Parc Naturel des Mangroves du Fleuve Cacheu, Seal Ledges Provincial Nature Reserve, Sebkhet Halk Elmanzel et Oued Essed, Sebkhet Soliman, Réserve Naturelle d'Intérêt Communautaire de la Somone, Songor Biosphere Reserve, Tanbi Wetland Complex and Watamu Marine National Park).

In Asia

As of 2022, it is estimated that 63 million people in the East and South Asia are already at risk from a 100-year flood, in large part due to inadequate coastal protection in many countries. This will be greatly exacerbated in the future: Asia has the largest population at risk from sea level and Bangladesh, China, India, Indonesia, Japan, Pakistan, the Philippines, Thailand and Vietnam alone account for 70% number of people exposed to sea level rise during the 21st century. This is entirely due to the region's densely populated coasts, as the rate of sea level rise in Asia is generally similar to the global average.

Exceptions include the Indo-Pacific region, where it had been around 10% faster since the 1990s, and the coast of China, where globally "extreme" sea level rise had been detected since the 1980s, and it is believed that the difference between and of global warming would have a disproportionate impact on flood frequency.

It is also estimated that future sea level rise along the Japanese Honshu Island would be up to 25 cm faster than the global average under RCP8.5, the intense climate change scenario. RCP8.5 is additionally associated with the loss of at least a third of the Japanese beaches and 57-72% of Thai beaches.

One estimate finds that Asia will suffer direct economic damages of US$167.6 billion at 0.47 meters of sea level rise, US$272.3 billion at 1.12 meters and US$338.1 billion at 1.75 meters (along with the indirect impact of 8.5, 24 or US$15 billion from population displacement at those levels), with China, India, the Republic of Korea, Japan, Indonesia and Russia experiencing the largest economic losses. Out of the 20 coastal cities expected to see the highest flood losses by 2050, 13 are in Asia.

For nine of those (Bangkok, Guangzhou, Ho Chi Minh City, Jakarta, Kolkata, Nagoya, Tianjin, Xiamen and Zhanjiang) sea level rise would be compounded by subsidence. By 2050, Guangzhou would see 0.2 meters of sea level rise and the estimated annual economic losses of US$254 million - the highest in the world. One estimate calculates that in the absence of adaptation, cumulative economic losses caused by sea level rise in Guangzhou under RCP8.5 would reach ~US$331 billion by 2050, ~US$660 billion by 2070 and US$1.4 trillion by 2100, while the impact of high-end ice sheet instability would increase these figures to ~US$420 billion, ~US$840 billion and ~US$1.8 trillion, respectively.

In Shanghai, coastal inundation amounts to ~0.03% of local GDP; but would increase to 0.8% (confidence interval of 0.4-1.4%) by 2100 even under the "moderate" RCP4.5 scenario in the absence of adaptation. Likewise, failing to adapt to sea level rise in Mumbai would result in the damages of 112-162 billion USD by 2050, which would nearly triple by 2070. As the result, efforts like the Mumbai Coastal Road are being implemented, although they are likely to affect coastal ecosystems and fishing livelihoods. Nations with extensive rice production along the coasts like Bangladesh, Vietnam and China are already seeing adverse impacts from saltwater intrusion.

It is estimated that sea level rise in Bangladesh may force the relocation of up to one-third of power plants as early as 2030, while a similar proportion would have to deal with the increased salinity of their cooling water by then.

Research from 2010s indicates that by 2050, between 0.9 and 2.1 million people would be displaced by sea level rise alone: this would likely necessitate the creation of ~594,000 additional jobs and ~197,000 housing units in the areas receiving the displaced persons, as well as to secure the supply of additional ~783 billion calories worth of food.

In 2021, another paper estimated that 816,000 would be directly displaced by sea level rise by 2050, but this would be increased to 1,3 million when the indirect effects are taken into account.

Both studies assume that the majority of the displaced people would travel to the other areas of Bangladesh, and attempt to estimate population changes in different localities.

Table 5.2: Net Variations in the Population Due to Sea Level Rise in 2050 in Selected Districts

District	Net Flux	Net Flux	Rank	Rank
Dhaka	207,373	−34, 060	1	11
Narayanganj	−95,003	−126,694	2	1
Shariatpur	−80,916	−124,444	3	3
Barisal	−80,669	64,252	4	6
Munshiganj	−77,916	−124,598	5	2
Madaripur	61,791	−937	6	60
Chandpur	−37,711	−70,998	7	4
Jhalakati	35,546	9,198	8	36
Satkhira	−32,287	−19,603	9	23
Khulna	−28,148	−9,982	10	33
Cox's Bazar	−25,680	−16,366	11	24
Bagherat	24,860	12,263		12

In an attempt to address these challenges, the Bangladesh Delta Plan 2100 has been launched in 2018. As of 2020, it was seen falling short of most of its initial targets. The progress is being monitored.

In 2019, the president of Indonesia, Joko Widodo, declared that the city of Jakarta is sinking to a degree that requires him to move the capital to another city. A study conducted between 1982 and 2010 found that some areas of Jakarta have been sinking by as much as 28 cm (11 inches) per year due to ground water drilling and the weight of its buildings, and the problem is now exacerbated by sea level rise. However, there are concerns that building in a new location will increase tropical deforestation. Other so called sinking cities, such as Bangkok or Tokyo, are vulnerable to these compounding subsidence with sea level rise.

In Australasia

In Australia, erosion and flooding of Queensland's Sunshine Coast beaches is projected to intensify by 60% by 2030, with severe impacts on tourism in the absence of adaptation. Adaptation costs to sea level rise under the high-emission RCP8.5 scenario are projected to be three times greater than the adaptation costs to low-emission RCP2.6 scenario. For 0.2- to 0.3-m sea level rise (set to occur by 2050), what is currently a 100-year flood would occur every year in New Zealand cities of Wellington and Christchurch. Under 0.5 m sea level rise, the current 100-year flood in Australia would be likely to occur several times a year, while in New Zealand, buildings with a collective worth of NZ$12.75 billion would become exposed to new 100-year floods. A metre or so of sea level rise would threaten assets in New Zealand with a worth of NZD$25.5 billion (with a disproportionate impact on Maori-owned holdings and cultural heritage objects), and Australian assets with a worth of AUD$164-226 billion (including many unsealed roads and railway lines). The latter represents a 111% rise in Australia's inundation costs between 2020 and 2100.

Central and South America

By 2100, a minimum of 3-4 million people in South America would be directly affected by coastal flooding and erosion. 6% of the population of Venezuela, 56% of the population of Guyana (including in the capital, Georgetown, much of which is already below the sea level) and 68% of the population of Suriname are already living in low-lying areas exposed to sea level rise. In Brazil, the coastal ecoregion of Caatinga is responsible for 99% of its shrimp production, yet its unique conditions are threatened by a combination of sea level rise, ocean warming and ocean acidification. The port complex of Santa Catarina had been interrupted by extreme wave or wind behavior 76 times in one 6-year period in 2010s, with a 25,000-50,000 USD loss for each idle day. In Port of Santos, storm surges were three times more frequent between 2000 and 2016 than between 1928 and 1999.

In Europe

Many sandy coastlines in Europe are vulnerable to erosion caused by sea level rise. In Spain, Costa del Maresme is anticipated to retreat by 16 meters by 2050 relative to 2010, and potentially by 52 meters by 2100 under RCP8.5 Other vulnerable coastlines include Tyrrhenian Sea coast of Italy's Calabria region, Barra-Vagueira coast in Portugal and Norlev Strand in Denmark.

In France, it was estimated that 8,000-10,000 people would be forced to migrate away from the coasts by 2080. The Italian city of Venice is located on islands. It is highly vulnerable to flooding and has already spent $6 billion on a barrier system. A quarter of the German state of Schleswig-Holstein, inhabited by over 350,000 people, is at low elevation and has been vulnerable to flooding since the preindustrial times. Many levees already exist, but to its complex geography, a flexible mix of hard and soft measures was chosen, which is intended to support a safety margin of >1 meter rise per century. In the United Kingdom, sea level at the end of the century would increase by 53 to 115 centimetres at the mouth of river Thames and 30 to 90 centimetres at Edinburgh. To address this reality, it has divided its coast into 22 areas, each covered by a Shoreline Management Plan. Those are further sub-divided into 2000 management units in total, spanning across three "epochs" (0-20 years, 20-50 and 50-100 years).

The Netherlands is a country that sits partially below sea level and is subsiding. It has responded by extending its Delta Works program. Drafted in 2008, the Delta Commission report said that the country must plan for a rise in the North Sea up to 1.3 m (4 ft 3 in) by 2100 and plan for a 2-4 m (7-13 ft) rise by 2200. It advised annual spending between €1.0 and €1.5 billion for measures such as broadening coastal dunes and strengthening sea and river dikes. Worst-case evacuation plans were also drawn up.

In North America

As of 2017, around 95 million Americans lived on the coast: for Canada and Mexico, this figure amounts to 6.5 million and 19 million people. Increased chronic nuisance flooding and king tide flooding is already an issue in the highly vulnerable state of Florida, as well as alongside the US East Coast. On average, the number of days with tidal flooding in the USA increased 2 times in the years 2000-2020, reaching 3-7 days per year. In some areas the increase was much stronger: 4 times in the Southeast Atlantic and 11 times in the Western Gulf. By the year 2030 the average number is expected to be 7-15 days, reaching 25-75 days by 2050. U.S. coastal cities have responded to that through beach nourishment or beach replenishment, where mined sand is trucked in and added, in addition to other adaptation measures such as zoning, restrictions on state funding, and building code standards. Along an estimated 15% of the US coastline, the majority of local groundwater levels are already below the sea level. This places those groundwater reservoirs at risk of sea water intrusion, which renders fresh water unusable once its concentration exceeds 2-3%. The damages are also widespread in Canada and will affect both major cities like Halifax and the more remote locations like Lennox Island, whose Mi'kmaq community is already considering relocation due to widespread coastal erosion. In Mexico, the damages from SLR to tourism hotspots like Cancun, Isla Mujeres, Playa del Carmen, Puerto Morelos and Cozumel could amount to 1.4-2.3 billion USD. The increase in storm surge due to sea level rise is also a problem. For example, due to this effect Hurricane Sandy caused additional 8 billion dollars in damage, impacted 36,000 more houses and 71,000 more people.

In the future, northern Gulf of Mexico, Atlantic Canada and the Pacific coast of Mexico would experience the greatest sea level rise. By 2030, flooding along the US Gulf Coast could cause economic losses of up to US$176 billion: around US$50 billion may be avoided through nature-based solutions like wetland restoration and oyster reef restoration. By 2050, the frequency of coastal flooding in the US is expected to rise tenfold to four "moderate" flooding events per year, even without storms or heavy rainfall. In the New York City, current 100-year flood would occur once in 19-68 years by 2050 and 4-60 years by 2080. By 2050, 20 million people in the greater New York City area would be threatened, as 40% of the existing water treatment facilities would be compromised and 60% of power plants will need to be relocated. By 2100, sea level rise of 0.9 m (3 ft) and 1.8 m (6 ft) would threaten 4.2 and 13.1 million people in the US, respectively. In California alone, 2 m (6+1?2 ft) of SLR could affect 600,000 people and threaten over US$150 billion in property with inundation, potentially representing over 6% of the state's GDP. In North Carolina, a meter of SLR inundates 42% of the Albemarle-Pamlico Peninsula, costing up to US$14 billion (at 2016 value of the currency). In nine southeast US states, the same level of sea level rise would claim up to 13,000 historical and archaeological sites, including over 1000 sites eligible for inclusion in the National Register for Historic Places.

In Island Nations

Small island states are nations whose populations are concentrated on atolls and other low islands. Atolls on average reach 0.9-1.8 m (3-6 ft) above sea level. This means that no other place is more vulnerable to coastal erosion, flooding and salt intrusion into soils and freshwater caused by sea level rise. The latter may render an island uninhabitable well before it is completely flooded. Already, children in small island states are encountering hampered access to food and water and are suffering an increased rate of mental and social disorders due to these stressors. At current rates, sea level would be high enough to make the Maldives uninhabitable by 2100, while five of the Solomon Islands have already disappeared due to the combined effects of sea level rise and stronger trade winds that were pushing water into the Western Pacific.

Adaptation to sea level rise is costly for small island nations as a large portion of their population lives in areas that are at risk. Nations like Maldives, Kiribati and Tuvalu are already forced to consider controlled international migration of their population in response to rising seas, since the alternative of uncontrolled migration threatens to exacerbate the humanitarian crisis of climate refugees. In 2014, Kiribati had purchased 20 square kilometers of land (about 2.5% of Kiribati's current area) on the Fijian island of Vanua Levu to relocate its population there once their own islands are lost to the sea.

While Fiji is also impacted by sea level rise, it is in a comparatively safer position, and its residents continue to rely on local adaptation like moving further inland and increasing sediment supply to combat erosion instead of relocating entirely. Fiji has also issued a green bond of $50 million to invest in green initiatives and use the proceeds to fund adaptation efforts, and it is restoring coral reefs and mangroves to protect itself flooding and erosion as a more cost-efficient alternative to building sea walls, with the nations of Palau and Tonga adopting similar efforts. At the same time, even when an island is not threatened with complete disappearance due to flooding, tourism and local economies may end up devastated. For instance, a sea level rise of 1.0 m (3 ft 3 in) would cause partial or complete inundation of 29% of coastal resorts in the Caribbean, while a further 49-60% of coastal resorts would be at risk from resulting coastal erosion.

Adaptation

Cutting greenhouse gas emissions can slow and stabilize the rate of sea level rise after 2050, greatly reducing its costs and damages, but cannot stop it outright. Thus, climate change adaptation to sea level rise is inevitable. The most straightforward approach is to first cease development in vulnerable areas and ultimately move the people and infrastructure away from them. Such retreat from sea level rise often results in the loss of livelihoods, and the displacement of newly impoverished people could burden their new homes and accelerate social tensions.

It is possible to avoid or at least delay the retreat from sea level rise with enhanced protections like dams, levees or improved natural defenses, or through accommodation like building standards updated to reduce damage from floods, addition of storm water valves to address more frequent and severe flooding at high tide, or cultivating crops more tolerant of saltwater mixing into the soil, even at an increased cost. These options can be further divided into hard and soft adaptation. The former generally involves large-scale changes to human societies and ecological systems, often through the construction of capital-intensive infrastructure. Soft adaptation involves strengthening natural defenses and local community adaptation, usually with simple, modular and locally owned technology. The two types of adaptation might be complementary or mutually exclusive. Adaptation options often require significant investment, but the costs of doing nothing are far greater. For instance, effective adaptation measures are predicted to reduce future annual costs of flooding in 136 of the world's largest coastal cities from $1 trillion by 2050 if no adaptation was done, to a little over $60 billion annually, while costing $50 billion per year. However, it has been suggested that in the case of very high sea level rise, retreat away from the coast would have a lower impact on the GDP of India and Southeast Asia then attempting to protect every coastline.

To be successful, adaptation needs to anticipate sea level rise well ahead of time. As of 2023, the global state of adaptation planning is mixed. A survey of 253 planners from 49 countries found that while 98% are aware of sea level rise projections, 26% have not yet formally integrated them into their policy documents. Only around a third of respondents from Asian and South American countries have done so, compared to 50% in Africa, and >75% in Europe, Australasia and North America. 56% of all surveyed planners have structured plans which account for 2050 and 2100 sea level rise, but 53% only plan using a single projection, rather than a range of two or three projections. Just 14% plan using four projections, including that of the "extreme" or "high-end" sea level rise. Another study found that while >75% of regional sea level rise assessments from the West and Northeastern United States included at least three estimates (usually RCP2.6, RCP4.5 and RCP8.5), and sometimes included extreme scenarios, 88% of projections from the American South had only a single estimate. Similarly, no assessment from the South went beyond 2100, while 14 assessments from the West went up to 2150, and three from the Northeast went to 2200. 56% of all localities were also found to underestimate the upper end of sea level rise relative to IPCC Sixth Assessment Report.

CHAPTER 6

PHYSICAL HYDROGEOLOGY

6.1 INTRODUCTION

Physical hydrogeology is a branch of geology that focuses on the study of groundwater, its occurrence, movement, and interactions with geological materials. Groundwater, stored beneath the Earth's surface in aquifers, plays a crucial role in sustaining ecosystems, providing drinking water, and influencing geological processes. Physical hydrogeology stands at the intersection of geology, hydrology, and environmental science, offering invaluable insights into the dynamics of groundwater systems. From the principles of groundwater flow and geological formations hosting aquifers to the practical applications in water resource management, hydrogeology plays a pivotal role in addressing water challenges. As we confront growing water scarcity, pollution, and climate change, continued research and advancements in physical hydrogeology will be essential for ensuring the sustainable use and protection of this vital freshwater resource. Through interdisciplinary collaboration, technological innovation, and informed decision-making, the field of physical hydrogeology will continue to shape our understanding of groundwater and contribute to the development of effective water management strategies.

6.1.1 Fundamental Principles of Physical Hydrogeology

1. Hydrologic Cycle: The hydrologic cycle, also known as the water cycle, is at the core of physical hydrogeology. It describes the continuous movement of water between the atmosphere, surface water bodies, and the subsurface. Precipitation, infiltration, evaporation, and transpiration are key processes that influence the distribution of water on Earth, with groundwater serving as a vital component of this cycle.
2. Porosity and Permeability: Understanding the properties of geological materials is crucial in hydrogeology. Porosity refers to the percentage of void space within a rock or sediment, indicating its ability to store water. Permeability, on the other hand, measures the ease with which water can flow through a material. Rocks with high porosity and permeability, such as sandstone and limestone, are often excellent aquifer materials.

3. Darcy's Law: Darcy's Law is a fundamental principle that governs groundwater flow. It establishes a relationship between the hydraulic conductivity of a material, the cross-sectional area through which water flows, the hydraulic gradient (the slope of the water table or potentiometric surface), and the rate of groundwater flow. This law forms the basis for understanding the movement of water through aquifers.
4. Groundwater Movement: Groundwater moves primarily through the process of advection, driven by differences in hydraulic head. Advection involves the bulk movement of water within the subsurface. Additionally, dispersion and diffusion play roles in groundwater movement, influencing the spreading and mixing of solutes within the groundwater system.

6.1.2 Geological Formations and Aquifers

1. Aquifer Types: Aquifers are geological formations that can store and transmit water. Confined aquifers are sandwiched between impermeable layers, while unconfined aquifers are open to the land surface. Aquitards and aquicludes, in contrast, are formations with low permeability that impede the flow of groundwater. The arrangement of these geological units influences groundwater storage and movement.
2. Karst Aquifers: Karst aquifers are a unique type characterized by the dissolution of soluble rocks such as limestone and gypsum. The dissolution creates conduits, caves, and sinkholes, forming a highly permeable subsurface network. Karst aquifers often exhibit rapid groundwater flow and are vulnerable to contamination due to their direct connection with the surface.
3. Glacial Aquifers: In regions shaped by glaciation, such as during the Pleistocene epoch, glacial aquifers play a significant role in groundwater storage. Glacial deposits, including till and outwash, form heterogeneous aquifer systems with varying degrees of permeability. Understanding the complex structure of these aquifers is essential for managing groundwater resources in glaciated landscapes.
4. Volcanic Aquifers: Volcanic rocks, such as basalt, can form aquifers with unique characteristics. The vesicular nature of some volcanic rocks creates pore spaces, while fractures and lava tubes enhance permeability. Volcanic aquifers can exhibit high hydraulic conductivity, facilitating rapid groundwater flow.

6.1.3 Applications and Challenges in Water Resource Management

1. Groundwater Exploration and Well Design: Physical hydrogeology is instrumental in the exploration and development of groundwater resources. Geophysical methods, hydrogeological mapping, and drilling

techniques help identify potential aquifers. Well design and construction take into account the geological characteristics of the subsurface to optimize water extraction.

2. Groundwater Contamination and Remediation: Understanding the movement of contaminants in the subsurface is crucial in addressing groundwater pollution. Physical hydrogeology principles aid in assessing the fate and transport of contaminants, developing conceptual models, and designing remediation strategies. Techniques such as pump-and-treat, in-situ bioremediation, and monitored natural attenuation leverage hydrogeological concepts for effective cleanup.
3. Land-Subsidence Mitigation: Over-extraction of groundwater can lead to land subsidence, a phenomenon where the land surface sinks. Physical hydrogeology helps identify regions susceptible to subsidence and develop strategies to mitigate its impacts. Artificial recharge, reducing pumping rates, and land-use planning are among the measures employed to address land subsidence.
4. Groundwater and Climate Change: Climate change can influence groundwater recharge patterns, alter precipitation regimes, and impact aquifer storage. Physical hydrogeology provides insights into how these changes may affect groundwater resources. Developing climate-resilient groundwater management strategies involves anticipating shifts in hydrological patterns and adapting to the evolving conditions.
5. Sustainable Groundwater Management: As groundwater resources face increasing pressures, sustainable groundwater management becomes imperative. Groundwater models, based on physical hydrogeological principles, assist in simulating groundwater flow, predicting aquifer response to pumping, and evaluating the sustainability of extraction rates. Integrated water resource management frameworks incorporate hydrogeological data to balance competing water needs.

6.1.4 Advancements in Hydrogeological Research

1. Numerical Modeling: Numerical modeling has revolutionized hydrogeological research. Computer models, such as MODFLOW, simulate complex groundwater flow scenarios, incorporating geological data, recharge rates, and extraction rates. These models aid in predicting aquifer behavior, optimizing well-field designs, and assessing the impacts of various management strategies.
2. Remote Sensing and GIS: Remote sensing technologies and Geographic Information Systems (GIS) play a crucial role in hydrogeological investigations. Satellite imagery, aerial photography, and geospatial data enhance the understanding of surface and subsurface features. These

tools assist in mapping aquifer extents, monitoring land-use changes, and evaluating the vulnerability of groundwater systems.

3. Isotope Hydrology: Isotope hydrology involves the use of isotopic tracers to study the origin, movement, and age of groundwater. Isotopes of hydrogen, oxygen, and noble gases provide valuable information about the sources of recharge, residence times, and the interaction between surface water and groundwater. Isotope hydrology has applications in understanding paleoclimate, identifying pollution sources, and improving water resource management.

6.1.5 Challenges and Future Directions

1. ***Anthropogenic Impacts:*** Anthropogenic activities, including urbanization, industrialization, and agriculture, pose significant challenges to groundwater sustainability. Over-extraction, contamination, and alteration of natural recharge patterns are consequences of human activities. Addressing these impacts requires interdisciplinary approaches that integrate hydrogeological knowledge with social, economic, and policy considerations.
2. ***Climate Change Uncertainties:*** The uncertainties associated with climate change pose challenges for hydrogeological predictions. Changing precipitation patterns, altered temperatures, and the frequency of extreme weather events can influence groundwater recharge and discharge dynamics. Ongoing research aims to enhance the predictive capabilities of hydrogeological models under evolving climate scenarios.
3. ***Transboundary Aquifer Management:*** Many aquifers extend across national borders, requiring international collaboration for effective management. Developing frameworks for transboundary aquifer cooperation, sharing data, and establishing joint monitoring programs are essential steps in addressing the challenges associated with shared groundwater resources.
4. ***Emerging Contaminants:*** The presence of emerging contaminants, such as pharmaceuticals and industrial chemicals, poses new challenges for hydrogeologists. Understanding the fate and transport of these contaminants in the subsurface, as well as developing effective remediation strategies, requires ongoing research and technological advancements.

6.2 HYDRAULIC CONDUCTIVITY

In science and engineering, hydraulic conductivity (K, in SI units of meters per second), is a property of porous materials, soils and rocks, that describes the ease with which a fluid (usually water) can move through the pore space, or fracture network. It depends on the intrinsic permeability (k, unit: m) of the material, the degree of saturation, and on the density and viscosity of the fluid. Saturated hydraulic conductivity, K_{sat}, describes water movement through saturated media. By definition, hydraulic conductivity is the ratio of volume flux to hydraulic gradient yielding a quantitative measure of a saturated soil's ability to transmit water when subjected to a hydraulic gradient.

6.2.1 Methods of Determination

There are two broad approaches for determining hydraulic conductivity:

1. In the *empirical* approach the hydraulic conductivity is correlated to soil properties like pore-size and particle-size (grain-size) distributions, and soil texture.
2. In the *experimental* approach the hydraulic conductivity is determined from hydraulic experiments that are interpreted using Darcy's law.

The experimental approach is broadly classified into:

1. Laboratory tests using soil samples subjected to hydraulic experiments.
2. *Field tests* (on site, in situ) that are differentiated into:
    a. small-scale field tests, using observations of the water level in cavities in the soil; and
    b. large-scale field tests, like pumping tests in wells or by observing the functioning of existing horizontal drainage systems.

The small-scale field tests are further subdivided into:

1. infiltration tests in cavities *above* the water table; and
2. slug tests in cavities *below* the water table.

The methods of determining hydraulic conductivity and other hydraulic properties are investigated by numerous researchers and include additional empirical approaches.

6.2.2 Estimation by Empirical Approach

Estimation from Grain Size

Allen Hazen derived an empirical formula for approximating hydraulic conductivity from grain-size analyses:

$$K = C(D_{10})^2$$

where

- C Hazen's empirical coefficient, which takes a value between 0.0 and 1.5 (depending on literature), with an average value of 1.0. A.F. Salarashayeri & M. Siosemarde indicate C is usually between 1.0 and 1.5, with D in mm and K in cm/s.

- is the diameter of the 10 percentile grain size of the material.

Pedotransfer Function

A pedotransfer function (PTF) is a specialized empirical estimation method, used primarily in the soil sciences, but increasingly used in hydrogeology. There are many different PTF methods, however, they all attempt to determine soil properties, such as hydraulic conductivity, given several measured soil properties, such as soil particle size, and bulk density.

6.2.3 Determination by Experimental Approach

There are relatively simple and inexpensive laboratory tests that may be run to determine the hydraulic conductivity of a soil: constant-head method and falling-head method.

Laboratory Methods

Constant-head Method

The constant-head method is typically used on granular soil. This procedure allows water to move through the soil under a steady state head condition while the volume of water flowing through the soil specimen is measured over a period of time. By knowing the volume ÄV of water measured in a time over a specimen of length L and cross-sectional area A, as well as the head h, the hydraulic conductivity (K) can be derived by simply rearranging Darcy's law:

$$K = \frac{\Delta V}{\Delta t} \frac{L}{Ah}$$

Proof: Darcy's law states that the volumetric flow depends on the pressure differential ÄP between the two sides of the sample, the permeability k and the viscosity m as:

$$\frac{\Delta V}{\Delta t} = -\frac{kA}{\mu L} \Delta P$$

In a constant head experiment, the head (difference between two heights) defines an excess water mass, ñ*Ah*, where ñ is the density of water. This mass weighs down on the side it is on, creating a pressure differential of ÄP = ñ*gh*, where g is the gravitational acceleration. Plugging this directly into the above gives

$$\frac{\Delta V}{\Delta t} = -\frac{k\rho g A}{\mu L} h$$

If the hydraulic conductivity is defined to be related to the hydraulic permeability as

$$K = \frac{k\rho g}{\mu},$$

this gives the result.

Falling-head Method

In the falling-head method, the soil sample is first saturated under a specific head condition. The water is then allowed to flow through the soil without adding any water, so the pressure head declines as water passes through the specimen. The advantage to the falling-head method is that it can be used for both fine-grained and coarse-grained soils.. If the head drops from h_i to h_f in a time Ät, then the hydraulic conductivity is equal to

$$K = \frac{L}{\Delta t} \ln \frac{h_f}{h_i}$$

Proof: As above, Darcy's law reads

$$\frac{\Delta V}{\Delta t} = -K \frac{A}{L} h$$

The decrease in volume is related to the falling head by ÄV = Ä*hA*. Plugging this relationship into the above, and taking the limit as Ät□0, the differential equation

$$\frac{dh}{dt} = -\frac{K}{L} h$$

has the solution

$$h(t) = h_i e^{-\frac{K}{L}(t-t_i)}.$$

Plugging in $h(t_f) = h_f$ and rearranging gives the result.

In-situ (Field) Methods

In compare to laboratory method, field methods gives the most reliable information about the permeability of soil with minimum disturbances. In laboratory methods, the degree of disturbances affect the reliability of value of permeability of the soil.

Pumping Test

Pumping test is the most reliable method to calculate the coefficient of permeability of a soil. This test is further classified into Pumping in test and pumping out test.

Augerhole Method

There are also in-situ methods for measuring the hydraulic conductivity in the field. When the water table is shallow, the augerhole method, a slug test, can be used for determining the hydraulic conductivity below the water table. The method was developed by Hooghoudt (1934) in The Netherlands and introduced in the US by Van Bavel en Kirkham (1948). The method uses the following steps:

1. an augerhole is perforated into the soil to below the water table;
2. water is bailed out from the augerhole;

3. the rate of rise of the water level in the hole is recorded; and
4. the K-value is calculated from the data as:

$$K = F\frac{H_o - H_t}{t}$$

where:

- K is the horizontal saturated hydraulic conductivity (m/day)
- H is the depth of the water level in the hole relative to the water table in the soil (cm):
  * $H_t = H$ at time t
  * $H_0 = H$ at time $t = 0$
- t is the time (in seconds) since the first measurement of H as H_0
- F is a factor depending on the geometry of the hole:

$$F = \frac{4000r}{h'}\left(20 + \frac{D}{r}\right)\left(2 - \frac{h'}{D}\right)$$

where:

- r is the radius of the cylindrical hole (cm)
- h' is the average depth of the water level in the hole relative to the water table in the soil (cm), found as $h' = \frac{H_o + H_t}{2}$
- D is the depth of the bottom of the hole relative to the water table in the soil (cm).

The picture shows a large variation of K-values measured with the augerhole method in an area of 100 ha. The ratio between the highest and lowest values is 25.

6.2.4 Related Magnitudes

Transmissivity

The transmissivity is a measure of how much water can be transmitted horizontally, such as to a pumping well.

1. *Transmissivity* should not be confused with the similar word transmittance used in optics, meaning the fraction of incident light that passes through a sample.

An aquifer may consist of n soil layers. The transmissivity T_i of a horizontal flow for the ith soil layer with a *saturated* thickness d_i and horizontal hydraulic conductivity K_i is:

$$T_i = K_i d_i$$

Transmissivity is directly proportional to horizontal hydraulic conductivity K_i and thickness d_i. Expressing K_i in m/day and d_i in m, the transmissivity T_i is found in units m/day. The total transmissivity T_t of the aquifer is the sum of every layer's transmissivity:

$$T_t = \sum T_i$$

The *apparent* horizontal hydraulic conductivity K_A of the aquifer is:

$$K_A = \frac{T_t}{D_t}$$

where D_t, the total thickness of the aquifer, is the sum of each layer's individual thickness: $D_t = \sum d_i$.

The transmissivity of an aquifer can be determined from pumping tests.

Influence of the Water Table

When a soil layer is above the water table, it is not saturated and does not contribute to the transmissivity. When the soil layer is entirely below the water table, its saturated thickness corresponds to the thickness of the soil layer itself. When the water table is inside a soil layer, the saturated thickness corresponds to the distance of the water table to the bottom of the layer.

As the water table may behave dynamically, this thickness may change from place to place or from time to time, so that the transmissivity may vary accordingly. In a semi-confined aquifer, the water table is found within a soil layer with a negligibly small transmissivity, so that changes of the total transmissivity (D_t) resulting from changes in the level of the water table are negligibly small.

When pumping water from an unconfined aquifer, where the water table is inside a soil layer with a significant transmissivity, the water table may be drawn down whereby the transmissivity reduces and the flow of water to the well diminishes.

Resistance

The *resistance* to vertical flow (R_i) of the *i*th soil layer with a *saturated* thickness d_i and vertical hydraulic conductivity K_{vi} is:

$$R_i = \frac{d_i}{K_{v_i}}$$

Expressing K_{vi} in m/day and d_i in m, the resistance (R_i) is expressed in days. The total resistance (R_t) of the aquifer is the sum of each layer's resistance:

$$R_t = \sum R_i = \sum \frac{d_i}{K_{v_i}}$$

The *apparent* vertical hydraulic conductivity (K_{vA}) of the aquifer is:

$$K_{v_A} = \frac{D_t}{R_t}$$

where D_t is the total thickness of the aquifer: $D_t = \sum d_i$.

The resistance plays a role in aquifers where a sequence of layers occurs with varying horizontal permeability so that horizontal flow is found mainly in the layers with high horizontal permeability while the layers with low horizontal permeability transmit the water mainly in a vertical sense.

6.2.5 Anisotropy

When the horizontal and vertical hydraulic conductivity of the soil layer differ considerably, the layer is said to be anisotropic with respect to hydraulic conductivity. When the *apparent* horizontal and vertical hydraulic conductivity differ considerably, the aquifer is said to be anisotropic with respect to hydraulic conductivity. An aquifer is called *semi-confined* when a saturated layer with a relatively small horizontal hydraulic conductivity (the semi-confining layer or aquitard) overlies a layer with a relatively high horizontal hydraulic conductivity so that the flow of groundwater in the first layer is mainly vertical and in the second layer mainly horizontal. The resistance of a semi-confining top layer of an aquifer can be determined from pumping tests. When calculating flow to drains or to a well field in an aquifer with the aim to control the water table, the anisotropy is to be taken into account, otherwise the result may be erroneous.

6.2.6 Relative Properties

Because of their high porosity and permeability, sand and gravel aquifers have higher hydraulic conductivity than clay or unfractured granite aquifers. Sand or gravel aquifers would thus be easier to extract water from (e.g., using a pumping well) because of their high transmissivity, compared to clay or unfractured bedrock aquifers.

Hydraulic conductivity has units with dimensions of length per time (e.g., m/s, ft/day and (gal/day)/ft^2); transmissivity then has units with dimensions of length squared per time. The following table gives some typical ranges (illustrating the many orders of magnitude which are likely) for *K* values.

Hydraulic conductivity (*K*) is one of the most complex and important of the properties of aquifers in hydrogeology as the values found in nature:

1. range over many orders of magnitude (the distribution is often considered to be lognormal),
2. vary a large amount through space (sometimes considered to be randomly spatially distributed, or stochastic in nature),
3. are directional (in general *K* is a symmetric second-rank tensor; e.g., vertical *K* values can be several orders of magnitude smaller than horizontal *K* values),

4. are scale dependent (testing a m^3 of aquifer will generally produce different results than a similar test on only a cm^3 sample of the same aquifer),
5. must be determined indirectly through field pumping tests, laboratory column flow tests or inverse computer simulation, (sometimes also from grain size analyses), and
6. are very dependent (in a non-linear way) on the water content, which makes solving the unsaturated flow equation difficult. In fact, the variably saturated K for a single material varies over a wider range than the saturated K values for all types of materials.

6.3 DARCY'S LAW

Darcy's Law stands as a foundational principle in hydrogeology, providing essential insights into the movement of groundwater within the subsurface. Named after the French engineer Henri Darcy, who first formulated the law in the mid-19th century, this fundamental concept has played a pivotal role in understanding the dynamics of groundwater flow.

Darcy's Law, a cornerstone of hydrogeology, has played a pivotal role in unraveling the mysteries of groundwater flow within the subsurface. From its formulation in the 19th century to its integration with modern numerical models, Darcy's Law has remained a fundamental tool for hydrogeologists worldwide. Its applications in diverse fields, including water resource management, environmental protection, and geothermal energy, underscore its versatility. As research in hydrogeology progresses, advancements in modeling techniques, field measurements, and uncertainty analysis contribute to refining the application of Darcy's Law in diverse hydrogeological settings. By addressing challenges and embracing emerging trends, the field continues to evolve, providing valuable insights into the complex interactions governing subsurface water movement. Darcy's Law remains a beacon, guiding hydrogeologists in their quest to sustainably manage and protect one of our planet's most vital resources - groundwater.

6.3.1 Formulation of Darcy's Law

Darcy's Law describes the rate of flow of groundwater through a porous medium. The basic form of the law can be expressed mathematically as:

$$Q = -KA\frac{dh}{dl}$$

Where:

- Q represents the volumetric flow rate of groundwater (m^3/s),
- K is the hydraulic conductivity of the porous medium (m/s),
- A is the cross-sectional area through which groundwater flows (m^2),

- $\frac{dh}{dl}$ dhdh denotes the hydraulic gradient, representing the change in hydraulic head per unit distance (m/m).

The negative sign in the equation signifies that groundwater flows from regions of higher hydraulic head to lower hydraulic head, following the natural direction of water movement within the subsurface.

6.3.2 Significance in Hydrogeology

1. ***Quantifying Groundwater Flow:*** Darcy's Law is instrumental in quantifying the rate at which groundwater moves through porous formations. By incorporating hydraulic conductivity, cross-sectional area, and hydraulic gradient, hydrogeologists can determine the volume of groundwater passing through a given section of the subsurface over time. This information is crucial for understanding the dynamics of aquifer systems.
2. ***Defining Hydraulic Conductivity:*** Hydraulic conductivity (K) is a key parameter in Darcy's Law, representing the ability of a porous medium to transmit water. It is influenced by factors such as the size and arrangement of grains in the aquifer material. Measuring hydraulic conductivity helps characterize the permeability of aquifers, providing insights into how easily water can flow through different geological formations.
3. ***Groundwater Velocity Calculation:*** Darcy's Law enables the calculation of groundwater velocity, providing information about the speed at which water moves through the subsurface. This velocity is crucial for understanding the time it takes for water to travel from recharge areas to discharge areas, influencing the management of groundwater resources and the assessment of contaminant transport.
4. ***Assessing Aquifer Characteristics:*** The application of Darcy's Law allows hydrogeologists to assess and compare the characteristics of different aquifers. By determining hydraulic conductivity and hydraulic gradients in various geological formations, researchers can identify aquifers that are more conducive to groundwater extraction or understand the factors influencing groundwater movement in a particular area.
5. ***Predicting Flow Paths:*** Understanding the hydraulic gradients and hydraulic conductivities of aquifers aids in predicting groundwater flow paths. Darcy's Law forms the basis for groundwater modeling, allowing hydrogeologists to simulate and predict the movement of water through subsurface formations under different scenarios. This predictive capability is valuable for water resource management and environmental assessments.

6.3.3 Applications of Darcy's Law

1. ***Well Design and Pumping Tests:*** Darcy's Law plays a crucial role in well design and the interpretation of pumping tests. Hydrogeologists use the law to estimate the hydraulic conductivity of aquifers, evaluate the efficiency of wells, and design extraction systems that optimize groundwater yield while minimizing potential negative impacts on the aquifer.
2. ***Contaminant Transport:*** The transport of contaminants in groundwater is governed by Darcy's Law. By understanding the rate of groundwater flow and the characteristics of the porous medium, hydrogeologists can model the movement of contaminants and assess the potential risks associated with pollution. This is essential for developing effective remediation strategies and protecting groundwater quality.
3. ***Managed Aquifer Recharge:*** Darcy's Law is applied in managed aquifer recharge projects. These projects involve intentionally recharging aquifers with water to enhance groundwater storage. By understanding hydraulic conductivity and hydraulic gradients, hydrogeologists can design and optimize recharge systems to replenish aquifers and mitigate the impacts of over-extraction.
4. ***Geothermal Energy:*** Darcy's Law is relevant in geothermal energy applications, where heat exchange with subsurface formations occurs. Understanding the movement of fluids in geothermal reservoirs is crucial for optimizing energy extraction. Darcy's Law helps assess the feasibility of geothermal projects and predicts the efficiency of heat transfer in subsurface systems.
5. ***Land Subsidence Mitigation:*** Over-extraction of groundwater can lead to land subsidence, a phenomenon where the land surface sinks. Darcy's Law is applied to evaluate the impacts of groundwater pumping on subsurface pressures and to design strategies for land subsidence mitigation, such as artificial recharge programs or sustainable pumping practices.

6.3.4 Challenges and Limitations

1. ***Assumptions in Homogeneity:*** Darcy's Law assumes homogeneity in the porous medium, meaning that the properties influencing groundwater flow are constant throughout the aquifer. In reality, geological formations may exhibit heterogeneity, leading to variations in hydraulic conductivity and affecting the accuracy of predictions.
2. ***Neglect of Chemical and Biological Processes:*** Darcy's Law primarily addresses the physical movement of water and does not consider chemical or biological processes in detail. In cases where groundwater flow interacts with contaminants or undergoes geochemical reactions, supplementary

models and considerations are required to comprehensively assess water quality.

3. ***Transient Conditions:*** Darcy's Law is applicable for steady-state conditions, where groundwater flow remains constant over time. In situations where flow conditions are transient, such as during pumping tests or changes in recharge rates, modifications to Darcy's Law are necessary to account for dynamic changes in groundwater flow.
4. ***Limited in Fractured Rock Environments:*** In fractured rock aquifers, where groundwater flow is influenced by fractures and faults, Darcy's Law may have limitations. The discrete nature of fractures and the complex connectivity within fractured rock formations require more sophisticated models to accurately represent groundwater movement.

6.3.5 Advancements and Future Directions

1. ***Coupling with Numerical Models:*** Advances in numerical modeling techniques have allowed hydrogeologists to couple Darcy's Law with other governing equations, enabling more comprehensive simulations of groundwater flow. Numerical models, such as MODFLOW, incorporate Darcy's Law to simulate complex hydrogeological scenarios and improve predictions under various conditions.
2. ***Incorporating Uncertainty Analysis:*** The integration of uncertainty analysis techniques is an emerging trend in hydrogeological research. Incorporating uncertainties in hydraulic conductivity measurements, geological heterogeneity, and other parameters enhances the reliability of predictions based on Darcy's Law. Probabilistic approaches contribute to a more realistic representation of groundwater flow systems.
3. ***Advancements in Field Measurement Techniques:*** Field measurement techniques, such as geophysical methods and advanced monitoring technologies, have advanced significantly. These tools enable more accurate determination of hydraulic conductivity and hydraulic gradients, providing hydrogeologists with high-resolution data for better calibration of models based on Darcy's Law.
4. ***Advances in Contaminant Transport Modeling:*** Recent developments in contaminant transport modeling integrate Darcy's Law with reactive transport models. This allows researchers to simulate the movement of contaminants in response to chemical reactions within the subsurface. Improved understanding of the coupling between physical and chemical processes enhances the assessment of groundwater quality.

6.4 HYDRAULIC PROPERTIES OF FRACTURED ROCKS

A fracture is any separation in a geologic formation, such as a *joint* or a *fault* that divides the rock into two or more pieces. A fracture will sometimes form a deep fissure or crevice in the rock. Fractures are commonly caused by stress exceeding the rock strength, causing the rock to lose cohesion along its weakest plane. Fractures can provide permeability for fluid movement, such as water or hydrocarbons. Highly fractured rocks can make good aquifers or hydrocarbon reservoirs, since they may possess both significant permeability and fracture porosity.

6.4.1 Brittle Deformation

Fractures are forms of brittle deformation. There are two types of primary brittle deformation processes. Tensile fracturing results in *joints*. *Shear fractures* are the first initial breaks resulting from shear forces exceeding the cohesive strength in that plane. After those two initial deformations, several other types of secondary brittle deformation can be observed, such as *frictional sliding* or *cataclastic flow* on reactivated joints or faults. Most often, fracture profiles will look like either a blade, ellipsoid, or circle.

6.4.2 Causes

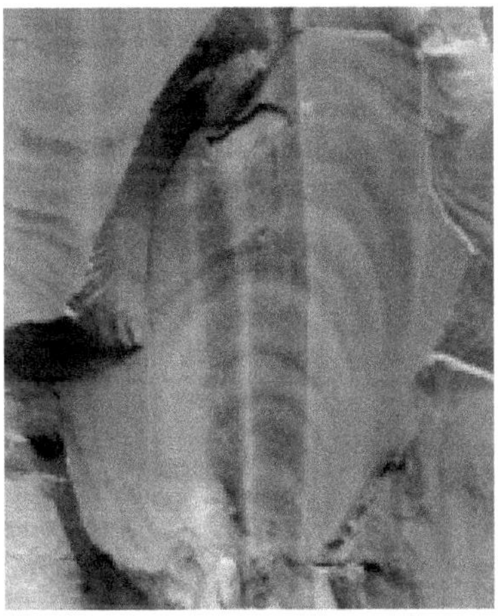

Fig. 6.1: The Concentric Circles in this Sandstone in Arizona are Plumose (Plume-like) Structures that can form during the Formation and Propagation of a Fracture

Fractures in rocks can be formed either due to compression or tension. Fractures due to compression include thrust faults. Fractures may also be a result from shear or tensile stress. Some of the primary mechanisms are discussed below.

Modes

First, there are three modes of fractures that occur (regardless of mechanism):

1. ***Mode I crack:*** Opening mode (a tensile stress normal to the plane of the crack)

2. ***Mode II crack:*** Sliding mode (a shear stress acting parallel to the plane of the crack and perpendicular to the crack front)
3. ***Mode III crack:*** Tearing mode (a shear stress acting parallel to the plane of the crack and parallel to the crack front)

Tensile Fractures

Rocks contain many pre-existing cracks where development of tensile fracture, or Mode I fracture, may be examined.

The first form is in *axial stretching.* In this case a remote tensile stress, σ_n, is applied, allowing microcracks to open slightly throughout the tensile region. As these cracks open up, the stresses at the crack tips intensify, eventually exceeding the rock strength and allowing the fracture to propagate. This can occur at times of rapid overburden erosion. Folding also can provide tension, such as along the top of an anticlinal fold axis. In this scenario the tensile forces associated with the stretching of the upper half of the layers during folding can induce tensile fractures parallel to the fold axis.

Another, similar tensile fracture mechanism is *hydraulic fracturing.* In a natural environment, this occurs when rapid sediment compaction, thermal fluid expansion, or fluid injection causes the pore fluid pressure, $ó_p$, to exceed the pressure of the least principal normal stress, $ó_n$. When this occurs, a tensile fracture opens perpendicular to the plane of least stress.

Tensile fracturing may also be induced by applied compressive loads, $ó_n$, along an axis such as in a Brazilian disk test. This applied compression force results in *longitudinal splitting.* In this situation, tiny tensile fractures form parallel to the loading axis while the load also forces any other microfractures closed. To picture this, imagine an envelope, with loading from the top. A load is applied on the top edge, the sides of the envelope open outward, even though nothing was pulling on them. Rapid deposition and compaction can sometimes induce these fractures.

Tensile fractures are almost always referred to as *joints*, which are fractures where no appreciable slip or shear is observed.

To fully understand the effects of applied tensile stress around a crack in a brittle material such a rock, fracture mechanics can be used. The concept of fracture mechanics was initially developed by A. A. Griffith during World War I. Griffith looked at the energy required to create new surfaces by breaking material bonds versus the elastic strain energy of the stretched bonds released. By analyzing a rod under uniform tension Griffith determined an expression for the critical stress at which a favorably orientated crack will grow. The critical stress at fracture is given by,

$$\sigma_f = \left(\frac{2E\gamma}{\pi a}\right)^{1/2}$$

where ã = surface energy associated with broken bonds, E = Young's modulus, and a = half crack length. Fracture mechanics has generalized to that ã represents energy dissipated in fracture not just the energy associated with creation of new surfaces

Linear Elastic Fracture Mechanics

Linear elastic fracture mechanics (LEFM) builds off the energy balance approach taken by Griffith but provides a more generalized approach for many crack problems. LEFM investigates the stress field near the crack tip and bases fracture criteria on stress field parameters. One important contribution of LEFM is the stress intensity factor, K, which is used to predict the stress at the crack tip. The stress field is given by

$$\sigma_{ij}(r,\theta) = \frac{K}{(2\pi r)^{1/2}} f_{ij}(\theta)$$

where *K* is the stress intensity factor for Mode I, II, or III cracking and is a dimensionless quantity that varies with applied load and sample geometry. As the stress field gets close to the crack tip, i.e., becomes a fixed function of . With knowledge of the geometry of the crack and applied far field stresses, it is possible to predict the crack tip stresses, displacement, and growth. Energy release rate is defined to relate K to the Griffith energy balance as previously defined. In both LEFM and energy balance approaches, the crack is assumed to be cohesionless behind the crack tip. This provides a problem for geological applications such a fault, where friction exists all over a fault. Overcoming friction absorbs some of the energy that would otherwise go to crack growth. This means that for Modes II and III crack growth, LEFM and energy balances represent local stress fractures rather than global criteria.

Crack Formation and Propagation

Cracks in rock do not form smooth path like a crack in a car windshield or a highly ductile crack like a ripped plastic grocery bag. Rocks are a polycrystalline material so cracks grow through the coalescing of complex microcracks that occur in front of the crack tip. This area of microcracks is called the brittle process zone. The shear crack, shown in blue, propagates when tensile cracks, shown in red, grow perpendicular to the direction of the least principal stresses.

The tensile cracks propagate a short distance then become stable, allowing the shear crack to propagate. This type of crack propagation should only be considered an example. Fracture in rock is a 3D process with cracks growing in all directions. It is also important to note that once the crack grows, the microcracks in the brittle process zone are left behind leaving a weakened section of rock. This weakened section is more susceptible to changes in pore pressure and dilatation or compaction.

Fig. 6.2: Rough Surfaces on a Piece of Fractured Granite

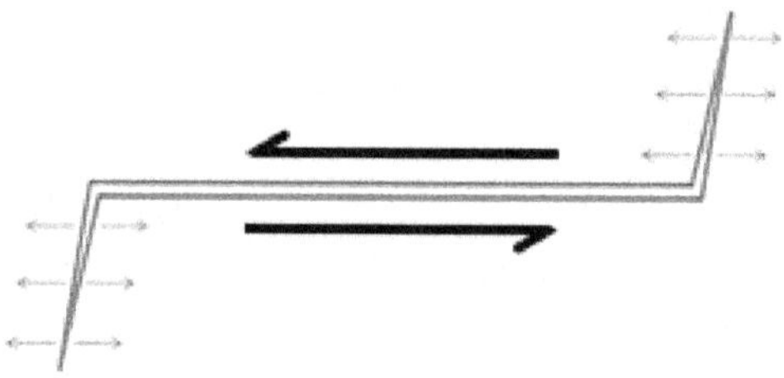

Fig. 6.3: Shear Fracture (Blue) under Shear Loading (Black Arrows) in Rock. Tensile Cracks, also Referred to as Wing Cracks (Red) Grow at an Angle from the Edges of the Shear Fracture Allowing the Shear Fracture to Propagate by the Coalescing of these Tensile Fractures

Note that this description of formation and propagation considers temperatures and pressures near the Earth's surface. Rocks deep within the earth are subject to very high temperatures and pressures.

This causes them to behave in the semi-brittle and plastic regimes which result in significantly different fracture mechanisms. In the plastic regime cracks acts like a plastic bag being torn. In this case stress at crack tips goes to two mechanisms, one which will drive propagation of the crack and the other which will blunt the crack tip.

In the brittle-ductile transition zone, material will exhibit both brittle and plastic traits with the gradual onset of plasticity in the polycrystalline rock. The main form of deformation is called cataclastic flow, which will cause fractures to fail and propagate due to a mixture of brittle-frictional and plastic deformations.

Joint Types

Describing joints can be difficult, especially without visuals. The following are descriptions of typical natural fracture joint geometries that might be encountered in field studies:

1. *Plumose Structures* are fracture networks that form at a range of scales, and spread outward from a *joint origin*. The joint origin represents a point at which the fracture begins. The *mirror zone* is the joint morphology closest to the origin that results in very smooth surfaces. *Mist zones* exist on the fringe of mirror zones and represent the zone where the joint surface slightly roughens. *Hackle zones* predominate after mist zones, where the joint surface begins to get fairly rough. This hackle zone severity designates *barbs*, which are the curves away from the *plume axis*.
2. *Orthogonal Joints* occur when the joints within the system occur at mutually perpendicular angles to each other.
3. *Conjugate Joints* occur when the joints intersect each other at angles significantly less than ninety degrees.
4. *Systematic Joints* are joint systems in which all the joints are parallel or subparallel, and maintain roughly the same spacing from each other.
5. *Columnar Joints* are joints that cut the formation vertically in (typically) hexagonal columns. These tend to be a result of cooling and contraction in hypabyssal intrusions or lava flows.
6. *Desiccation cracks* are joints that form in a layer of mud when it dries and shrinks. Like columnar joints, these tend to be hexagonal in shape.
7. *Sigmoidal Joints* are joints that run parallel to each other, but are cut by sigmoidal (stretched S) joints in between.
8. *Sheeting joints* are joints that often form near surface, and as a result form parallel to the surface. These can also be recognized in *exfoliation joints*.
9. In joint systems where relatively long joints cut across the outcrop, the throughgoing joints act as *master joints* and the short joints that occur in between are *cross joints*.
10. *Poisson effect* is the creation of vertical contraction fractures that are a result of the relief of overburden over a formation.
11. *Pinnate joints* are joints that form immediately adjacent to and parallel to the shear face of a fault. These joints tend to merge with the faults at an angle between 35 and 45 degrees to the fault surface.
12. *Release joints* are tensile joints that form as a change in geologic shape results in the manifestation of local or regional tension that can create Mode I tensile fractures.
13. Concurrent joints that display a *ladder pattern* are interior regions with one set of joints that are fairly long, and the conjugate set of joints for the pattern remain relatively short, and terminate at the long joint.
14. Sometimes joints can also display *grid patterns*, which are fracture sets that have mutually crosscutting fractures.

15. An *en echelon* or *stepped* array represents a set of tensile fractures that form within a fault zone parallel to each other.

Faults and Shear Fractures

Faults are another form of fracture in a geologic environment. In any type of faulting, the active fracture experiences shear failure, as the faces of the fracture slip relative to each other.

As a result, these fractures seem like large scale representations of Mode II and III fractures, however that is not necessarily the case. On such a large scale, once the shear failure occurs, the fracture begins to curve its propagation towards the same direction as the tensile fractures.

In other words, the fault typically attempts to orient itself perpendicular to the plane of least principal stress. This results in an out-of-plane shear relative to the initial reference plane.

Therefore, these cannot necessarily be qualified as Mode II or III fractures. An additional, important characteristic of shear-mode fractures is the process by which they spawn *wing cracks*, which are tensile cracks that form at the propagation tip of the shear fractures. As the faces slide in opposite directions, tension is created at the tip, and a mode I fracture is created in the direction of the $\sigma_{h\text{-}max}$, which is the direction of maximum principal stress.

Shear-failure criteria is an expression that attempts to describe the stress at which a shear rupture creates a crack and separation. This criterion is based largely off of the work of Charles Coulomb, who suggested that as long as all stresses are compressive, as is the case in shear fracture, the shear stress is related to the normal stress by:

$$\sigma_s = C + \mu(\sigma_n - \sigma_f),$$

where C is the *cohesion* of the rock, or the shear stress necessary to cause failure given the normal stress across that plane equals 0. μ is the coefficient of internal friction, which serves as a constant of proportionality within geology. σ_n is the normal stress across the fracture at the instant of failure, σ_f represents the pore fluid pressure. It is important to point out that pore fluid pressure has a significant impact on shear stress, *especially* where pore fluid pressure approaches *lithostatic pressure*, which is the normal pressure induced by the weight of the overlying rock.

This relationship serves to provide the *coulomb failure envelope* within the Mohr-Coulomb Theory.

Frictional sliding is one aspect for consideration during shear fracturing and faulting. The shear force parallel to the plane must overcome the frictional force to move the faces of the fracture across each other. In fracturing, frictional sliding typically only has significant effects on the reactivation on existing shear fractures.

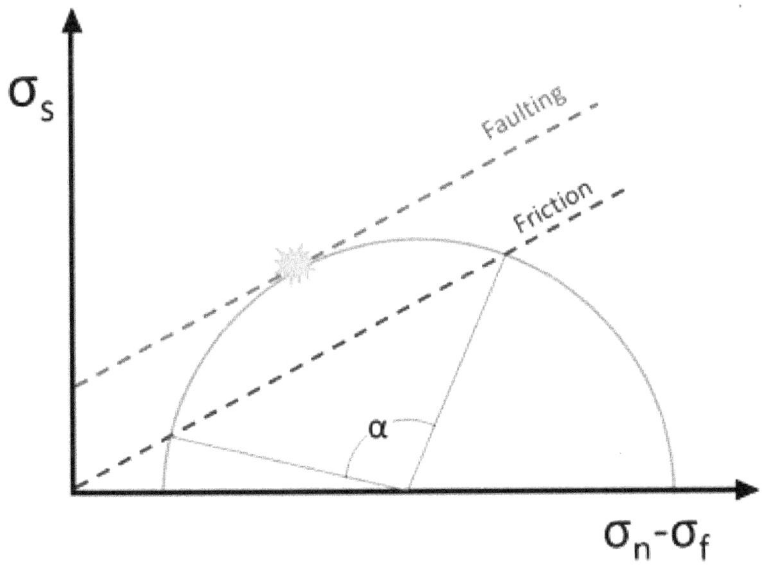

Fig. 6.4: 2D Mohr's Diagram Showing the Different Failure Criteria for Frictional Sliding vs Faulting

The shear force required to slip fault is less than force required to fracture and create new faults as shown by the Mohr-Coulomb diagram. Since the earth is full of existing cracks and this means for any applied stress, many of these cracks are more likely to slip and redistribute stress than a new crack is to initiate. The Mohr's Diagram shown, provides a visual example. For a given stress state in the earth, if an existing fault or crack exists orientated anywhere from “$a/4$ to + $a/4$, this fault will slip before the strength of the rock is reached and a new fault is formed. While the applied stresses may be high enough to form a new fault, existing fracture planes will slip before fracture occurs.

One important idea when evaluating the friction behavior within a fracture is the impact of *asperities*, which are the irregularities that stick out from the rough surfaces of fractures. Since both faces have bumps and pieces that stick out, not all of the fracture face is actually touching the other face. The cumulative impact of asperities is a reduction of the *real area of contact*', which is important when establishing frictional forces.

Subcritical Crack Growth

Sometimes, it is possible for fluids within the fracture to cause fracture propagation with a much lower pressure than initially required. The reaction between certain fluids and the minerals the rock is composed of can lower the stress required for fracture below the stress required throughout the rest of the rock. For instance, water and quartz can react to form a substitution of OH molecules for the O molecules in the quartz mineral lattice near the fracture tip. Since the OH bond is much lower than that with O, it effectively reduces the necessary tensile stress required to extend the fracture.

6.4.3 Engineering Considerations

In geotechnical engineering a fracture forms a discontinuity that may have a large influence on the mechanical behavior (strength, deformation, etc.) of soil and rock masses in, for example, tunnel, foundation, or slope construction.

Fractures also play a significant role in minerals exploitation. One aspect of the upstream energy sector is the production from naturally fractured reservoirs. There are a good number of naturally fractured reservoirs in the United States, and over the past century, they have provided a substantial boost to the nation's net hydrocarbon production.

The key concept is while low porosity, brittle rocks may have very little natural storage or flow capability, the rock is subjected to stresses that generate fractures, and these fractures can actually store a very large volume of hydrocarbons, capable of being recovered at very high rates. One of the most famous examples of a prolific naturally fractured reservoir was the Austin Chalk formation in South Texas. The chalk had very little porosity, and even less permeability. However, tectonic stresses over time created one of the most extensive fractured reservoirs in the world. By predicting the location and connectivity of fracture networks, geologists were able to plan horizontal wellbores to intersect as many fracture networks as possible. Many people credit this field for the birth of true horizontal drilling in a developmental context. Another example in South Texas is the Georgetown and Buda limestone formations.

Furthermore, the recent uprise in prevalence of unconventional reservoirs is actually, in part, a product of natural fractures. In this case, these microfractures are analogous to Griffith Cracks, however they can often be sufficient to supply the necessary productivity, especially after completions, to make what used to be marginally economic zones commercially productive with repeatable success.

However, while natural fractures can often be beneficial, they can also act as potential hazards while drilling wells. Natural fractures can have very high permeability, and as a result, any differences in hydrostatic balance down the well can result in well control issues. If a higher pressured natural fracture system is encountered, the rapid rate at which formation fluid can flow into the wellbore can cause the situation to rapidly escalate into a blowout, either at surface or in a higher subsurface formation. Conversely, if a lower pressured fracture network is encountered, fluid from the wellbore can flow very rapidly into the fractures, causing a loss of hydrostatic pressure and creating the potential for a blowout from a formation further up the hole.

6.4.4 Fracture Modeling

Since the mid-1980s, 2D and 3D computer modeling of fault and fracture networks has become common practice in Earth Sciences. This technology became known as "DFN" (discrete fracture network") modeling, later modified into "DFFN" (discrete fault and fracture network") modeling. The technology consists of defining the statistical variation of various parameters such as size, shape, and orientation and modeling the fracture network in space in a semi-probabilistic way in two or three dimensions. Computer algorithms and speed of calculation have become sufficiently capable of capturing and simulating the complexities and geological variabilities in three dimensions, manifested in what became known as the "DMX Protocol".

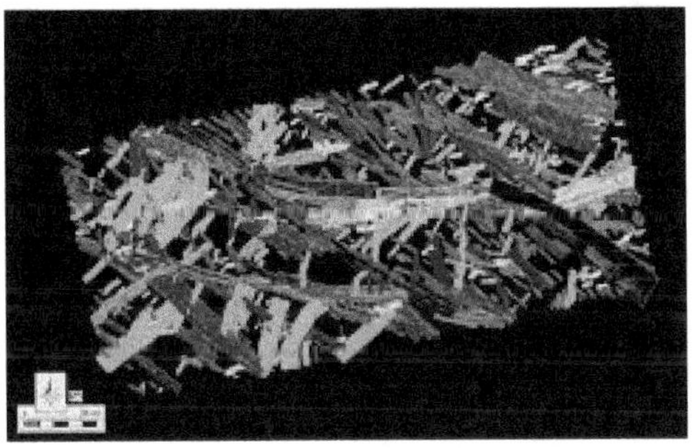

Fig. 6.5: Three Dimensional Computer Model of a Fracture and Fault Network (DFN/DFFN)

6.5 KARST AQUIFER PROPERTIES

Karst aquifers represent a distinctive type of subsurface formation characterized by soluble rocks, such as limestone, gypsum, and dolomite. These geological wonders give rise to intricate landscapes featuring sinkholes, caves, and underground conduits. The properties of karst aquifers set them apart from other aquifer types, significantly influencing water storage, movement, and the challenges associated with their management. Karst aquifers, with their unique properties and intricate geological features, present both challenges and opportunities for water resource management. The high permeability, conduits, and dynamic behavior of karst aquifers necessitate specialized approaches for investigation and sustainable use. Understanding the hydrogeological characteristics of karst aquifers is essential for protecting water quality, mitigating contamination risks, and ensuring the long-term viability of these vital groundwater resources. As advancements in hydrogeological research continue, the management of karst aquifers will benefit from innovative technologies, improved modeling techniques, and interdisciplinary efforts to balance human needs with the preservation of these geological marvels.

6.5.1 Hydrogeological Characteristics of Karst Aquifers

1. High Permeability and Porosity: One of the defining features of karst aquifers is their high permeability and porosity. The dissolution of soluble rocks by water creates conduits and cavities, resulting in a network of interconnected fractures and channels. This intricate structure enhances the flow of water through the aquifer, often making karst aquifers more permeable than other types.
2. Conduits and Caves: The dissolution of rock in karst environments leads to the formation of conduits and caves. These conduits act as preferential pathways for groundwater flow, allowing water to move rapidly through the subsurface. Caves and conduits can range in size from small fractures to extensive underground caverns, forming an intricate labyrinth beneath the Earth's surface.
3. Swallow Holes and Sinkholes: Karst aquifers are often marked by surface features such as swallow holes and sinkholes. Swallow holes are points where surface water infiltrates into the subsurface, initiating the dissolution of the underlying rock. Sinkholes, resulting from the collapse of underground caverns or dissolution of the overlying material, are common in karst landscapes and provide direct access to the underlying aquifer.
4. Vadose Zone and Phreatic Zone: Karst aquifers typically consist of two main zones: the vadose zone and the phreatic zone. The vadose zone is the unsaturated zone above the water table, where water infiltrates through fractures and cavities. The phreatic zone, below the water table, is saturated with groundwater, and water flows through conduits and interconnected fractures, defining the dynamic nature of karst aquifers.

6.5.2 Challenges Associated with Karst Aquifers

1. Vulnerability to Contamination: The rapid movement of water through conduits in karst aquifers increases their vulnerability to contamination. Surface pollutants, such as agricultural runoff or improperly managed waste disposal, can rapidly reach the phreatic zone, affecting the quality of groundwater. Contaminants can move swiftly through the conduits, making source protection challenging.
2. Surface-Subsurface Connectivity: The interconnected nature of karst aquifers creates a strong connectivity between surface and subsurface water systems. Changes in surface water levels, such as during heavy rainfall or flooding events, can impact groundwater levels and flow. Conversely, contamination at the surface can easily infiltrate and affect the quality of groundwater in the aquifer.

3. Spatial Variability: Karst aquifers exhibit significant spatial variability in terms of hydraulic conductivity and flow patterns. The presence of conduits, caves, and fractures creates a heterogeneous subsurface environment. This variability poses challenges for accurately characterizing and modeling the aquifer, requiring sophisticated techniques to capture the complex geometry of flow paths.
4. Sinkhole Formation and Subsidence: The formation of sinkholes in karst areas can lead to subsidence and pose risks to infrastructure and land use. As underground voids collapse, sinkholes can form suddenly, impacting roads, buildings, and agricultural areas. Subsidence can affect the stability of the land surface, necessitating careful land-use planning and monitoring.

6.5.3 Key Properties Influencing Karst Aquifer Behavior

1. Conduit Flow: Conduit flow is a dominant feature of karst aquifers, allowing water to move rapidly through well-defined conduits and channels. The presence of conduits influences the overall behavior of the aquifer, impacting groundwater velocities, recharge rates, and the response to precipitation events.
2. Transient Flow Conditions: Karst aquifers often exhibit transient flow conditions, with rapid responses to changes in recharge and discharge. The dynamic nature of these aquifers makes them susceptible to fluctuations in water levels, especially during storm events, where conduits convey large volumes of water to the aquifer rapidly.
3. Seasonal Variability: Seasonal variations in precipitation have a pronounced impact on karst aquifers. During periods of increased rainfall, recharge rates rise, and water levels in conduits and fractures increase. Conversely, in dry seasons, recharge diminishes, and water levels may decline. Understanding these seasonal variations is crucial for sustainable management.
4. Speleogenesis: Speleogenesis refers to the formation and development of caves and conduits within karst aquifers. The intricate network of conduits results from the dissolution of soluble rocks over geological timescales. Speleogenesis influences the overall structure of the aquifer, determining the pathways through which water moves and the storage capacity of the system.

6.5.4 Methods for Investigating Karst Aquifer Properties

1. Geophysical Surveys: Geophysical methods, such as electrical resistivity imaging and ground-penetrating radar, are employed to characterize subsurface structures in karst aquifers. These techniques provide insights

into the distribution of fractures, conduits, and other geological features, aiding in the development of conceptual models.

2. Tracer Tests: Tracer tests involve introducing a known quantity of a tracer substance into the aquifer and monitoring its movement. The tracking of tracers helps identify flow paths, residence times, and connectivity within the karst aquifer. Tracer tests are valuable for understanding the complex behavior of conduits and fractures.
3. Hydrogeological Modeling: Hydrogeological models, often numerical, are used to simulate the behavior of karst aquifers. These models incorporate data on hydraulic conductivity, porosity, and the geometry of conduits to simulate groundwater flow under different scenarios. Modeling aids in predicting responses to various stressors and supports sustainable management practices.
4. Dye Tracing: Dye tracing involves introducing a visible dye into a known point within the aquifer and monitoring its emergence at springs or other outlets. This method provides information on flow paths, velocity, and connectivity. Dye tracing is particularly effective in delineating the extent of conduits and understanding the dynamics of karst aquifers.

6.5.5 Sustainable Management and Protection

1. Source Water Protection: Given the vulnerability of karst aquifers to contamination, source water protection measures are essential. Implementing protective measures in recharge areas, regulating land use, and monitoring potential contaminant sources contribute to safeguarding the quality of groundwater in karst aquifers.
2. Land-Use Planning: Land-use planning in karst regions should consider the unique properties of these aquifers. Avoiding high-density development in sinkhole-prone areas, protecting recharge zones, and implementing best management practices for agriculture are vital components of sustainable land-use planning.
3. Monitoring and Early Warning Systems: Continuous monitoring of water levels, spring discharge, and water quality is crucial for understanding the behavior of karst aquifers. Early warning systems can provide alerts about potential changes in groundwater conditions, such as rising water levels during heavy rainfall, helping mitigate risks associated with flooding or contamination events.
4. Public Awareness and Education: Public awareness and education programs play a crucial role in promoting responsible behavior in karst regions. Informing communities about the vulnerabilities of karst aquifers, the importance of source water protection, and the potential risks associated

with land use can foster a sense of responsibility and participation in sustainable practices.

6.6 SINKHOLES AND LAND SUBSIDENCE

6.6.1 Sinkhole

A sinkhole is a depression or hole in the ground caused by some form of collapse of the surface layer. A cenote is a type of sinkhole that exposes groundwater underneath. Sink and stream sink are more general terms for sites that drain surface water, possibly by infiltration into sediment or crumbled rock. Most sinkholes are caused by karst processes – the chemical dissolution of carbonate rocks, collapse or suffosion processes. Sinkholes are usually circular and vary in size from tens to hundreds of meters both in diameter and depth, and vary in form from soil-lined bowls to bedrock-edged chasms. Sinkholes may form gradually or suddenly, and are found worldwide.

Formation

Natural Processes

Sinkholes may capture surface drainage from running or standing water, but may also form in high and dry places in specific locations. Sinkholes that capture drainage can hold it in large limestone caves. These caves may drain into tributaries of larger rivers.

The formation of sinkholes involves natural processes of erosion or gradual removal of slightly soluble bedrock (such as limestone) by percolating water, the collapse of a cave roof, or a lowering of the water table. Sinkholes often form through the process of suffosion. For example, groundwater may dissolve the carbonate cement holding the sandstone particles together and then carry away the lax particles, gradually forming a void.

Occasionally a sinkhole may exhibit a visible opening into a cave below. In the case of exceptionally large sinkholes, such as the Minyé sinkhole in Papua New Guinea or Cedar Sink at Mammoth Cave National Park in Kentucky, an underground stream or river may be visible across its bottom flowing from one side to the other.

Sinkholes are common where the rock below the land surface is limestone or other carbonate rock, salt beds, or in other soluble rocks, such as gypsum, that can be dissolved naturally by circulating ground water. Sinkholes also occur in sandstone and quartzite terrains.

As the rock dissolves, spaces and caverns develop underground. These sinkholes can be dramatic, because the surface land usually stays intact until there is not enough support. Then, a sudden collapse of the land surface can occur.

Space and Planetary Bodies

On 2 July 2015, scientists reported that active pits, related to sinkhole collapses and possibly associated with outbursts, were found on the comet 67P/Churyumov-Gerasimenko by the Rosetta space probe.

Artificial Processes

Collapses, commonly incorrectly labeled as sinkholes, also occur due to human activity, such as the collapse of abandoned mines and salt cavern storage in salt domes in places like Louisiana, Mississippi and Texas, in the United States of America. More commonly, collapses occur in urban areas due to water main breaks or sewer collapses when old pipes give way. They can also occur from the overpumping and extraction of groundwater and subsurface fluids.

Sinkholes can also form when natural water-drainage patterns are changed and new water-diversion systems are developed. Some sinkholes form when the land surface is changed, such as when industrial and runoff-storage ponds are created; the substantial weight of the new material can trigger a collapse of the roof of an existing void or cavity in the subsurface, resulting in development of a sinkhole.

Human Accelerated Sinkholes

The U.S. Geological Survey notes that "It is a frightening thought to imagine the ground below your feet or house suddenly collapsing and forming a big hole in the ground." Human activities can accelerate collapses of karst sinkholes, causing collapse within a few years that would normally evolve over thousands of years under natural conditions. Soil-collapse sinkholes, which are characterized by the collapse of cavities in soil that have developed where soil falls down into underlying rock cavities, pose the most serious hazards to life and property. Fluctuation of the water level accelerates this collapse process. When water rises up through fissures in the rock, it reduces soil cohesion. Later, as the water level moves downward, the softened soil seeps downwards into rock cavities. Flowing water in karst conduits carries the soil away, preventing soil from accumulating in rock cavities and allowing the collapse process to continue.

Induced sinkholes occur where human activity alters how surface water recharges groundwater. Many human-induced sinkholes occur where natural diffused recharge is disturbed and surface water becomes concentrated. Activities that can accelerate sinkhole collapses include timber removal, ditching, laying pipelines, sewers, water lines, storm drains, and drilling. These activities can increase the downward movement of water beyond the natural rate of groundwater recharge. The increased runoff from the impervious surfaces of roads, roofs, and parking lots also accelerate man-induced sinkhole collapses.

Some induced sinkholes are preceded by warning signs, such as cracks, sagging, jammed doors, or cracking noises, but others develop with little or no warning. However, karst development is well understood, and proper site characterization can avoid karst disasters. Thus most sinkhole disasters are predictable and preventable rather than "acts of God". The American Society of Civil Engineers has declared that the potential for sinkhole collapse must be a part of land-use planning in karst areas. Where sinkhole collapse of structures could cause loss of life, the public should be made aware of the risks.

The most likely locations for sinkhole collapse are areas where there is already a high density of existing sinkholes. Their presence shows that the subsurface contains a cave system or other unstable voids. Where large cavities exist in the limestone large surface collapses can occur, such the Winter Park, Florida sinkhole collapse. Recommendations for land uses in karst areas should avoid or minimize alterations of the land surface and natural drainage.

Since water level changes accelerate sinkhole collapse, measures must be taken to minimize water level changes. The areas most susceptible to sinkhole collapse can be identified and avoided. In karst areas the traditional foundation evaluations (bearing capacity and settlement) of the ability of soil to support a structure must be supplemented by geotechnical site investigation for cavities and defects in the underlying rock. Since the soil/rock surface in karst areas are very irregular the number of subsurface samples (borings and core samples) required per unit area is usually much greater than in non-karst areas.

In 2015, the U.S. Geological Survey estimated the cost for repairs of damage arising from karst-related processes as at least $300 million per year over the preceding 15 years, but noted that this may be a gross underestimate based on inadequate data. The greatest amount of karst sinkhole damage in the United States occurs in Florida, Texas, Alabama, Missouri, Kentucky, Tennessee, and Pennsylvania. The largest recent sinkhole in the USA is possibly one that formed in 1972 in Montevallo, Alabama as a result of man-made lowering of the water level in a nearby rock quarry. This "December Giant" or "Golly Hole" sinkhole measures 130 m (425 ft) long, 105 m (350 ft) wide and 45 m (150 ft) deep.

Other areas of significant karst hazards include the Ebro Basin in northern Spain; the island of Sardinia; the Italian peninsula; the Chalk areas in southern England; Sichuan, China; Jamaica; France;Croatia; Bosnia and Herzegovina; Slovenia; and Russia, where one-third of the total land area is underlain by karst.

Occurrence

Sinkholes tend to occur in karst landscapes. Karst landscapes can have up to thousands of sinkholes within a small area, giving the landscape a pock-marked appearance. These sinkholes drain all the water, so there are only subterranean rivers in these areas.

Examples of karst landscapes with numerous massive sinkholes include Khammouan Mountains (Laos) and Mamo Plateau (Papua New Guinea). The largest known sinkholes formed in sandstone are Sima Humboldt and Sima Martel in Venezuela.

Some sinkholes form in thick layers of homogeneous limestone. Their formation is facilitated by high groundwater flow, often caused by high rainfall; such rainfall causes formation of the giant sinkholes in the Nakanaï Mountains, on the New Britain island in Papua New Guinea. Powerful underground rivers may form on the contact between limestone and underlying insoluble rock, creating large underground voids.

In such conditions, the largest known sinkholes of the world have formed, like the 662-metre-deep (2,172 ft) Xiaozhai Tiankeng (Chongqing, China), giant sótanos in Querétaro and San Luis Potosí states in Mexico and others.

Unusual processes have formed the enormous sinkholes of Sistema Zacatón in Tamaulipas (Mexico), where more than 20 sinkholes and other karst formations have been shaped by volcanically heated, acidic groundwater. This has produced not only the formation of the deepest water-filled sinkhole in the world—Zacatón—but also unique processes of travertine sedimentation in upper parts of sinkholes, leading to sealing of these sinkholes with travertine lids.

The U.S. state of Florida in North America is known for having frequent sinkhole collapses, especially in the central part of the state. Underlying limestone there is from 15 to 25 million years old. On the fringes of the state, sinkholes are rare or non-existent; limestone there is around 120,000 years old.

The Murge area in southern Italy also has numerous sinkholes. Sinkholes can be formed in retention ponds from large amounts of rain.

On the Arctic seafloor, methane emissions have caused large sinkholes to form.

Human Uses

Sinkholes have been used for centuries as disposal sites for various forms of waste. A consequence of this is the pollution of groundwater resources, with serious health implications in such areas.

The Maya civilization sometimes used sinkholes in the Yucatán Peninsula (known as cenotes) as places to deposit precious items and human sacrifices.

When sinkholes are very deep or connected to caves, they may offer challenges for experienced cavers or, when water-filled, divers. Some of the most spectacular are the Zacatón cenote in Mexico (the world's deepest water-filled sinkhole), the Boesmansgat sinkhole in South Africa, Sarisariñama tepuy in Venezuela, the Sótano del Barro in Mexico, and in the town of Mount Gambier, South Australia. Sinkholes that form in coral reefs and islands that collapse to enormous depths are known as blue holes and often become popular diving spots.

Local Names

Large and visually unusual sinkholes have been well known to local people since ancient times. Nowadays sinkholes are grouped and named in site-specific or generic names. Some examples of such names are listed below.

1. Aven: In the south of France this name means pit cave in the Occitan language.
2. Black holes (not to be confused with cosmic black holes): This term refers to a group of unique, round, water-filled pits in the Bahamas. These formations seem to be dissolved in carbonate mud from above, by the sea water. The dark color of the water is caused by a layer of phototropic microorganisms concentrated in a dense, purple colored layer at 15 to 20 m (49 to 66 ft) depth; this layer "swallows" the light. Metabolism in the layer of microorganisms causes heating of the water. One of them is the Black Hole of Andros.
3. Blue holes: This name was initially given to the deep underwater sinkholes of the Bahamas but is often used for any deep water-filled pits formed in carbonate rocks. The name originates from the deep blue color of water in these sinkholes, which is created by the high clarity of the water and the great depth of the sinkholes; only the deep blue color of the visible spectrum can penetrate such depth and return after reflection.
4. Cenotes: This refers to the characteristic water-filled sinkholes in the Yucatán Peninsula, Belize and some other regions. Many cenotes have formed in limestone deposited in shallow seas created by the Chicxulub meteorite's impact.
5. Dolina: Slavic toponym (from Slovene language dòla:"valley" or "dale"). The name is widely used in various slavic regions of Eastern Europe for sinkholes, including Romania.
6. Foiba: Friulan Italian dialect word (from the Latin fOvea: "pit" or "chasm"). The name, is given to sinkholes in the frontier zone between the Italian region of Friuli Venezia Giulia, Croatia and Slovenia, in the Karst Plateau.
7. Sótanos: This name is given to several giant pits in several states of Mexico.
8. Tiankengs: These are extremely large sinkholes, typically deeper and wider than 250 m (820 ft), with mostly vertical walls, most often created by the collapse of caverns. The term means sky holes in Chinese; many of this largest type of sinkhole are located in China.
9. Tomo: This term is used in New Zealand karst country to describe sinkholes.

Piping Pseudokarst

The 2010 Guatemala City sinkhole formed suddenly in May of that year; torrential rains from Tropical Storm Agatha and a bad drainage system were blamed for its creation. It swallowed a three-story building and a house; it measured approximately 20 m (66 ft) wide and 30 m (98 ft) deep. A similar hole had formed nearby in February 2007.

This large vertical hole is not a true sinkhole, as it did not form via the dissolution of limestone, dolomite, marble, or any other water-soluble rock. Instead, they are examples of "piping pseudokarst", created by the collapse of large cavities that had developed in the weak, crumbly Quaternary volcanic deposits underlying the city. Although weak and crumbly, these volcanic deposits have enough cohesion to allow them to stand in vertical faces and to develop large subterranean voids within them. A process called "soil piping" first created large underground voids, as water from leaking water mains flowed through these volcanic deposits and mechanically washed fine volcanic materials out of them, then progressively eroded and removed coarser materials. Eventually, these underground voids became large enough that their roofs collapsed to create large holes.

Crown Hole

A crown hole is subsidence due to subterranean human activity, such as mining and military trenches. Examples have included, instances above World War I trenches in Ypres, Belgium; near mines in Nitra, Slovakia; a limestone quarry in Dudley, England; and above an old gypsum mine in Magheracloone, Ireland.

6.6.2 Subsidence

Subsidence is a general term for downward vertical movement of the Earth's surface, which can be caused by both natural processes and human activities. Subsidence involves little or no horizontal movement, which distinguishes it from slope movement.

Processes that lead to subsidence include dissolution of underlying carbonate rock by groundwater; gradual compaction of sediments; withdrawal of fluid lava from beneath a solidified crust of rock; mining; pumping of subsurface fluids, such as groundwater or petroleum; or warping of the Earth's crust by tectonic forces. Subsidence resulting from tectonic deformation of the crust is known as tectonic subsidence and can create accommodation for sediments to accumulate and eventually lithify into sedimentary rock.

Ground subsidence is of global concern to geologists, geotechnical engineers, surveyors, engineers, urban planners, landowners, and the public in general. Pumping of groundwater or petroleum has led to subsidence of as much as 9 meters (30 ft) in many locations around the world and incurring costs measured in hundreds of millions of US dollars.

Causes

Dissolution of Limestone

Subsidence frequently causes major problems in karst terrains, where dissolution of limestone by fluid flow in the subsurface creates voids (i.e., caves). If the roof of a void becomes too weak, it can collapse and the overlying rock and earth will fall into the space, causing subsidence at the surface. This type of subsidence can cause sinkholes which can be many hundreds of meters deep.

Mining

Several types of sub-surface mining, and specifically methods which intentionally cause the extracted void to collapse (such as pillar extraction, longwall mining and any metalliferous mining method which uses "caving" such as "block caving" or "sub-level caving") will result in surface subsidence. Mining-induced subsidence is relatively predictable in its magnitude, manifestation and extent, except where a sudden pillar or near-surface tunnel collapse occurs (usually very old workings). Mining-induced subsidence is nearly always very localized to the surface above the mined area, plus a margin around the outside.

The vertical magnitude of the subsidence itself typically does not cause problems, except in the case of drainage (including natural drainage)–rather, it is the associated surface compressive and tensile strains, curvature, tilts and horizontal displacement that are the cause of the worst damage to the natural environment, buildings and infrastructure.

Where mining activity is planned, mining-induced subsidence can be successfully managed if there is co-operation from all of the stakeholders. This is accomplished through a combination of careful mine planning, the taking of preventive measures, and the carrying out of repairs post-mining.

Extraction of Petroleum and Natural Gas

If natural gas is extracted from a natural gas field the initial pressure (up to 60 MPa (600 bar)) in the field will drop over the years. The pressure helps support the soil layers above the field. If the gas is extracted, the overburden pressure sediment compacts and may lead to earthquakes and subsidence at the ground level.

Since exploitation of the Slochteren (Netherlands) gas field started in the late 1960s the ground level over a 250 km area has dropped by a current maximum of 30 cm.

Extraction of petroleum likewise can cause significant subsidence. The city of Long Beach, California, has experienced 9 meters (30 ft) over the course of 34 years of petroleum extraction, resulting in damage of over $100 million to infrastructure in the area. The subsidence was brought to a halt when secondary recovery wells pumped enough water into the oil reservoir to stabilize it.

Earthquake

Land subsidence can occur in various ways during an earthquake. Large areas of land can subside drastically during an earthquake because of offset along fault lines. Land subsidence can also occur as a result of settling and compacting of unconsolidated sediment from the shaking of an earthquake.

The Geospatial Information Authority of Japan reported immediate subsidence caused by the 2011 Tôhoku earthquake. In Northern Japan, subsidence of 0.50 m (1.64 ft) was observed on the coast of the Pacific Ocean in Miyako, Tôhoku, while Rikuzentakata, Iwate measured 0.84 m (2.75 ft). In the south at Sôma, Fukushima, 0.29 m (0.95 ft) was observed. The maximum amount of subsidence was 1.2 m (3.93 ft), coupled with horizontal diastrophism of up to 5.3 m (17.3 ft) on the Oshika Peninsula in Miyagi Prefecture.

Groundwater-related Subsidence

Groundwater-related subsidence is the subsidence (or the sinking) of land resulting from groundwater extraction. It is a growing problem in the developing world as cities increase in population and water use, without adequate pumping regulation and enforcement. One estimate has 80% of serious land subsidence problems associated with the excessive extraction of groundwater, making it a growing problem throughout the world. Groundwater fluctuations can also indirectly affect the decay of organic material. The habitation of lowlands, such as coastal or delta plains, requires drainage. The resulting aeration of the soil leads to the oxidation of its organic components, such as peat, and this decomposition process may cause significant land subsidence. This applies especially when groundwater levels are periodically adapted to subsidence, in order to maintain desired unsaturated zone depths, exposing more and more peat to oxygen. In addition to this, drained soils consolidate as a result of increased effective stress. In this way, land subsidence has the potential of becoming self-perpetuating, having rates up to 5 cm/yr. Water management used to be tuned primarily to factors such as crop optimization but, to varying extents, avoiding subsidence has come to be taken into account as well.

Faulting Induced

When differential stresses exist in the Earth, these can be accommodated either by geological faulting in the brittle crust, or by ductile flow in the hotter and more fluid mantle. Where faults occur, absolute subsidence may occur in the hanging wall of normal faults. In reverse, or thrust, faults, relative subsidence may be measured in the footwall.

Isostatic Subsidence

The crust floats buoyantly in the asthenosphere, with a ratio of mass below the "surface" in proportion to its own density and the density of the asthenosphere. If mass is added to a local area of the crust (e.g., through deposition), the crust subsides to compensate and maintain isostatic balance.

The opposite of isostatic subsidence is known as isostatic rebound–the action of the crust returning (sometimes over periods of thousands of years) to a state of isostacy, such as after the melting of large ice sheets or the drying-up of large lakes after the last ice age. Lake Bonneville is a famous example of isostatic rebound. Due to the weight of the water once held in the lake, the earth's crust subsided nearly 200 feet (61 m) to maintain equilibrium. When the lake dried up, the crust rebounded. Today at Lake Bonneville, the center of the former lake is about 200 feet (61 m) higher than the former lake edges.

Seasonal Effects

Many soils contain significant proportions of clay. Because of the very small particle size, they are affected by changes in soil moisture content. Seasonal drying of the soil results in a lowering of both the volume and the surface of the soil. If building foundations are above the level reached by seasonal drying, they move, possibly resulting in damage to the building in the form of tapering cracks.

Trees and other vegetation can have a significant local effect on seasonal drying of soils. Over a number of years, a cumulative drying occurs as the tree grows. That can lead to the opposite of subsidence, known as heave or swelling of the soil, when the tree declines or is felled. As the cumulative moisture deficit is reversed, which can last up to 25 years, the surface level around the tree will rise and expand laterally. That often damages buildings unless the foundations have been strengthened or designed to cope with the effect.

Weight of Buildings

High buildings can create land subsidence by pressing the soil beneath with their weight. The problem is already felt in New York City, San Francisco Bay Area, Lagos.

Impacts

Sinking Cities

Sinking cities are urban environments that are in danger of disappearing due to their rapidly changing landscapes. The largest contributors to these cities becoming unlivable are the combined effects of climate change (manifested through sea level rise, intensifying storms, and storm surge), land subsidence, and accelerated urbanization. Many of the world's largest and most rapidly growing cities are located along rivers and coasts, exposing them to natural disasters. As countries continue to invest people, assets, and infrastructure into these cities, the loss potential in these areas also increases. Sinking cities must overcome substantial barriers to properly prepare for today's dynamic environmental climate.

6.7 GROUNDWATER POTENTIAL AND HYDRAULIC HEAD

Groundwater potential and hydraulic head are integral components of hydrogeological understanding, shaping our ability to manage and sustainably utilize subsurface water resources. The complex interplay between geological, hydrological, and anthropogenic factors underscores the importance of ongoing research and advancements in monitoring technologies. As we navigate challenges such as climate change and increasing human demands on water resources, a comprehensive understanding of groundwater potential and hydraulic head will remain essential for informed decision-making, sustainable water resource management, and the preservation of this invaluable freshwater source.

6.7.1 Groundwater Potential

Groundwater potential refers to the capacity of the subsurface to yield water under specific hydrogeological conditions. It represents the ability of an aquifer to release water to wells or springs, and it is influenced by factors such as geological properties, aquifer characteristics, and recharge rates. Assessing groundwater potential is crucial for identifying suitable locations for water abstraction, understanding aquifer behavior, and managing water resources sustainably.

Factors Influencing Groundwater Potential

1. Geological Properties: The geological characteristics of an area, including the type of rocks and sediments present, influence groundwater potential. Permeable formations such as sandstone and limestone generally have higher groundwater potential than impermeable formations like clay.
2. Aquifer Properties: The properties of the aquifer, such as porosity and hydraulic conductivity, significantly impact groundwater potential. Aquifers with high porosity can store more water, while high hydraulic conductivity facilitates greater water movement within the subsurface.
3. Recharge Rates: The rate at which water replenishes the aquifer, known as recharge, is a critical factor in determining groundwater potential. Areas with higher rates of natural recharge often exhibit greater groundwater potential.
4. Topography: The topography of the land surface influences the movement of groundwater. Groundwater potential tends to be higher in areas with lower elevations and where topographic gradients favor the movement of water.

5. Human Activities: Anthropogenic activities, such as excessive groundwater pumping or improper land use practices, can impact groundwater potential negatively. Over-extraction can lead to lowered water tables and reduced groundwater potential in affected areas.

Methods for Assessing Groundwater Potential

1. Geophysical Surveys: Geophysical methods, including resistivity surveys and seismic surveys, are employed to characterize subsurface structures and assess the potential for groundwater occurrence. These surveys provide insights into the geological formations and help identify potential aquifer zones.
2. Drilling and Well Testing: Drilling wells and conducting pump tests are direct methods for assessing groundwater potential. By monitoring the response of wells to pumping, hydrogeologists can estimate aquifer characteristics, such as transmissivity and storativity, which contribute to understanding groundwater potential.
3. Hydrogeological Modeling: Numerical models, such as MODFLOW, simulate groundwater flow based on hydrogeological parameters. These models can help predict groundwater potential under various scenarios, considering factors like pumping rates, recharge, and aquifer properties.
4. Isotope Hydrology: Isotope studies involve analyzing the stable isotopes of water to trace the origin and movement of groundwater. Isotope hydrology contributes to understanding groundwater potential by providing insights into recharge sources and groundwater age.

6.7.2 Hydraulic Head

Hydraulic head is a key parameter in hydrogeology that represents the potential energy of groundwater at a specific point in the subsurface. It is measured as the height of a column of water above a reference point, typically the water table. Hydraulic head is crucial for describing and understanding groundwater flow, as it governs the direction and velocity of subsurface water movement.

Components of Hydraulic Head

1. Elevation Head (Z): Elevation head represents the potential energy associated with the elevation of groundwater above a reference datum. It is calculated as the height of the water table or piezometric surface above a chosen reference point.
2. Pressure Head (P): Pressure head accounts for the pressure exerted by the weight of the water column above a specific point in the aquifer. It is influenced by the depth of water and is measured as the head equivalent to the pressure exerted by the overlying water.

3. Velocity Head (V): Velocity head represents the kinetic energy associated with the movement of groundwater. In cases where groundwater is flowing, velocity head becomes a significant component of the total hydraulic head.
4. Total Hydraulic Head (H): The total hydraulic head at a point in the aquifer is the sum of elevation head, pressure head, and velocity head. Mathematically, it is expressed as $H=Z+P+V$.

Significance of Hydraulic Head

1. Groundwater Flow Direction: Hydraulic head governs the direction of groundwater flow. Groundwater naturally moves from areas of higher hydraulic head to areas of lower hydraulic head. Understanding hydraulic head gradients is essential for predicting flow paths and designing effective groundwater extraction systems.
2. Aquifer Response to Pumping: Pumping wells lower the hydraulic head in the vicinity, creating a cone of depression. Monitoring changes in hydraulic head during pumping tests helps assess the aquifer's response, estimate transmissivity, and evaluate the sustainable yield of the aquifer.
3. Influence on Contaminant Transport: Hydraulic head gradients impact the movement of contaminants in groundwater. Contaminants tend to migrate from areas of higher hydraulic head to lower hydraulic head, emphasizing the role of hydraulic head in understanding contaminant transport and designing remediation strategies.
4. Aquifer Recharge and Discharge: Changes in hydraulic head influence aquifer recharge and discharge. Recharge occurs in areas with higher hydraulic head, replenishing the aquifer, while discharge areas have lower hydraulic head, resulting in the outflow of groundwater to springs, rivers, or wells.

6.7.3 Measurement and Monitoring

1. Piezometers and Monitoring Wells: Piezometers and monitoring wells are installed in aquifers to measure hydraulic head at specific locations. These wells allow hydrogeologists to observe changes in water levels over time, assess the response to pumping, and monitor seasonal variations in hydraulic head.
2. Pressure Transducers and Data Loggers: Pressure transducers and data loggers provide continuous monitoring of hydraulic head. These devices can be installed in piezometers or wells to record water levels at regular intervals. The collected data offer insights into variations in hydraulic head under different conditions.

3. Field Surveys and GPS: Field surveys, coupled with Global Positioning System (GPS) technology, aid in accurately determining the elevation head component of hydraulic head. Combining field measurements with spatial data allows for comprehensive mapping of hydraulic head gradients across an aquifer.
4. Remote Sensing Technologies: Remote sensing technologies, including satellite imagery and aerial photography, contribute to monitoring changes in land surface features related to groundwater. These technologies can help identify areas of potential recharge or discharge by observing changes in vegetation patterns, land cover, or surface water features.

6.7.4 Implications for Water Resource Management

1. Sustainable Groundwater Extraction: Understanding groundwater potential and hydraulic head is essential for sustainable groundwater extraction. Properly designed wells, based on hydraulic head gradients, ensure efficient water abstraction while avoiding over-pumping and adverse impacts on aquifer sustainability.
2. Contaminant Management: Hydraulic head gradients influence the movement of contaminants in groundwater. Knowledge of hydraulic head variations allows for effective monitoring of contaminant plumes, assessment of potential risks, and the implementation of remediation measures to protect water quality.
3. Aquifer Recharge Strategies: Managing groundwater potential involves considering aquifer recharge strategies. Identifying areas with higher hydraulic head for artificial recharge can enhance the replenishment of aquifers, contributing to long-term water resource sustainability.
4. Land-Use Planning: Hydraulic head gradients play a crucial role in land-use planning, helping to delineate areas suitable for sustainable development, agriculture, or natural habitat conservation. Proper land-use planning considers the interplay between land surface features, groundwater potential, and hydraulic head variations.

6.7.5 Challenges and Future Directions

1. Complexity of Aquifer Systems: The complexity of aquifer systems, including heterogeneity in geological formations and variations in hydraulic conductivity, poses challenges in accurately characterizing groundwater potential and hydraulic head. Advances in geophysical methods and modeling techniques aim to address these challenges.
2. Climate Change Impacts: Climate change introduces uncertainties in precipitation patterns and recharge rates, influencing groundwater potential. Integrating climate change projections into hydrogeological

models will be crucial for anticipating shifts in hydraulic head and adapting to changing hydrological conditions.

3. Anthropogenic Pressures: Anthropogenic pressures, such as urbanization, agricultural expansion, and excessive groundwater pumping, can alter hydraulic head gradients. Sustainable groundwater management requires addressing these pressures through regulatory measures, efficient water use practices, and community engagement.
4. Advancements in Monitoring Technologies: Future advancements in monitoring technologies, such as the use of advanced sensors and real-time data analytics, will enhance our ability to continuously observe changes in hydraulic head. This will contribute to more adaptive and responsive groundwater management strategies.

6.8 GROUNDWATER FLOW DIRECTION

In hydrogeology, groundwater flow is defined as the "part of streamflow that ha infiltrated the ground, entered the phreatic zone, and has been (or is at a particular time) discharged into a stream channel or springs; and seepage water." It is governed by the groundwater flow equation. Groundwater is water that is found underground in cracks and spaces in the soil, sand and rocks. Where water has filled these spaces is the phreatic (also called) saturated zone. Groundwater is stored in and moves slowly (compared to surface runoff in temperate conditions and watercourses) through layers or zones of soil, sand and rocks: aquifers. The rate of groundwater flow depends on the permeability (the size of the spaces in the soil or rocks and how well the spaces are connected) and the hydraulic head (water pressure). In polar regions groundwater flow may be obstructed by permafrost.

6.9 WATER TABLE AND POTENTIOMETRIC SURFACE MAPS

6.9.1 Water Table

The water table is the upper surface of the zone of saturation. The zone of saturation is where the pores and fractures of the ground are saturated with groundwater, which may be fresh, saline, or brackish, depending on the locality. It can also be simply explained as the depth below which the ground is saturated.

The water table is the surface where the water pressure head is equal to the atmospheric pressure (where gauge pressure = 0). It may be visualized as the "surface" of the subsurface materials that are saturated with groundwater in a given vicinity.

The groundwater may be from precipitation or from groundwater flowing into the aquifer. In areas with sufficient precipitation, water infiltrates through pore spaces in the soil, passing through the unsaturated zone. At increasing depths, water fills in more of the pore spaces in the soils, until a zone of saturation is reached. Below the water table, in the phreatic zone (zone of saturation), layers of permeable rock that yield groundwater are called aquifers. In less permeable soils, such as tight bedrock formations and historic lakebed deposits, the water table may be more difficult to define.

"Water table" and "water level" are not synonymous. If a deeper aquifer has a lower permeable unit that confines the upward flow, then the water level in this aquifer may rise to a level that is greater or less than the elevation of the actual water table. The elevation of the water in this deeper well is dependent upon the pressure in the deeper aquifer and is referred to as the potentiometric surface, not the water table.

Form

The water table may vary due to seasonal changes such as precipitation and evapotranspiration. In undeveloped regions with permeable soils that receive sufficient amounts of precipitation, the water table typically slopes toward rivers that act to drain the groundwater away and release the pressure in the aquifer. Springs, rivers, lakes and oases occur when the water table reaches the surface. Groundwater entering rivers and lakes accounts for the base-flow water levels in water bodies.

Surface Topography

Within an aquifer, the water table is rarely horizontal, but reflects the surface relief due to the capillary effect (capillary fringe) in soils, sediments and other porous media. In the aquifer, groundwater flows from points of higher pressure to points of lower pressure, and the direction of groundwater flow typically has both a horizontal and a vertical component.

The slope of the water table is known as the "hydraulic gradient", which depends on the rate at which water is added to and removed from the aquifer and the permeability of the material. The water table does not always mimic the topography due to variations in the underlying geological structure (e.g., folded, faulted, fractured bedrock).

Perched Water Tables

A perched water table (or perched aquifer) is an aquifer that occurs above the regional water table. This occurs when there is an impermeable layer of rock or sediment (aquiclude) or relatively impermeable layer (aquitard) above the main water table/aquifer but below the land surface. If a perched aquifer's flow intersects the surface, at a valley wall, for example, the water is discharged as a spring.

Fluctuations

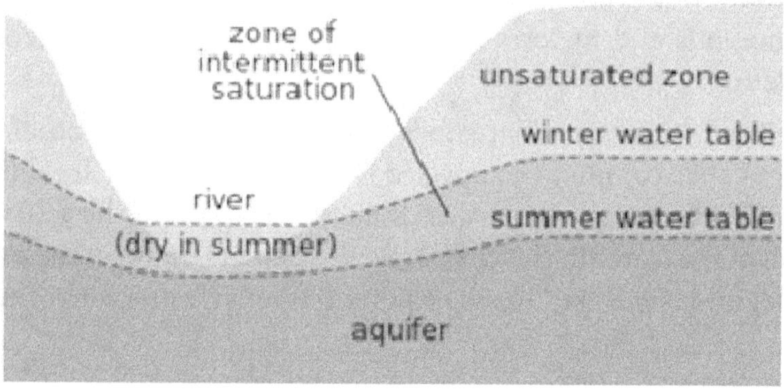

Fig. 6.6: Seasonal Fluctuations in the Water Table may Cause River Beds to Dry up during the Dry Season

Tidal

On low-lying oceanic islands with porous soil, freshwater tends to collect in lenticular pools on top of the denser seawater intruding from the sides of the islands. Such an island's freshwater lens, and thus the water table, rises and falls with the tides.

Seasonal

In some regions, for example, Great Britain or California, winter precipitation is often higher than summer precipitation and so the groundwater storage is not fully recharged in summer. Consequently, the water table is lower during the summer. This disparity between the level of the winter and summer water table is known as the "zone of intermittent saturation", wherein the water table will fluctuate in response to climatic conditions.

Long-term

Fossil water is groundwater that has remained in an aquifer for several millennia and occurs mainly in deserts. It is non-renewable by present-day rainfall due to its depth below the surface, and any extraction causes a permanent change in the water table in such regions.

Effects on Crop Yield

Most crops need a water table at a minimum depth. For some important food and fiber crops a classification was made because at shallower depths the crop suffers a yield decline.

Table 6.1: Effects

Crop and Location	DWT Tolerance	Classification	Explanation
Wheat, Nile Delta, Egypt	45	Very tolerant	Resists shallow water tables
Sugar cane, Australia	60	Tolerant	The water table should be deeper than 60 cm
Banana, Surinam	70	Slightly	Yield declines at water tables < 70 cm deep sensitive
Cotton, Nile Delta	90	Sensitive	Cotton needs dry feet, water table should be deep

Effects on Construction

A water table close to the surface affects excavation, drainage, foundations, wells and leach fields (in areas without municipal water and sanitation), and more.

When excavation occurs near enough to the water table to reach its capillary action, groundwater must be removed during construction. This is conspicuous in Berlin, which is built on sandy, marshy ground, and the water table is generally 2 meters below the surface. Pink and blue pipes can often be seen carrying groundwater from construction sites into the Spree river (or canals).

6.9.2 Potentiometric Surface

A potentiometric surface is the imaginary plane where a given reservoir of fluid will "equalize out to" if allowed to flow. A potentiometric surface is based on hydraulic principles. For example, two connected storage tanks with one full and one empty will gradually fill/drain to the same level. This is because of atmospheric pressure and gravity. This idea is heavily used in city water supplies - a tall water tower containing the water supply has a great enough potentiometric surface to provide flowing water at a decent pressure to the houses it supplies.

For groundwater "potentiometric surface" is a synonym of "piezometric surface" which is an imaginary surface that defines the level to which water in a confined aquifer would rise were it completely pierced with wells. If the potentiometric surface lies above the ground surface, a flowing artesian well results. Contour maps and profiles of the potentiometric surface can be prepared from the well data.

6.10 TYPES OF GROUNDWATER CONDITIONS

Groundwater, hidden beneath the Earth's surface, is a critical component of the global water cycle. The conditions under which groundwater exists are diverse and dynamic, influenced by geological, hydrological, and climatic factors. Understanding the types of groundwater conditions is essential for effective water resource management, sustainable extraction practices, and addressing challenges such as contamination and depletion. The diversity of groundwater conditions reflects the intricate interplay of geological, hydrological, and climatic factors. Each type of groundwater condition presents unique challenges and opportunities for water resource management. Understanding the characteristics, influences, and implications of these conditions is crucial for sustainable extraction practices, protection against contamination, and the preservation of groundwater resources. As we navigate the complexities of groundwater systems, ongoing research, monitoring, and innovative management strategies will play a pivotal role in ensuring the resilience and longevity of this vital component of the Earth's water cycle.

6.10.1 Unconfined Aquifers

Unconfined aquifers, also known as water-table aquifers, are characterized by an open water table that directly interacts with the atmosphere. The water in these aquifers is not confined by impermeable layers above, allowing direct contact between the aquifer and the atmosphere. Rainfall and other forms of precipitation directly contribute to recharge, and the water table fluctuates in response to seasonal variations and pumping.

1. Characteristics:
    a. Open water table in direct contact with the atmosphere.
    b. Vulnerable to surface contamination due to the direct connection with the land surface.
    c. Water table responds rapidly to changes in precipitation, recharge, or pumping.
2. Influences:
    a. Local topography affects the shape of the water table.
    b. High susceptibility to contamination from surface activities.
    c. Sensitive to changes in land use and vegetation cover.
3. Implications:
    a. Common in alluvial and porous formations.
    b. Quick response to changes in precipitation, making them susceptible to droughts and over-extraction.

6.10.2 Confined Aquifers

Confined aquifers, in contrast to unconfined aquifers, are sandwiched between impermeable layers, limiting the interaction between the aquifer and the atmosphere. These impermeable layers, often composed of clay or silt, confine the groundwater within the aquifer, creating pressure within the confined space. Confined aquifers are characterized by distinct boundaries and may exhibit artesian conditions, where groundwater can flow to the surface without the need for pumping.

1. Characteristics:
   a. Bounded above and below by impermeable layers.
   b. Higher pressure within the confined aquifer, potentially leading to artesian conditions.
   c. Recharge occurs through specific recharge zones or boundaries.
2. Influences:
   a. Impermeable layers determine the lateral and vertical extent of the confined aquifer.
   b. Hydraulic head gradients within the aquifer influence groundwater movement.
3. Implications:
   a. Artesian conditions may result in natural springs or flowing wells.
   b. More protection from surface contamination compared to unconfined aquifers.

6.10.3 Perched Aquifers

Perched aquifers are localized, elevated groundwater bodies that exist above the regional water table. These aquifers typically form in areas where an impermeable layer creates a barrier to downward percolation. The perched water table is often found in hillsides or elevated terrain, and its characteristics can vary significantly from the regional water table.

1. Characteristics:
   a. Separated from the regional water table by an impermeable layer.
   b. Occurs in localized, elevated positions, often on hillsides.
   c. Recharge is limited, and the perched aquifer may be small in size.
2. Influences:
   a. Topography and the presence of impermeable layers determine the formation of perched aquifers.
   b. Limited recharge potential compared to unconfined or confined aquifers.
3. Implications:
   a. Susceptible to rapid depletion during dry periods.

b. Vulnerable to contamination due to proximity to the land surface.

6.10.4 Leaky Confined Aquifers

Leaky confined aquifers, also known as semi-confined aquifers, have characteristics of both confined and unconfined aquifers. They are confined by impermeable layers but possess zones where water can leak through, allowing some interaction with the atmosphere. These aquifers often exhibit a combination of confined and unconfined conditions, with the water table fluctuating in response to both natural and human-induced factors.

1. Characteristics:
   a. Confined by impermeable layers with zones of permeability.
   b. Water table fluctuates, influenced by both confined and unconfined conditions.
   c. Recharge occurs through permeable zones.
2. Influences:
   a. Presence of leaky zones determines the interaction with the atmosphere.
   b. Hydraulic conductivity variations within the aquifer influence groundwater movement.
3. Implications:
   a. Vulnerable to contamination in leaky zones.
   b. May respond to changes in pumping rates and precipitation differently than fully confined aquifers.

6.10.5 Karst Aquifers

Karst aquifers are unique groundwater systems formed in soluble rocks such as limestone, gypsum, or dolomite. The dissolution of these rocks creates distinctive landscapes characterized by sinkholes, caves, and underground conduits. Karst aquifers often have high permeability due to the presence of conduits, allowing rapid movement of water through the subsurface.

1. Characteristics:
   a. Formed in soluble rocks with distinctive surface features.
   b. High permeability due to the presence of conduits and caves.
   c. Rapid response to changes in precipitation.
2. Influences:
   a. Geological processes such as dissolution and speleogenesis shape the aquifer.

b. Vulnerable to surface contamination due to the direct connection between surface features and groundwater.

3. Implications:
   a. Unique hydrogeological features such as sinkholes and caves.
   b. Challenges in contaminant transport due to rapid groundwater movement.

6.10.6 Coastal Aquifers

Coastal aquifers are situated near coastlines and are influenced by both freshwater and seawater interactions. These aquifers may extend beneath the land surface and the ocean floor. The interface between freshwater and seawater is dynamic and influenced by factors such as tides, pumping rates, and changes in land use.

1. Characteristics:
   a. Proximity to coastlines with potential saltwater intrusion.
   b. Dynamic interface between freshwater and seawater.
   c. Vulnerable to sea level rise and changes in coastal hydrodynamics.
2. Influences:
   a. Tidal fluctuations and changes in groundwater levels affect the movement of the interface.
   b. Human activities, such as groundwater extraction, can induce saltwater intrusion.
3. Implications:
   a. Risk of saltwater intrusion impacting freshwater resources.
   b. Requires careful management to prevent over-extraction and maintain freshwater availability.

6.10.7 Alluvial Aquifers

Alluvial aquifers are formed in unconsolidated sediments such as gravel, sand, and silt deposited by rivers or streams. These aquifers are often associated with river valleys and floodplains. Alluvial aquifers have high hydraulic conductivity, allowing for rapid groundwater movement.

1. Characteristics:
   a. Formed in unconsolidated sediments deposited by rivers.
   b. High hydraulic conductivity due to the nature of the sediments.
   c. Frequently associated with river valleys and floodplains.
2. Influences:

   a. River dynamics, sedimentation, and flood events contribute to the formation of alluvial aquifers.
   b. Recharge often occurs during periods of high river flow and flooding.
3. Implications:
   a. High potential for groundwater extraction due to the permeable nature of the sediments.
   b. Vulnerable to contamination from surface activities, especially during flood events.

6.10.8 Mountain Front Recharge

Mountain front recharge occurs in mountainous regions where precipitation infiltrates the ground at the base of the mountains, recharging the underlying aquifer. This process is influenced by the elevation and topography of the mountains, creating conditions favorable for groundwater recharge.

1. Characteristics:
   a. Precipitation infiltrates at the base of mountains.
   b. Recharge occurs due to the elevation gradient and topography.
   c. Groundwater flows from mountainous areas towards lower elevations.
2. Influences:
   a. Topography and elevation gradient influence the movement of groundwater.
   b. Vegetation cover and land use impact recharge rates.
3. Implications:
   a. Critical for sustaining base flow in downstream rivers.
   b. Vulnerable to changes in precipitation patterns and land use in mountainous regions.

CHAPTER 7

GROUNDWATER: TIPPING POINTS

7.1 DEPLETION AND AQUIFER COLLAPSE

7.1.1 Water Extraction

Water extraction is the process of taking water from any source, either temporarily or permanently, for flood control or to obtain water for, for example, irrigation. The extracted water could also be used as drinking water after suitable treatment.

Depending on the environmental legislation in the country, controls may be placed on extraction to limit the amount of water that can be removed. The over-extraction of water can lead to dry rivers or declining groundwater levels.

The science of hydrogeology is used to determine safe water extraction levels. Water can go through dams that are used to regulate or stop water from coming though, creating hydroelectricity.

Effects of Overextraction

Saltwater Intrusion

Saltwater intrusion is the movement of saline water into freshwater aquifers, which can lead to groundwater quality degradation, including drinking water sources, and other consequences. Saltwater intrusion can naturally occur in coastal aquifers, owing to the hydraulic connection between groundwater and seawater. Because saline water has a higher mineral content than freshwater, it is denser and has a higher water pressure. As a result, saltwater can push inland beneath the freshwater. In other topologies, submarine groundwater discharge can push fresh water into saltwater.

Certain human activities, especially groundwater pumping from coastal freshwater wells, have increased saltwater intrusion in many coastal areas. Water extraction drops the level of fresh groundwater, reducing its water pressure and allowing saltwater to flow further inland. Other contributors to saltwater intrusion include navigation channels or agricultural and drainage channels, which provide conduits for saltwater to move inland. Sea level rise caused by climate change also contributes to saltwater intrusion. Saltwater intrusion can also be worsened by extreme events like hurricane storm surges.

Groundwater Contamination

Groundwater contamination of water is seen vastly through the high needs of irrigation, drinking, and to support organic life. Keeping contaminants at a minimum is at a high demand and treated using arsenic, chloride and other chemicals to extract the pollutants. Humans can be a direct cause of these pollutants through over extraction.

Certain leading causes for groundwater contamination comes from lowering water tables due to the over extraction of water and the water table not being able to recharge as quickly as needed. With this being said, polluted water from the surface (rivers and streams), makes it way into the groundwater more quickly and easily and results in a water quality problem due to the surface water pollution.

Land Subsidence

Land Subsidence is another effect linked to the over extraction of groundwater. When large amounts of groundwater is extracted from aquifers beneath, surrounding areas above. When water from the aquifer is extracted at a large amount, the sediment, certain rock types, is separated due to the lack of water being used to make sure the sediment stays tightly together. The over extraction of groundwater is a human caused activity that causes these ground failures that create pore spaces where water once was occupying. The sudden sinking of the soils surface causes infrastructure damage and a higher risk of flood damage due to the displacement of the Earth's surface.

Groundwater Extraction Laws

Public Rights and Laws Established towards Groundwater Extraction

Groundwater laws contain the information revolving around the rights of water extraction and the withdrawals from aquifers in the United States that is measured. The most of the groundwater that is mainly withdrawn or extracted from aquifers consist of primarily irrigation towards the Southwest and the West with close to 85 to 90 withdrawn. With an expected increase of the demand of water for domestic usage in the future, systems are to be regulated and land rights vary on the consumption of groundwater rights. When requirements haven't been met through the water extraction, States water resources control and take steps on authorization of the requirements for groups of individuals or corporations.

7.1.2 Salinization

Salinity is the saltiness or amount of salt dissolved in a body of water, called saline water. It is usually measured in g/L or g/kg (grams of salt per liter/kilogram of water; the latter is dimensionless and equal to ‰). Salinity is an important factor in determining many aspects of the chemistry of natural waters and of biological processes within it, and is a thermodynamic state variable that, along with temperature and pressure, governs physical characteristics like the density and heat capacity of the water. A contour line of constant salinity is called an isohaline, or sometimes isohale.

Definitions

Salinity in rivers, lakes, and the ocean is conceptually simple, but technically challenging to define and measure precisely. Conceptually the salinity is the quantity of dissolved salt content of the water. Salts are compounds like sodium chloride, magnesium sulfate, potassium nitrate, and sodium bicarbonate which dissolve into ions. The concentration of dissolved chloride ions is sometimes referred to as chlorinity. Operationally, dissolved matter is defined as that which can pass through a very fine filter (historically a filter with a pore size of 0.45 ìm, but nowadays usually 0.2 ìm). Salinity can be expressed in the form of a mass fraction, i.e., the mass of the dissolved material in a unit mass of solution. Seawater typically has a mass salinity of around 35 g/kg, although lower values are typical near coasts where rivers enter the ocean. Rivers and lakes can have a wide range of salinities, from less than 0.01 g/kg to a few g/kg, although there are many places where higher salinities are found. The Dead Sea has a salinity of more than 200 g/kg. Precipitation typically has a TDS of 20 mg/kg or less.

Whatever pore size is used in the definition, the resulting salinity value of a given sample of natural water will not vary by more than a few percent (%). Physical oceanographers working in the abyssal ocean, however, are often concerned with precision and intercomparability of measurements by different researchers, at different times, to almost five significant digits. A bottled seawater product known as IAPSO Standard Seawater is used by oceanographers to standardize their measurements with enough precision to meet this requirement.

Composition

Measurement and definition difficulties arise because natural waters contain a complex mixture of many different elements from different sources (not all from dissolved salts) in different molecular forms. The chemical properties of some of these forms depend on temperature and pressure. Many of these forms are difficult to measure with high accuracy, and in any case complete chemical analysis is not practical when analyzing multiple samples. Different practical definitions of salinity result from different attempts to account for these problems, to different levels of precision, while still remaining reasonably easy to use.

For practical reasons salinity is usually related to the sum of masses of a subset of these dissolved chemical constituents (so-called solution salinity), rather than to the unknown mass of salts that gave rise to this composition (an exception is when artificial seawater is created). For many purposes this sum can be limited to a set of eight major ions in natural waters, although for seawater at highest precision an additional seven minor ions are also included. The major ions dominate the inorganic composition of most (but by no means all) natural waters. Exceptions include some pit lakes and waters from some hydrothermal springs.

The concentrations of dissolved gases like oxygen and nitrogen are not usually included in descriptions of salinity. However, carbon dioxide gas, which when dissolved is partially converted into carbonates and bicarbonates, is often included. Silicon in the form of silicic acid, which usually appears as a neutral molecule in the pH range of most natural waters, may also be included for some purposes (e.g., when salinity/density relationships are being investigated).

Seawater

The term 'salinity' is, for oceanographers, usually associated with one of a set of specific measurement techniques. As the dominant techniques evolve, so do different descriptions of salinity. Salinities were largely measured using titration-based techniques before the 1980s. Titration with silver nitrate could be used to determine the concentration of halide ions (mainly chlorine and bromine) to give a chlorinity. The chlorinity was then multiplied by a factor to account for all other constituents. The resulting 'Knudsen salinities' are expressed in units of parts per thousand (ppt or ‰).

The use of electrical conductivity measurements to estimate the ionic content of seawater led to the development of the scale called the practical salinity scale 1978 (PSS-78). Salinities measured using PSS-78 do not have units. The suffix psu or PSU (denoting practical salinity unit) is sometimes added to PSS-78 measurement values. The addition of PSU as a unit after the value is "formally incorrect and strongly discouraged".

In 2010 a new standard for the properties of seawater called the thermodynamic equation of seawater 2010 (TEOS-10) was introduced, advocating absolute salinity as a replacement for practical salinity, and conservative temperature as a replacement for potential temperature. This standard includes a new scale called the reference composition salinity scale. Absolute salinities on this scale are expressed as a mass fraction, in grams per kilogram of solution. Salinities on this scale are determined by combining electrical conductivity measurements with other information that can account for regional changes in the composition of seawater. They can also be determined by making direct density measurements.

A sample of seawater from most locations with a chlorinity of 19.37 ppt will have a Knudsen salinity of 35.00 ppt, a PSS-78 practical salinity of about 35.0, and a TEOS-10 absolute salinity of about 35.2 g/kg. The electrical conductivity of this water at a temperature of 15 °C is 42.9 mS/cm. On the global scale, it is extremely likely that human-caused climate change has contributed to observed surface and subsurface salinity changes since the 1950s, and projections of surface salinity changes throughout the 21st century indicate that fresh ocean regions will continue to get fresher and salty regions will continue to get saltier.

Salinity is serving as a tracer of different masses. Surface water is pulled in to replace the sinking water, which in turn eventually becomes cold and salty enough to sink. Salinity distribution contributes to shape the oceanic circulation.

Lakes and Rivers

Limnologists and chemists often define salinity in terms of mass of salt per unit volume, expressed in units of mg/L or g/L. It is implied, although often not stated, that this value applies accurately only at some reference temperature because solution volume varies with temperature. Values presented in this way are typically accurate to the order of 1%. Limnologists also use electrical conductivity, or “reference conductivity”, as a proxy for salinity. This measurement may be corrected for temperature effects, and is usually expressed in units of ìS/cm.

A river or lake water with a salinity of around 70 mg/L will typically have a specific conductivity at 25 °C of between 80 and 130 ìS/cm. The actual ratio depends on the ions present. The actual conductivity usually changes by about 2% per degree Celsius, so the measured conductivity at 5 °C might only be in the range of 50–80 ìS/cm.

Direct density measurements are also used to estimate salinities, particularly in highly saline lakes. Sometimes density at a specific temperature is used as a proxy for salinity. At other times an empirical salinity/density relationship developed for a particular body of water is used to estimate the salinity of samples from a measured density.

Table 7.1: Water Salinity

Fresh Water	Brackish Water	Saline Water	Brine
< 0.05%	0.05–3%	3–5%	> 5%
< 0.5 ‰	0.5–30 ‰	30–50 ‰	> 50 ‰

Classification of Water Bodies based upon Salinity

Marine waters are those of the ocean, another term for which is euhaline seas. The salinity of euhaline seas is 30 to 35 ‰. Brackish seas or waters have salinity in the range of 0.5 to 29 ‰ and metahaline seas from 36 to 40 ‰. These waters are all regarded as thalassic because their salinity is derived from the ocean and defined as homoiohaline if salinity does not vary much over time (essentially constant).

In contrast to homoiohaline environments are certain poikilohaline environments (which may also be thalassic) in which the salinity variation is biologically significant.

Poikilohaline water salinities may range anywhere from 0.5 to greater than 300 ‰. The important characteristic is that these waters tend to vary in salinity over some biologically meaningful range seasonally or on some other roughly comparable time scale. Put simply, these are bodies of water with quite variable salinity.

Highly saline water, from which salts crystallize (or are about to), is referred to as brine.

Environmental Considerations

Salinity is an ecological factor of considerable importance, influencing the types of organisms that live in a body of water. As well, salinity influences the kinds of plants that will grow either in a water body, or on land fed by a water (or by a groundwater). A plant adapted to saline conditions is called a halophyte. A halophyte which is tolerant to residual sodium carbonate salinity are called glasswort or saltwort or barilla plants. Organisms (mostly bacteria) that can live in very salty conditions are classified as extremophiles, or halophiles specifically. An organism that can withstand a wide range of salinities is euryhaline.

Salts are expensive to remove from water, and salt content is an important factor in water use, factoring into potability and suitability for irrigation. Increases in salinity have been observed in lakes and rivers in the United States, due to common road salt and other salt de-icers in runoff.

The degree of salinity in oceans is a driver of the world's ocean circulation, where density changes due to both salinity changes and temperature changes at the surface of the ocean produce changes in buoyancy, which cause the sinking and rising of water masses. Changes in the salinity of the oceans are thought to contribute to global changes in carbon dioxide as more saline waters are less soluble to carbon dioxide. In addition, during glacial periods, the hydrography is such that a possible cause of reduced circulation is the production of stratified oceans. In such cases, it is more difficult to subduct water through the thermohaline circulation.

Not only is salinity a driver of ocean circulation, but changes in ocean circulation also affect salinity, particularly in the subpolar North Atlantic where from 1990 to 2010 increased contributions of Greenland meltwater were counteracted by increased northward transport of salty Atlantic waters. However, North Atlantic waters have become fresher since the mid-2010s due to increased Greenland meltwater flux.

7.2 ECOLOGICAL DISRUPTION

7.2.1 Habitat Destruction

Habitat destruction (also termed habitat loss and habitat reduction) is the process by which a natural habitat becomes incapable of supporting its native species. The organisms that previously inhabited the site are displaced or dead, thereby reducing biodiversity and species abundance. Habitat destruction is the leading cause of biodiversity loss. Fragmentation and loss of habitat have become one of the most important topics of research in ecology as they are major threats to the survival of endangered species.

Activities such as harvesting natural resources, industrial production and urbanization are human contributions to habitat destruction. Pressure from agriculture is the principal human cause. Some others include mining, logging, trawling, and urban sprawl. Habitat destruction is currently considered the primary cause of species extinction worldwide. Environmental factors can contribute to habitat destruction more indirectly. Geological processes, climate change, introduction of invasive species, ecosystem nutrient depletion, water and noise pollution are some examples. Loss of habitat can be preceded by an initial habitat fragmentation.

Attempts to address habitat destruction are in international policy commitments embodied by Sustainable Development Goal 15 "Life on Land" and Sustainable Development Goal 14 "Life Below Water". However, the United Nations Environment Programme report on "Making Peace with Nature" released in 2021 found that most of these efforts had failed to meet their internationally agreed upon goals.

Impacts on Organisms

When a habitat is destroyed, the carrying capacity for indigenous plants, animals, and other organisms is reduced so that populations decline, sometimes up to the level of extinction.

Habitat loss is perhaps the greatest threat to organisms and biodiversity. Temple (1986) found that 82% of endangered bird species were significantly threatened by habitat loss. Most amphibian species are also threatened by native habitat loss, and some species are now only breeding in modified habitat. Endemic organisms with limited ranges are most affected by habitat destruction, mainly because these organisms are not found anywhere else in the world, and thus have less chance of recovering. Many endemic organisms have very specific requirements for their survival that can only be found within a certain ecosystem, resulting in their extinction.

Extinction may also take place very long after the destruction of habitat, a phenomenon known as extinction debt. Habitat destruction can also decrease the range of certain organism populations. This can result in the reduction of genetic diversity and perhaps the production of infertile youths, as these organisms would have a higher possibility of mating with related organisms within their population, or different species. One of the most famous examples is the impact upon China's giant panda, once found in many areas of Sichuan. Now it is only found in fragmented and isolated regions in the southwest of the country, as a result of widespread deforestation in the 20th century.

As habitat destruction of an area occurs, the species diversity offsets from a combination of habitat generalists and specialists to a population primarily consisting of generalist species. Invasive species are frequently generalists that are able to survive in much more diverse habitats. Habitat destruction leading to climate change offsets the balance of species keeping up with the extinction threshold leading to a higher likelihood of extinction.

Habitat loss is one of the main environmental causes of the decline of biodiversity on local, regional, and global scales. Many believe that habitat fragmentation is also a threat to biodiversity however some believe that it is secondary to habitat loss. The reduction of the amount of habitat available results in specific landscapes that are made of isolated patches of suitable habitat throughout a hostile environment/ matrix. This process is generally due to pure habitat loss as well as fragmentation effects. Pure habitat loss refers to changes occurring in the composition of the landscape that causes a decrease in individuals. Fragmentation effects refer to an addition of effects occurring due to the habitat changes. Habitat loss can result in negative effects on the dynamic of species richness. The order Hymenoptera is a diverse group of plant pollinators who are highly susceptible to the negative effects of habitat loss, this could result in a domino effect between the plant-pollinator interactions leading to major conservation implications within this group.

Destruction of Populations

Habitat fragmentation has a major impact on animal specie populations because it deprives species of what they are naturally accustomed to. This makes the species isolated, reduces the area where they can live, and creates new ecological boundaries. Some studies have shown that changes in the abiotic and biotic parameters have caused a greater impact on the ecology than the reduction in habitat size itself. They concluded that crowding a species into one space will eventually lead to the extinction of that species.

The destruction and fragmentation of natural habitats are currently the leading factors in species extinction. This is because the loss and fragmentation of habitats results in much smaller populations. Reduced population sizes ends up creating higher chances of extinction.

Studies have shown that there is no relationship between habitat patch and species number when it comes to habitat specialist plants species located in fragmented landscapes. This could potentially be due to drastic declines of plant species areas due to changes in the surrounding land.

Predators Affecting the Population of the Prey

In recent times the destruction of habitat has been the cause of the loss of many species. Sometimes the area may be small of destruction but as time goes by slowly that will cause an increase in extinction. Loss of habitat is not always the direct cause of extinction; there are other reasons causes for extinction that connect back to the loss of habitat. For example, if the sole predator in an ecosystem were to go extinct, prey populations would increase, which could possibly result in overpopulation. A higher amount of any species that can cause them to use too much of their resources. Since many species depend on limited natural resources, with the overuse they will eventually run out degrade their habitat.

Habitat destruction and fragmentation are the two most important factors in species extinction. The negative effects of decreasing size and increasing isolation of habitat are misinterpreted by fragmentation, but in reality they are much more larger effects on the population. Fragmentation generally has either no effect or a negative effect on population survival. Since habitat loss of fragmentation typically occurs together it is still not clear which process has a larger effect on extinction. Increasing isolation and habitat loss with fragmentation are all connected in a way that has negatively affected the environment.

Geography

Biodiversity hotspots are chiefly tropical regions that feature high concentrations of endemic species and, when all hotspots are combined, may contain over half of the world's terrestrial species. These hotspots are suffering from habitat loss and destruction. Most of the natural habitat on islands and in areas of high human population density has already been destroyed (WRI, 2003). Islands suffering extreme habitat destruction include New Zealand, Madagascar, the Philippines, and Japan. South and East Asia—especially China, India, Malaysia, Indonesia, and Japan—and many areas in West Africa have extremely dense human populations that allow little room for natural habitat. Marine areas close to highly populated coastal cities also face degradation of their coral reefs or other marine habitat. Forest City, a township in southern Malaysia built on Environmentally Sensitive Area (ESA) Rank 1 wetland is one such example, with irreversible reclamation proceeding prior to environmental impact assessment and approvals. Other such areas include the eastern coasts of Asia and Africa, northern coasts of South America, and the Caribbean Sea and its associated islands.

Regions of unsustainable agriculture or unstable governments, which may go hand in-hand, typically experience high rates of habitat destruction. South Asia, Central America, Sub-Saharan Africa, and the Amazonian tropical rainforest areas of South America are the main regions with unsustainable agricultural practices and/or government mismanagement.

Areas of high agricultural output tend to have the highest extent of habitat destruction. In the U.S., less than 25% of native vegetation remains in many parts of the East and Midwest. Only 15% of land area remains unmodified by human activities in all of Europe.

Currently, changes occurring in different environments around the world are changing the specific geographical habitats that are suitable for plants to grow. Therefore, the ability for plants to migrate to suitable environment areas will have a strong impact on the distribution of plant diversity. However, at the moment, the rates of plant migration that are influenced by habitat loss and fragmentation are not as well understood as they could be.

Ecosystems

Tropical rainforests have received most of the attention concerning the destruction of habitat. From the approximately 16 million square kilometers of tropical rainforest habitat that originally existed worldwide, less than 9 million square kilometers remain today. The current rate of deforestation is 160,000 square kilometers per year, which equates to a loss of approximately 1% of original forest habitat each year.

Other forest ecosystems have suffered as much or more destruction as tropical rainforests. Deforestation for farming and logging have severely disturbed at least 94% of temperate broadleaf forests; many old growth forest stands have lost more than 98% of their previous area because of human activities. Tropical deciduous dry forests are easier to clear and burn and are more suitable for agriculture and cattle ranching than tropical rainforests; consequently, less than 0.1% of dry forests in Central America's Pacific Coast and less than 8% in Madagascar remain from their original extents.

Plains and desert areas have been degraded to a lesser extent. Only 10–20% of the world's drylands, which include temperate grasslands, savannas, and shrublands, scrub, and deciduous forests, have been somewhat degraded. But included in that 10–20% of land is the approximately 9 million square kilometers of seasonally drylands that humans have converted to deserts through the process of desertification.

The tallgrass prairies of North America, on the other hand, have less than 3% of natural habitat remaining that has not been converted to farmland.

Wetlands and marine areas have endured high levels of habitat destruction. More than 50% of wetlands in the U.S. have been destroyed in just the last 200 years. Between 60% and 70% of European wetlands have been completely destroyed. In the United Kingdom, there has been an increase in demand for coastal housing and tourism which has caused a decline in marine habitats over the last 60 years. The rising sea levels and temperatures have caused soil erosion, coastal flooding, and loss of quality in the UK marine ecosystem. About one-fifth (20%) of marine coastal areas have been highly modified by humans. One-fifth of coral reefs have also been destroyed, and another fifth has been severely degraded by overfishing, pollution, and invasive species; 90% of the Philippines' coral reefs alone have been destroyed. Finally, over 35% of the mangrove ecosystems worldwide have been destroyed.

Natural Causes

Habitat destruction through natural processes such as volcanism, fire, and climate change is well documented in the fossil record. One study shows that habitat fragmentation of tropical rainforests in Euramerica 300 million years ago led to a great loss of amphibian diversity, but simultaneously the drier climate spurred on a burst of diversity among reptiles.

Human Causes

Habitat destruction caused by humans includes land conversion from forests, etc. to arable land, urban sprawl, infrastructure development, and other anthropogenic changes to the characteristics of land. Habitat degradation, fragmentation, and pollution are aspects of habitat destruction caused by humans that do not necessarily involve over destruction of habitat, yet result in habitat collapse. Desertification, deforestation, and coral reef degradation are specific types of habitat destruction for those areas (deserts, forests, coral reefs).

Geist and Lambin (2002) assessed 152 case studies of net losses of tropical forest cover to determine any patterns in the proximate and underlying causes of tropical deforestation. Their results, yielded as percentages of the case studies in which each parameter was a significant factor, provide a quantitative prioritization of which proximate and underlying causes were the most significant. The proximate causes were clustered into broad categories of agricultural expansion (96%), infrastructure expansion (72%), and wood extraction (67%). Therefore, according to this study, forest conversion to agriculture is the main land use change responsible for tropical deforestation. The specific categories reveal further insight into the specific causes of tropical deforestation: transport extension (64%), commercial wood extraction (52%), permanent cultiation (48%), cattle ranching (46%), shifting (slash and burn) cultivation (41%), subsistence agriculture (40%), and fuel wood extraction for domestic use (28%). One result is that shifting cultivation is not the primary cause of deforestation in all world regions, while transport extension (including the construction of new roads) is the largest single proximate factor responsible for deforestation.

Global Warming

Rising global temperatures, caused by the greenhouse effect, contribute to habitat destruction, endangering various species, such as the polar bear. Melting ice caps promote rising sea levels and floods which threaten natural habitats and species globally.

Drivers

While the above-mentioned activities are the proximal or direct causes of habitat destruction in that they actually destroy habitat, this still does not identify why humans destroy habitat. The forces that cause humans to destroy habitat are known as drivers of habitat destruction. Demographic, economic, sociopolitical, scientific and technological, and cultural drivers all contribute to habitat destruction.

Demographic drivers include the expanding human population; rate of population increase over time; spatial distribution of people in a given area (urban versus rural), ecosystem type, and country; and the combined effects of poverty, age, family planning, gender, and education status of people in certain areas. Most of the exponential human population growth worldwide is occurring in or close to biodiversity hotspots. This may explain why human population density accounts for 87.9% of the variation in numbers of threatened species across 114 countries, providing indisputable evidence that people play the largest role in decreasing biodiversity. The boom in human population and migration of people into such species-rich regions are making conservation efforts not only more urgent but also more likely to conflict with local human interests. The high local population density in such areas is directly correlated to the poverty status of the local people, most of whom lacking an education and family planning.

According to the Geist and Lambin (2002) study, the underlying driving forces were prioritized as follows (with the percent of the 152 cases the factor played a significant role in): economic factors (81%), institutional or policy factors (78%), technological factors (70%), cultural or socio-political factors (66%), and demographic factors (61%). The main economic factors included commercialization and growth of timber markets (68%), which are driven by national and international demands; urban industrial growth (38%); low domestic costs for land, labor, fuel, and timber (32%); and increases in product prices mainly for cash crops (25%). Institutional and policy factors included formal pro-deforestation policies on land development (40%), economic growth including colonization and infrastructure improvement (34%), and subsidies for land-based activities (26%); property rights and land-tenure insecurity (44%); and policy failures such as corruption, lawlessness, or mismanagement (42%). The main technological factor was the poor application of technology in the wood industry (45%), which leads to wasteful logging practices. Within the broad category of cultural and sociopolitical factors are public attitudes and values (63%), individual/household behavior (53%), public unconcern toward forest environments (43%), missing basic values (36%), and unconcern by individuals (32%). Demographic factors were the in-migration of colonizing settlers into sparsely populated forest areas (38%) and growing population density—a result of the first factor—in those areas (25%).

There are also feedbacks and interactions among the proximate and underlying causes of deforestation that can amplify the process. Road construction has the largest feedback effect, because it interacts with—and leads to—the establishment of new settlements and more people, which causes a growth in wood (logging) and food markets. Growth in these markets, in turn, progresses the commercialization of agriculture and logging industries. When these industries become commercialized, they must become more efficient by utilizing larger or more modern machinery that often has a worse effect on the habitat than traditional farming and logging methods. Either way, more land is cleared more rapidly for commercial markets. This common feedback example manifests just how closely related the proximate and underlying causes are to each other.

Impact on Human Population

Fig. 7.1: The Draining and Development of Coastal Wetlands that Previously Protected the Gulf Coast Contributed to Severe Flooding in New Orleans

Habitat destruction can vastly increase an area's vulnerability to natural disasters like flood and drought, crop failure, spread of disease, and water contamination. On the other hand, a healthy ecosystem with good management practices can reduce the chance of these events happening, or will at least mitigate adverse impacts. Eliminating swamps—the habitat of pests such as mosquitoes—has contributed to the prevention of diseases such as malaria. Completely depriving an infectious agent (such as a virus) of its habitat—by vaccination, for example—can result in eradicating that infectious agent.

Agricultural land can suffer from the destruction of the surrounding landscape. Over the past 50 years, the destruction of habitat surrounding agricultural land has degraded approximately 40% of agricultural land worldwide via erosion, salinization, compaction, nutrient depletion, pollution, and urbanization. Humans also lose direct uses of natural habitat when habitat is destroyed. Aesthetic uses such as birdwatching, recreational uses like hunting and fishing, and ecotourism usually rely upon relatively undisturbed habitat. Many people value the complexity of the natural world and express concern at the loss of natural habitats and of animal or plant species worldwide.

Probably the most profound impact that habitat destruction has on people is the loss of many valuable ecosystem services. Habitat destruction has altered nitrogen, phosphorus, sulfur, and carbon cycles, which has increased the frequency and severity of acid rain, algal blooms, and fish kills in rivers and oceans and contributed tremendously to global climate change.

One ecosystem service whose significance is becoming better understood is climate regulation. On a local scale, trees provide windbreaks and shade; on a regional scale, plant transpiration recycles rainwater and maintains constant annual rainfall; on a global scale, plants (especially trees in tropical rainforests) around the world counter the accumulation of greenhouse gases in the atmosphere by sequestering carbon dioxide through photosynthesis.

Other ecosystem services that are diminished or lost altogether as a result of habitat destruction include watershed management, nitrogen fixation, oxygen production, pollination, waste treatment (i.e., the breaking down and immobilization of toxic pollutants), and nutrient recycling of sewage or agricultural runoff.

The loss of trees from tropical rainforests alone represents a substantial diminishing of Earth's ability to produce oxygen and to use up carbon dioxide. These services are becoming even more important as increasing carbon dioxide levels is one of themain contributors to global climate change. The loss of biodiversity may not directly affect humans, but the indirect effcts of losing many species as well as the diversity of ecosystems in general are enormous.

When biodiversity is lost, the environment loses many species that perform valuable and unique roles in the ecosystem. The environment and all its inhabitants rely on biodiversity to recover from extreme environmental conditions. When too much biodiversity is lost, a catastrophic event such as an earthquake, flood, or volcanic eruption could cause an ecosystem to crash, and humans would obviously suffer from that. Loss of biodiversity also means that humans are losing animals that could have served as biological-control agents and plants that could potentially provide higher-yielding crop varieties, pharmaceutical drugs to cure existing or future diseases (such as cancer), and new resistant crop-varieties for agricultural species susceptible to pesticide-resistant insects or virulent strains of fungi, viruses, and bacteria.

The negative effects of habitat destruction usually impact rural populations more directly than urban populations. Across the globe, poor people suffer the most when natural habitat is destroyed, because less natural habitat means fewer natural resources per capita, yet wealthier people and countries can simply pay more to continue to receive more than their per capita share of natural resources.

Another way to view the negative effects of habitat destruction is to look at the opportunity cost of destroying a given habitat. In other words, what do people lose out on with the removal of a given habitat? A country may increase its food supply by converting forest land to row-crop agriculture, but the value of the same land may be much larger when it can supply natural resources or services such as clean water, timber, ecotourism, or flood regulation and drought control.

Outlook

The rapid expansion of the global human population is increasing the world's food requirement substantially. Simple logic dictates that more people will require more food.

In fact, as the world's population increases dramatically, agricultural output will need to increase by at least 50%, over the next 30 years. In the past, continually moving to new land and soils provided a boost in food production to meet the global food demand. That easy fix will no longer be available, however, as more than 98% of all land suitable for agriculture is already in use or degraded beyond repair.

The impending global food crisis will be a major source of habitat destruction. Commercial farmers are going to become desperate to produce more food from the same amount of land, so they will use more fertilizers and show less concern for the environment to meet the market demand. Others will seek out new land or will convert other land-uses to agriculture. Agricultural intensification will become widespread at the cost of the environment and its inhabitants. Species will be pushed out of their habitat either directly by habitat destruction or indirectly by fragmentation, degradation, or pollution. Any efforts to protect the world's remaining natural habitat and biodiversity will compete directly with humans' growing demand for natural resources, especially new agricultural lands.

Solutions

Tropical deforestation: In most cases of tropical deforestation, three to four underlying causes are driving two to three proximate causes. This means that a universal policy for controlling tropical deforestation would not be able to address the unique combination of proximate and underlying causes of deforestation in each country. Before any local, national, or international deforestation policies are written and enforced, governmental leaders must acquire a detailed understanding of the complex combination of proximate causes and underlying driving forces of deforestation in a given area or country. This concept, along with many other results of tropical deforestation from the Geist and Lambin study, can easily be applied to habitat destruction in general.

Shoreline erosion: Coastal erosion is a natural process as storms, waves, tides and other water level changes occur. Shoreline stabilization can be done by barriers between land and water such as seawalls and bulkheads. Living shorelines are gaining attention as a new stabilization method. These can reduce damage and erosion while simultaneously providing ecosystem services such as food production, nutrient and sediment removal, and water quality improvement to society

Fig. 7.2: Example of Human Caused Habitat Destruction likely Capable of Reversing if Further Disturbance is Halted, Uganda

Fig. 7.3: Natural Vegetation along this Coastal Shoreline in North Carolina, US, is being Used to Reduce the Effects of Shoreline Erosion while Providing other Benefits to the Natural Ecosystem and the Human Community

Preventing an area from losing its specialist species to generalist invasive species depends on the extent of the habitat destruction that has already taken place. In areas where the habitat is relatively undisturbed, halting further habitat destruction may be enough. In areas where habitat destruction is more extreme (fragmentation or patch loss), restoration ecology may be needed.

Education of the general public is possibly the best way to prevent further human habitat destruction. Changing the dull creep of environmental impacts from being viewed as acceptable to being seen a reason for change to more sustainable practices. Education about the necessity of family planning to slow population growth is important as greater population leads to greater human caused habitat destruction.

The preservation and creation of habitat corridors can link isolated populations and increase pollination. Corridors are also known to reduce the negative impacts of habitat destruction.

The biggest potential to solving the issue of habitat destruction comes from solving the political, economical and social problems that go along with it such as, individual and commercial material consumption, sustainable extraction of resources, conservation areas, restoration of degraded land and addressing climate change.

Governmental leaders need to take action by addressing the underlying driving forces, rather than merely regulating the proximate causes. In a broader sense, governmental bodies at a local, national, and international scale need to emphasize:

1. Considering the irreplaceable ecosystem services provided by natural habitats.
2. Protecting remaining intact sections of natural habitat.
3. Finding ecological ways to increase agricultural output without increasing the total land in production.
4. Reducing human population and expansion. Apart from improving access to contraception globally, furthering gender equality also has a great benefit. When women have the same education (decision-making power), this generally leads to smaller families.

It is argued that the effects of habitat loss and fragmentation can be counteracted by including spatial processes in potential restoration management plans. However, even though spatial dynamics are incredibly important in the conservation and recovery of species, a limited amount of management plans are taking the spatial effects of habitat restoration and conservation into consideration.

7.2.2 Biodiversity Loss

Biodiversity loss includes the worldwide extinction of different species, as well as the local reduction or loss of species in a certain habitat, resulting in a loss of biological diversity. The latter phenomenon can be temporary or permanent, depending on whether the environmental degradation that leads to the loss is reversible through ecological restoration/ecological resilience or effectively permanent (e.g., through land loss). The current global extinction (frequently called the sixth mass extinction or Anthropocene extinction), has resulted in a biodiversity crisis being driven by human activities which push beyond the planetary boundaries and so far has proven irreversible.

The main direct threats to conservation (and thus causes for biodiversity loss) fall in eleven categories: Residential and commercial development; farming activities; energy production and mining; transportation and service corridors; biological resource usages; human intrusions and activities that alter, destroy, disturb habitats and species from exhibiting natural behaviors; natural system modification; invasive and problematic species, pathogens and genes; pollution; catastrophic geological events, climate change, and so on.

Numerous scientists and the IPBES Global Assessment Report on Biodiversity and Ecosystem Services assert that human population growth and overconsumption are the primary factors in this decline. However other scientists have criticized this, saying that loss of habitat is caused mainly by "the growth of commodities for export" and that population has very little to do with overall consumption, due to country wealth disparities.

Climate change is another threat to global biodiversity. For example, coral reefs – which are biodiversity hotspots – will be lost within the century if global warming continues at the current rate. However, habitat destruction e.g., for the expansion of agriculture, is currently the more significant driver of contemporary biodiversity loss, not climate change.

International environmental organizations have been campaigning to prevent biodiversity loss for decades, public health officials have integrated it into the One Health approach to public health practice, and increasingly preservation of biodiversity is part of international policy, as part of the response to the Triple planetary crisis. For example, the UN Convention on Biological Diversity is focused on preventing biodiversity loss and proactive conservation of wild areas. The international commitment and goals for this work is currently embodied by Sustainable Development Goal 15 "Life on Land" and Sustainable Development Goal 14 "Life Below Water". However, the United Nations Environment Programme report on "Making Peace with Nature" released in 2020 found that most of these efforts had failed to meet their international goals. Of the 20 biodiversity goals laid out by the Aichi Biodiversity Targets in 2010, only 6 were "partially achieved" by the deadline of 2020.

Loss Rate

Biodiversity is commonly defined as the variety of life on Earth in all its forms, including the diversity of species, their genetic variations, and the interaction of these lifeforms.

The most unique feature of Earth is the existence of life, and the most extraordinary feature of life is its diversity. The current rate of global diversity loss is estimated to be 100 to 1000 times higher than the (naturally occurring) background extinction rate, faster than at any other time in human history, and expected to still grow in the upcoming years. These rapidly rising extinction trends impacting numerous animal groups including mammals, birds, reptiles, amphibians and ray-finned fishes have prompted scientists to declare a contemporary biodiversity crisis, in both terrestrial and marine ecosystems.

Locally bounded loss rates can be measured using species richness and its variation over time. Raw counts may not be as ecologically relevant as relative or absolute abundances. Absolute abundance is expressed as a species population size or density while relative abundance is the percentage occurrence of the individuals of a species relative to other species (also called evenness). Taking into account the relative frequencies, many biodiversity indexes have been developed. Besides richness, evenness and heterogeneity are considered to be the main dimensions along which diversity can be measured.

As with all diversity measures, it is essential to accurately classify the spatial and temporal scope of the observation. "Definitions tend to become less precise as the complexity of the subject increases and the associated spatial and temporal scales widen." Biodiversity itself is not a single concept but can be split up into various scales (e.g., ecosystem diversity vs. habitat diversity or even biodiversity vs. habitat diversity) or different subcategories (e.g., phylogenetic diversity, species diversity, genetic diversity, nucleotide diversity). The question of net loss in confined regions is often a matter of debate but longer observation times are generally thought to be beneficial to loss estimates.

To compare rates between different geographic regions, latitudinal gradients in species diversity should also be considered.

In 2006, many more species were formally classified as rare or endangered or threatened; moreover, scientists have estimated that millions more species are at risk which have not been formally recognized.

In 2021, about 28 percent of the 134,400 species assessed using the IUCN Red List criteria are now listed as threatened with extinction—a total of 37,400 species compared to 16,119 threatened species in 2006.

A 2022 study, which surveyed more than 3,000 experts, states that "global biodiversity loss and its impacts may be greater than previously thought", and estimates that roughly 30% of species "have been globally threatened or driven extinct since the year 1500."

The World Wildlife Fund in 2022 reports an average population decline of 68% between 1970 and 2016 for 4,400 animal species around the world encompassing nearly 21,000 monitored populations.

Of 70,000 monitored species, some 48% are experiencing population declines from anthropogenic pressures (as of 2023), whereas only 3% have increasing populations.

Terrestrial Invertebrate Loss

In 2017, various publications described the dramatic reduction in absolute insect biomass and number of species in Germany and North America over a period of 27 years. As possible reasons for the decline, the authors highlight neonicotinoids and other agrochemicals. Writing in the journal PLOS One, Hallman et al. (2017) conclude that "the widespread insect biomass decline is alarming".

For example, the critical decline of earthworms (over 80% on average) has been recorded under non-ecological agricultural practices. Earthworms play an important role in ecosystem function. For example, they help with biological processing in soil, water, and even green house gas balancing. The decline of earthworm populations are said to be due to five reasons; soil degradation and destruction of habitat, climate change, biological invasion of nonnative species, poor soil management, and pollutant loading. Factors like tillage practices and intensive land use decimate the soil and plant roots that earthworms use to create their biomass, causing carbon and nitrogen cycles to be impacted negatively. Knowledge of earthworm species diversity is quite limited as not even 50% of them have been described. More studies upon earthworms and how they provide their ecosystem services must be done to gain a better understanding of going about preserving their diversity. With earthworm populations dwindling, this has caused for the Secretariat of the Convention on Biological Diversity to take action and promote the restoration and maintenance of the many diverse species of earthworms.

Birds Loss

Certain types of pesticide, neonicotinoids, probably contribute to the decline of certain bird species. A study funded by BirdLife International confirms that 51 species of birds are critically endangered and 8 could be classified as extinct or in danger of extinction. Nearly 30% of extinction is due to hunting and trapping for the exotic pet trade. Deforestation, caused by unsustainable logging and agriculture, could be the next extinction driver, because birds lose their habitat and their food. The biologist Luisa Arnedo said: "as soon as the habitat is gone, they're gone too".

Within the Amazon rainforest there is an area called Bele'm and it is an area of endemism. In Bele'm 76% of the land has already been stripped of its natural resources, including the trees of the forest. Within the area bird species are strongly affected by the deforestation, due to being put in that situation 56% of the birds are now in danger of going into extinction. With the climate changing as well as their habitat, the population of the birds will continue to decline. Even with protected areas of land, the efficiency in which birds are conserved are low.

Modern bird hunting and trapping is a common practice in South America. Some cultures in Brazil encourage bird hunting and trapping for commercial reasons. Some reasons include, selling the wild birds as pets, breeding the birds and selling the young, selling the birds for food, and selling them for religious and medicinal purposes.

Another increasingly abundant threat to bird populations is collisions and electrocutions due to power lines. Migratory species are at a higher risk of collision accidents and up to 1 billion birds are killed due to colliding with buildings each year in the United States.

Freshwater Species Loss

Freshwater ecosystems ranging from swamps, deltas, to rivers make up to 1% of earths surface. Although making up such little proportion of the earth, freshwater ecosystems are important because these kind of habitats are home to approximately one third of vertebrate species. Freshwater species are beginning to decline at twice the rate of other species such as those located on land or within the ocean, this rapid loss has already placed 27% of 29,500 species dependent on freshwater upon the IUCN Red List. With freshwater species declining so quickly, it is due to the poor systems in place that do not provide any protection to their biodiversity.

A study by 16 global conservation organizations found that the biodiversity crisis is most acute in freshwater ecosystems, with a rate of decline double that of oceans and forests. Global populations of freshwater fish are collapsing from anthropogenic impacts such as pollution and overfishing. Migratory fish populations have declined by 76% since 1970, and large "megafish" populations have fallen by 94% with 16 species declared extinct in 2020.

Native Species Richness Loss

Humans have altered plant richness in regional landscapes worldwide, transforming more than 75% of the terrestrial biomes to "anthropogenic biomes". This is seen through loss of native species being replaced and out competed by agriculture. Models indicate that about half of the biosphere has seen a "substantial net anthropogenic change" in species richness.

Trees

Scientists have warned, in a follow-up paper to their 2021 study, that a third of tree species are threatened with extinction, showing how this will significantly alter the world's ecosystems and could get averted with "urgent actions". They find that "Large-scale extinction of tree species will lead to major biodiversity losses in other species groups and substantially alter the cycling of carbon, water and nutrients in the world's ecosystems" and may "undermine the livelihoods of [...] billions". The GTA (global tree assessment) has determined that there are 17510 or 29.9% are considered threatened with extinction and there are 142 tree species recorded as extinct or extinct in the wild. Based on different forest types and their locations and conditions, different silvicultural methods of forest management can be taken to promote tree biodiversity, such as selective logging, thinning or crop tree management, and clear cutting and coppicing. It is important to make sure that any sort of sustainability intervention "should be minimal and mimic as much as possible small-scale natural disturbances" to avoid any further damage to the ecosystem.

Marine Species Richness Loss

Marine biodiversity encompasses any living organism which resides in the ocean, and describes various complex relationships within marine ecosystems. On a local and regional scale, marine communities are better understood compared to marine ecosystems on a global scale. In 2018, approximately 240,000 marine species had been documented, but many marine species - estimates range between 178,000 and 10 million oceanic species - remain to be described. Given the paucity of data on most marine species, it is likely that a number of 'rare' species not seen for decades in the world Ocean have already disappeared or are on the brink of extinction, unnoticed.

With anthropogenic pressure, this results in human activities having the strongest influences on marine biodiversity, with main drivers of global extinction being habitat loss, pollution, invasive species, and overexploitation. Greater pressure is placed on marine ecosystems with human settlements near coastal areas. Other indirect factors that have resulted in marine species to decline include climate change and change to oceanic biochemistry.

Overexploitation has resuled in the extinction of over 25 described marine species, which includes seabirds, marine mammals, algae, and fishes. Examples of extinct marine species include the Steller's sea cow (Hydrodamalis gigas) and the Caribbean monk seal (Monachus tropicalis). However, not all extinctions are because of humans. For example, in the 1930s, the eelgrass limpet (Lottia alveus) became extinct in the NW Atlantic area once the Zostera marina seagrass population declined upon exposure to a disease. The Lottia alveus were greatly impacted as the Zostera marina were their sole habitats.

Causes

Major factors for biotic stress and the ensuing accelerating loss rate are, amongst other threats:

1. Habitat loss, fragmentation and degradation.
2. Land use intensification (and ensuing land loss/habitat loss) has been identified to be a significant factor in loss of ecological services due to direct effects as well as biodiversity loss. Habitat fragmentation for commercial and agricultural uses (specifically monoculture farming) is another factor.

3. Excessive nutrient load and other forms of pollution.
4. Over-exploitation and unsustainable use (e.g., unsustainable fishing methods).
5. Armed conflict, which disrupts human livelihoods and institutions, contributes to habitat loss, and intensifies over-exploitation of economically valuable species, leading to population declines and local extinctions.
6. Invasive alien species that effectively compete for a niche, replacing indigenous species.
7. Climate change through heat stress and drought stress.

Invasive species and other disturbances have become more common in forests in the last several decades. These tend to be directly or indirectly connected to climate change and have negative consequences for forest ecosystems. Jared Diamond describes an "Evil Quartet" of habitat destruction, overkill, introduced species and secondary extinctions. Edward O. Wilson suggested the acronym HIPPO for the main causes of biodiversity loss, standing for Habitat destruction, Invasive species, Pollution, human over-Population and Over-harvesting.

According to the IUCN the main direct threats to conservation fall in eleven categories:

1. Residential & commercial development: Housing & urban areas, commercial & industrial areas, tourism & recreational areas.
2. Farming activities: Agriculture, aquaculture.
3. Energy production and mining: Renewable energy production, including hydroelectric dams, non-renewable energy production (oil and gas drilling), mining (fuel and minerals).
4. Transportation & service corridors: Service corridors (for example electrical & phone wires, oil & gas pipelines), transport corridors (roads, railroads, shipping lanes, and flight paths), collisions with the vehicles using the corridors, associated accidents and catastrophes (oil spills, electrocution, fire).
5. Biological resource usages: Hunting, persecution (predator control and pest control, superstitions), plant destruction or removal (human consumption, free-range livestock foraging), logging or wood harvesting, fishing.
6. Human intrusions & activities that alter, destroy, disturb habitats and species from exhibiting natural behaviors: Recreational activities, war, civil unrest, & military exercises, illegal activities (smuggling, vandalism), newly built housing.

7. Natural system modifications: Fire suppression or creation, water management (for example dam construction, groundwater pumping), other modifications (for example land reclamation), removing/reducing human maintenance (for example lack of indigenous management of key ecosystems, ceasing supplemental feeding of condors).
8. Invasive & problematic species, pathogens & genes: Invasive species, problematic native species, introduced genetic material (for example pesticide-resistant crops, genetically modified insects for biocontrol), pathogens & microbes.
9. Pollution: Sewage, military & industrial effluents, forestry and agricultural effluents, garbage & solid waste, air-borne pollutants, excess energy (for example noise pollution and heat pollution).
10. Catastrophic geological events (natural hazards): Earthquakes, tsunamis, avalanches, landslides, & volcanic eruptions and gas emissions.
11. Effects of climate change: Ecosystem encroachment (for example inundation of shoreline ecosystems & drowning of coral reefs from sea level rise), changes in geochemical regimes (for example ocean acidification), changes in temperature regimes (for example heat waves, marine heatwaves), changes in precipitation & hydrological regimes, severe weather events.

Habitat Destruction

Fig. 7.4: Deforestation and Increased Road-building in the Amazon Rainforest in Bolivia

Habitat destruction has played a key role in extinctions, especially in relation to tropical forest destruction. Factors contributing to habitat loss include: overconsumption, overpopulation, land use change, deforestation, pollution (air pollution, water pollution, soil contamination) and global warming or climate change.

Habitat size and numbers of species are systematically related. Physically larger species and those living at lower latitudes or in forests or oceans are more sensitive to reduction in habitat area. Conversion to "trivial" standardized ecosystems (e.g., monoculture following deforestation) effectively destroys habitat for the more diverse species that preceded the conversion. Even the simplest forms of agriculture affect diversity – through clearing/draining the land, discouraging weeds and "pests", and encouraging just a limited set of domesticated plant and animal species. In some countries, property rights or lax law/regulatory enforcement are associated with deforestation and habitat loss.

Co-extinctions are a form of habitat destruction. Co-extinction occurs when the extinction or decline in one species accompanies similar processes in another, such as in plants and beetles.

A 2019 report has revealed that bees and other pollinating insects have been wiped out of almost a quarter of their habitats across the United Kingdom. The population crashes have been happening since the 1980s and are affecting biodiversity. The increase in industrial farming and pesticide use, combined with diseases, invasive species, and climate change is threatening the future of these insects and the agriculture they support.

In 2019, research found that insects are destroyed by human activities like habitat destruction, pesticide poisoning, invasive species and climate change at a rate that will cause the collapse of ecological systems in the next 50 years if it cannot be stopped.

Human Overpopulation and Overconsumption

The world's population numbered nearly 7.6 billion as of mid-2017 and is forecast to peak towards the end of the 21st century at between 10–12 billion people. Scholars have argued that population size and growth, along with overconsumption, are significant factors in biodiversity loss and soil degradation.

Review articles, including the 2019 IPBES report, have also noted that human population growth and overconsumption are significant drivers of species decline. A 2022 study warned that conservation efforts will continue to fail if the primary drivers of biodiversity loss continue to be ignored, including population size and growth. In December 2022 Inger Andersen, the executive director of the United Nations Environment Programme, stated as delegates were meeting for COP 15 that "the more people there are, the more we put the Earth under heavy pressure. As far as biodiversity is concerned, we are at war with nature."

However, other scientists have criticized the assertion that population growth is a key driver for biodiversity loss. A scientific perspective, published in the journal Biological Conservation argues the main driver is the loss of habitat is caused by "the growth of commodities for export, particularly soybean and oil-palm, primarily for livestock feed or biofuel consumption in higher income economies." Because of the wealth disparities between countries, the paper shows a negative correlation between a countries' total population and its per capita footprint. While there is a strong positive correlation between country GDP and footprint. The study argues that population as a metric is ultimately unhelpful and counterproductive for tackling environmental challenges.

Human populations drive of biodiversity loss include habitat alteration, pollution, and overexploitation of resources. The resulting habitat destruction, which can come in many forms, is the global leader in loss of biodiversity

Land-use Change

Land use change describes the various changes facilitated by humans to the land, include deforestation, intensive monoculture, and urbanization. The 2019 IPBES Global Assessment Report on Biodiversity and Ecosystem Services asserts that industrial agriculture is the primary driver collapsing biodiversity. The UN's Global Biodiversity Outlook 2014 estimates that 70 percent of the projected loss of terrestrial biodiversity are caused by agriculture use. Moreover, more than 1/3 of the planet's land surface is utilised for crops and grazing of livestock. Agriculture destroys biodiversity by converting natural habitats to intensely managed systems and by releasing pollutants, including greenhouse gases. Food value chains further amplify impacts including through energy use, transport and waste.

More than 17,000 species are at risk of losing habitat by 2050 as agriculture continues to expand to meet future food needs (as of 2020). Greater agricultural efficiency in the developing world and large scale transitions to healthier, plant-based diets could help reduce habitat loss. A global shift towards largely plant-based diets would free up land to allow for the restoration of ecosystems and biodiversity, because in the 2010s over 80% of all global farmland was used to rear animals.

As of 2022 at least 64 million square kilometers (24.7 million square miles)—44% of terrestrial area—require conservation attention (ranging from protected areas to land-use policies) in order to secure important biodiversity areas, ecologically intact areas, and optimal locations for representation of species ranges and ecoregions.

The direct effects of urban growth on habitat loss are well understood: building construction often results in habitat destruction and fragmentation. The rise of urbanization greatly reduced biodiversity when large areas of natural habitat are fragmented, leading to selection for species that are adapted to urban environments. Small habitat patches are unable to support the same level of genetic or taxonomic diversity as they formerly could while some of the more sensitive species may become locally extinct. Species abundance populations are reduced due to the reduced fragmented area of habitat, this causes an increase of species isolation and forces species towards edge habitats and adapt to foraging elsewhere. Human caused habitat fragmentation tends to create barriers to dispersal which prevent species from moving with its ideal environment as its shifted by climate change. While the negative effects of fragmentation tend to be well known, the risk of fragmentation tends to have smaller effects on biodiversity, and can even change and strengthen certain inter-species relationships.

Infrastructure development in Key Biodiversity Areas (KBA) is a major driver of biodiversity loss, with infrastructure being present in roughly 80% of KBAs. Infrastructure development "drives conversion and fragmentation of natural habitat, pollution, disturbance, direct mortality through collisions with vehicles and structures, and can have impacts beyond the infrastructure site."

Pollution

Air Pollution

Air pollution adversely affects biodiversity and is considered the world's largest environmental health threat. Four greenhouse gases that are commonly studied and monitored are water vapor, carbon dioxide, methane, and nitrous oxide. In the past 250 years, concentrations of carbon dioxide and methane have increased, along with the introduction of purely anthropogenic emissions such as hydrofluorocarbons, perfluorocarbons, and sulfur hexafluoride into the atmosphere. These pollutants are emitted into the atmosphere by the burning of fossil fuels and biomass, deforestation, and agricultural practices which amplify the effects of climate change. As larger concentrations of greenhouse gases are released into the atmosphere, this causes the Earth's surface temperature to increase. This is because greenhouse gases are capable of absorbing, emitting, and trapping heat from the Sun and into the Earth's atmosphere. With the increase in temperature expected from increasing greenhouse gases, there will be higher levels of air pollution, greater variability in weather patterns, intensification of climate change effects, and changes in the distribution of vegetation in the landscape.

Other pollutants that are released from industrial and agricultural activity are sulfur dioxide and nitrogen oxides. Once sulfur dioxide and nitrogen oxide are introduced into the atmosphere, they can react with cloud droplets (cloud condensation nuclei), raindrops, or snowflakes, forming sulfuric acid and nitric acid. With the interaction between water droplets and sulfuric and nitric acids, wet deposition occurs and creates acid rain. As a result, these acids would be displaced into various environments and vegetation during precipitation, having significant aerial distance (hundreds of kilometres) from the emission source. Sulfur dioxide and nitrogen oxide can also be displaced onto vegetations through dry deposition.

Sulfur dioxide and nitrous oxide concentration has many implication on aquatic ecosystems, including acidity change, increased nitrogen and aluminum content, and altering biogeochemical processes. Typically, sulfur dioxide and nitrous oxide do not have direct physiological effects upon exposure; most effects are developed by accumulation and prolonged exposure of these gases in the environment, modifying soil and water chemistry. Consequently, sulfur largely contributes to lake and ocean acidification, and nitrogen initiates eutrophication of inland and coastal water bodies that lack nitrogen. Both of these phenomena alter the native aquatic biota composition and influence the original food web with higher acidity level, minimizing aquatic and marine biodiversity.

Nitrogen deposition has also affected terrestrial ecosystems, including forests, grasslands, alpine regions, and bogs. The influx of nitrogen has altered the natural biogeochemical cycle and promoted soil acidification. As a result, it is likely that plant and animal species composition and ecosystem functionality will decline with increased soil sensitivity; contribute to slower forest growth, tree damage at higher elevations, and replacement of native biota with nitrogen-loving species. Additionally, sulfate and nitrate can be leached from the soil, removing essential nutrients such as calcium and magnesium, and be deposited into freshwater, coastal, and oceanic environments, promoting eutrophication.

Noise Pollution

Noise generated by traffic, ships, vehicles, and aircraft can affect the survivability of wildlife species and can reach undisturbed habitats. Noise pollution is common in marine ecosystems, affecting at least 55 marine species. One study discovered that as seismic noises and naval sonar increases in marine ecosystems, cetacean, such as whales and dolphins, diversity decreases. Multiple studies have noticed that fewer fishes, such as cod, haddock, rockfish, herring, sand seal, and blue whiting, have been spotted in areas with seismic noises, with catch rates declining by 40–80%.

Noise pollution has also altered avian communities and diversity. Anthropogenic noises have a similar effect on bird population as seen in marine ecosystems, where noises reduce reproductive success; cannot detect predators due to interferences of anthropogenic noises, minimize nesting areas, increase stress response, and species abundances and richness declining. Noise pollution can alter the distribution and abundance of prey species, which can then impact predator populations.

Pollution from Fossil Fuel Extraction

Fossil fuel extraction and associated oil and gas pipelines have had major impacts on the biodiversity of terrestrial, freshwater, coastal and marine environments due to land conversion, habitat loss and degradation, contamination and pollution. An example is the Western Amazon region. Exploitation of fossil fuels has had significant impacts on biodiversity also through climate change following combustion of these fuels. Protected areas with rich biodiversity have been mapped out and many are located in areas containing unexploited fossil fuel reserves worth between 3 and 15 trillion USD (2018).

Invasive Species

Invasive species have major implications on biodiversity loss and have degraded various ecosystems worldwide. Invasive species are migrant species that have outcompeted and displaced native species, altered species richness and food webs, and changed ecosystems' functions and services. Invasive species are considered one of the top five factors which result in biodiversity loss. In the past half century, biological invasions have increased immensely worldwide due to economic globalization, resulting in biodiversity loss. Ecosystems that are vulnerable to biological invasions include coastal areas, freshwater ecosystems, islands, and places with a Mediterranean climate. One study conducted a meta-analysis on the impacts of invasive species on Mediterranean-type ecosystems, and observed a significant loss in native species richness.

Invasive species are introduced to new habitat, either intentionally or unintentionally, by human activities. The most common methods for the introduction of aquatic invasive species are by ballast water, on the hulls of ships, and attached to equipment such as fishing nets. Some invasive species may be better able to tolerate and adapt to changing climate conditions, giving them a competitive advantage over native species.

Climate change has changed typical conditions in various environments, allowing greater migration and distribution of species dependent on warm climate. This phenomenon could either result in greater biodiversity (new species being introduced to new environments), or reduce biodiversity (promotion of invasive species). A biological invasion is deemed successful if the invasive species can adapt and survive in the new environment, reproduce, disperse, nd compete with native communities. Some invasive species are known to have high dispersal rates and have major implications on a regional scale. For example, in 2010, muskrat, raccoon dog, thrips, and Chinese mitten crab were identified to have affected 20 to 50 regions in Europe.

Overexploitation

Overexploitation, also called overharvesting, refers to harvesting a renewable resource to the point of diminishing returns. Continued overexploitation can lead to the destruction of the resource, as it will be unable to replenish. The term applies to natural resources such as water aquifers, grazing pastures and forests, wild medicinal plants, fish stocks and other wildlife.

In ecology, overexploitation describes one of the five main activities threatening global biodiversity. Ecologists use the term to describe populations that are harvested at an unsustainable rate, given their natural rates of mortality and capacities for reproduction. This can result in extinction at the population level and even extinction of whole species. In conservation biology, the term is usually used in the context of human economic activity that involves the taking of biological resources, or organisms, in larger numbers than their populations can withstand. The term is also used and defined somewhat differently in fisheries, hydrology and natural resource management.

Overexploitation can lead to resource destruction, including extinctions. However, it is also possible for overexploitation to be sustainable, as discussed below in the section on fisheries. In the context of fishing, the term overfishing can be used instead of overexploitation, as can overgrazing in stock management, overlogging in forest management, overdrafting in aquifer management, and endangered species in species monitoring. Overexploitation is not an activity limited to humans. Introduced predators and herbivores, for example, can overexploit native flora and fauna.

Overexploitation occurs when a resource is consumed at an unsustainable rate. This occurs on land in the form of overhunting, excessive logging, poor soil conservation in agriculture and the illegal wildlife trade. Overexploitation can lead to resource destruction, including extinction. The overkill hypothesis, a pattern of large animal extinctions connected with human migration patterns, can be used to explain why megafaunal extinctions can occur within a relatively short time period.

Overfishing

Human demands and consumption have resulted in overfishing, which leads to a loss in biodiversity with reduction of fish species richness and of population abundances, and to depletion of large predatory fishes at the top of marine food webs. As of 2007, About 25% of world fisheries are overfished to the point where their current biomass is less than the level that maximizes their sustainable yield.

Reduction in global fish populations were first noticed during the 1990s. Currently, many commercial fishes have been overharvested: a 2020 report by FAO classified as overfished 34% of the fish stocks of the world's marine fisheries. By the same period, global fish populations were reduced by 38% compared to 1970. Regional examples abound: in the United States approximately 27% of exploited fish stocks are considered overfished. In Tasmania, over 50% of major fisheries species, such as the eastern gemfish, the southern rock lobster, southern bulkefin tuna, jack mackerel, or trumpeter, have declined over the past 75 years due to overfishing. The depletion of large predatory fishes at the top of marine food webs due to overfishing can have cascading effects on entire ecosystems. the loss of large predatory fish species can result in an increase in smaller predator populations, which in turn can lead to a decrease in herbivore populations, ultimately leading to a loss of kelp forests and other important habitats. Fishery methods, such as bottom trawling and longline fishing have caused habitat destruction, causing spatial diversity and regional species richness to decline. These methods contribute to unreported bycatch. Unwanted bycatch species commonly die while in captivity or after being released. Overexploitation of species removed from their ecosystems has impacts on trophic levels and the food web. Some studies, including the 2019 Intergovernmental Science-Policy Platform on Biodiversity and Ecosystem Services report, found that overfishing is the main driver of mass species extinction in the oceans. Overfishing has reduced fish and marine mammal biomass by 60% since the 1800s, and is currently driving over one-third of sharks and rays to extinction.

Climate Change

Fig. 7.5: Rainforest Ecosystems are Rich in Biodiversity

This is the Gambia River in Senegal's Niokolo-Koba National Park. Climate change has adversely affected terrestrial and marine ecosystems, including tundras, mangroves, coral reefs, and caves. Increasing global temperature, more frequent occurrence of extreme weather, and rising sea level are examples of the most impactful effects of climate change. Possible consequences of these effects include species decline and extinction and overall significant loss of biodiversity, change within ecosystems, increased prevalence of invasive species, loss of habitats, forests converting from carbon sinks to carbon sources, ocean acidification, disruption of the water cycle, increased occurrence and severity of natural disasters like wildfires and flooding, and lasting effects on species adaptation.

Some contemporary studies have suggested that addressing climate change alone will not resolve the biodiversity crisis.

Extinction Risks

There are several plausible pathways that could lead to an increased extinction risk from climate change. This is because every plant and animal species has evolved to exist within a certain ecological niche, and as climate change represents the long-term alteration of temperature and average weather patterns, it can push climatic conditions outside of the species' niche, which will ultimately render it extinct. Normally, species faced with changing conditions can either adapt in place through microevolution or move to another habitat with suitable conditions. However, the speed of recent climate change is so unprecedented, that even under "mid-range" scenarios of future warming, only 5% of current ectotherm locations (a category which includes amphibians, reptiles and all invertebrates) are within 50 km of a place which could serve as an equally suitable habitat at the end of this century.

Climate change also increases both the frequency and intensity of extreme weather events, which can directly wipe out regional populations of species. Those species occupying coastal and low-lying island habitats can also be rendered extinct by sea level rise; this has already happened with Bramble Cay melomys in Australia. Finally, climate change has been linked with the increased prevalence and global spread of certain diseases affecting wildlife. This includes Batrachochytrium dendrobatidis, a fungus identified as one of the main drivers of the worldwide decline in amphibian populations.

Effect on Plants

Pre-species barriers for plants are also the indirect effects of climate change due to human activities. First, the reduction in the number of birds and insects used to help pollinate plants will reduce the chance of reproduction between plants. Second, extended fire weather seasons may result in more severe burn conditions and shorter burn intervals, which can threaten the biodiversity of native vegetation. Besides, species habitat changes or migrations under changing weather conditions can cause non-native plants and pests to impact native vegetation diversity, making the latter less structurally functional and more vulnerable to external damage, leading to biodiversity loss.

Plant and animal populations are interconnected. There are a number of examples in nature that display this dependency. For example, pollinator-reliant plant species display an observable sensitivity to pollinator activity.

While plants are essential for human survival, they have not received the same attention as the conservation of animals. It is estimated that a third of all land plant species are at risk of extinction and 94% have yet to be evaluated in terms of their conservation status. Plants existing at the lowest trophic level require increased conservation in order to reduce negative impacts at higher trophic levels.

Changes in long term environmental conditions that can be collectively coined climate change are known to have had enormous impacts on current plant biodiversity patterns; further impacts are expected in the future. Environmental conditions play a key role in defining the function and geographic distributions of plants, in combination with other factors, thereby modifying patterns of biodiversity. It is predicted that climate change will remain one of the major drivers of biodiversity patterns in the future. Climate change is thought to be one of several factors causing biodiversity loss (human-triggered mass extinction), which is changing the distribution and abundance of many plants.

Predicting the effects that climate change will have on plant biodiversity can be achieved using various models, however bioclimatic models are most commonly used.

Effects on Aquatic Macroinvertebrates and Microbes

Many scientists have studied the effects of climate change on the community structures and behaviors of aquatic macroinvertebrates and microbes - which are the prominent foundation of nutrient cycling in aquatic systems. These organisms are responsible for breaking down organic matter into essential carbon and nutrients that get cycled throughout the system and maintain health and production of the entire habitat. However, there have been numerous studies (through experimental warming) that have shown increases in microbial respiration of carbon out of the system, with a simultaneous decrease in leaf litter breakdown caused by temperature-sensitive macroinvertebrates. As temperatures are expected to increase largely due to anthropogenic influence, the abundance, type, and efficiency of macroinvertebrate and microbial organisms in aquatic systems will likely be dramatically altered.

Impacts

Ecological Effects of Biodiversity Loss

Biodiversity loss also threatens the structure and proper functioning of ecosystems. Although ecosystems are able to adapt to the stresses associated with reductions in biodiversity to some degree, biodiversity loss reduces an ecosystem's complexity, as roles once played by multiple interacting species or individuals are played by fewer or none. The effects of species loss or changes in composition, and the mechanisms by which the effects manifest themselves, can differ among ecosystem properties, types, and pathways of potential community change. At higher levels of extinction (40 to 60 percent of species), the effects of species loss ranked with those of many other major drivers of environmental change, such as ozone pollution, acid deposition on forests and nutrient pollution. Finally, the effects are also seen on human needs such as clean water, air and food production over-time. For example, studies have demonstrated that more biologically diverse ecosystems are more productive. As a result, there has been growing concern that the very high rates of modern extinctions – due to habitat loss, overharvesting and other human-caused environmental changes – could reduce nature's ability to provide goods and services like food, clean water and a stable climate.

Biodiversity and genetic diversity are codependent. Diversity among species requires diversity within a species and vice versa. "If any one type is removed from the system, the cycle can break down, and the community becomes dominated by a single species."

An October 2020 analysis by Swiss Re found that one-fifth of all countries are at risk of ecosystem collapse as the result of anthropogenic habitat destruction and increased wildlife loss. If these losses are not reversed, this could trigger a total ecosystem collapse.

Regional Changes in Species Composition

Even though permanent global species loss is a more dramatic and tragic phenomenon than regional changes in species composition, even minor changes from a healthy stable state can have a dramatic influence on the food web and the food chain insofar as reductions in only one species can adversely affect the entire chain (coextinction), leading to an overall reduction in biodiversity, possible alternative stable states of an ecosystem notwithstanding. Ecological effects of biodiversity are usually counteracted by its loss. Reduced biodiversity in particular leads to reduced ecosystem services and eventually poses an immediate danger for food security, but also can have more lasting public health consequences for humans.

Impact on Food and Agriculture

In 2019, the UN's Food and Agriculture Organization produced its first report on The State of the World's Biodiversity for Food and Agriculture, which warned that "Many key components of biodiversity for food and agriculture at genetic, species and ecosystem levels are in decline."

The report states that "major global trends such as changes in climate, international markets and demography give rise to more immediate drivers such as land-use change, pollution and overuse of external inputs, overharvesting and the proliferation of invasive species. Interactions between drivers often exacerbate their effects on biodiversity for food and agriculture (BFA)." It further states that "loss and degradation of forest and aquatic ecosystems and, in many production systems, transition to intensive production of a reduced number of species, breeds and varieties, remain major drivers of loss of BFA and ecosystem services." The health of humans is largely dependent on the product of an ecosystem. With biodiversity loss, a huge impact on human health comes as well. Biodiversity makes it possible for humans to have a sustainable level of soils and the means to have the genetic factors to have food.

Many activists and scholars have suggested that there is a connection between plant patent protection and the loss of crop biodiversity, although such claims are contested.

Human Health

The decrease in biodiversity has several implications for human health. One such implication is the loss of medicinal plants. The use of plants for medicinal purposes is extensive, with ~70 to 80% of individuals worldwide relying solely on plant-based medicine as their primary source of healthcare. This dependency on plants for medicinal purposes is especially rife in developing countries. Local knowledge surrounding medicinal plants is useful for screening for new herbal medicines that may be useful for treating disease. Villages and communities which reside continually in a single geographic area over time, create, transmit and apply widespread information surrounding the medicinal resources in the area. Formal scientific methods have been useful in identifying the active ingredients used in ethnopharmacy and applying them to modern medicines. However, it is important that medicinal resources are managed appropriately as they become globally traded in order to prevent species endangerment. Changes to local ecosystems (such as access to food and clean water) can indirectly impact the local economy, and society (livihood and social interaction between people living in the impacted area). Therefore impacting the health of the people. The October 2020 "Era of Pandemics" report by IPBES asserted that the same human activities which are the underlying drivers of climate change and biodiversity loss are also the same drivers of pandemics, including the COVID-19 pandemic.

According to the biodiversity hypothesis, reduced contact of people with natural environment and biodiversity may adversely affect the human commensal microbiota and its immunomodulatory capacity. The hypothesis is based on the observation that two dominant socio-ecological trends – the loss of biodiversity and increasing incidence of inflammatory diseases – are interconnected.

Urbanization and fragmentation of habitats increasingly lead to loss of connection between human and natural environment. Furthermore, immunological non-communicable diseases have become increasingly common in recent decades especially in urbanized communities.

Others

The ongoing global biodiversity crisis not only involves biological extinctions, but also the loss of experience and the gradual fading of cultural knowledge and collective memory of species.

Proposed Solutions

There are many conservation challenges when dealing with biodiversity loss that a joint effort needs to be made through public policies, economic solutions, monitoring and education by governments, NGOs, conservationists etc. Incentives are required to protect species and conserve their natural habitat and disincentivize habitat loss and degradation (e.g., implementing sustainable development including targets of SDG 15). Other ways to achieve this goal are enforcing laws that prevent poaching wildlife, protect species from overhunting and overfishing and keep the ecosystems they rely on intact and secure from species invasions and land use conversion. Furthermore, conservation based models like the Global Safety Net are continuously being developed to consider the ecological connections that need to be addressed to effectively mitigate biodiversity loss. Action to protect biodiversity is very cost effective because it reduces the risk of pandemics due to pathogens from wildlife.

Conservationists and sustainable research scientists around the world have also developed systems-based approaches to help mitigate biodiversity loss. This methodology allows scientists to create contextual frameworks that consider the many nuances and linkages of environmental conservation like ecological footprints, planetary boundaries, ecological economics, etc. Considering all the many ways in which the natural and human world intersect can help researchers understand the intricacies that lead to biodiversity loss and find patterns that can be applied to similar situations. One example of these type of frameworks is the triple bottom line, which has been adopted by many businesses and organizations to evaluate their impact and progress towards the marriage of social, environmental, and economic success.

In September 2020 scientists recommend measures to reduce biodiversity loss, such as for addressing drivers of land-use change, and for increasing the extent of land under conservation management, efficiency in agriculture and the shares of plant-based diets.

Frans Timmermans, Vice-President of the European Commission, stated in 2022 that people are less aware of the threat of biodiversity loss than they are of the threat of climate change. A 2021 collaborative report by scientists from the IPBES and the IPCC says that biodiversity loss and climate change must be addressed simultaneously, as they are inexorably linked and have similar effects on human well-being.

According to a research produced by the International Union for Conservation of Nature and UNESCO, even though they cover less than 1% of the planet, nearly three-quarters of bird species, two thirds of mammals and more than half of hard corals have been recorded at World Heritage Sites. The report's recommendations encourage countries with World Heritage Sites to include them in their national biodiversity strategies and action plans.

International Action

There are many organizations devoted to the cause of prioritizing conservation efforts such as the Red List of Threatened Species from the International Union for Conservation of Nature and Natural Resources (IUCN) and the United States Endangered Species Act. British environmental scientist Norman Myers and his colleagues have identified 25 terrestrial biodiversity hotspots that could serve as priorities for habitat protection.

Many governments in the world have conserved portions of their territories under the Convention on Biological Diversity (CBD), a multilateral treaty signed in 1992–3. The 20 Aichi Biodiversity Targets, part of the CBD's Strategic Plan 2011–2020, were published in 2010. Aichi Target 11 aimed by 2020 to protect 17 percent of terrestrial and inland water areas and 10 percent of coastal and marine areas.

In 2019 the Intergovernmental Science-Policy Platform on Biodiversity and Ecosystem Services (IPBES), an international organization formed to serve a similar role to the Intergovernmental Panel on Climate Change (IPCC), published the Global Assessment Report on Biodiversity and Ecosystem Services which said that up to a million plant and animal species are facing extinction because of human activities. An October 2020 report by IPBES stated that the same human activities which are the underlying drivers of climate change and biodiversity loss, such as the destruction of wildlife and wild habitats, are also the same drivers of pandemics, including the COVID-19 pandemic. In 2022, IPBES listed some of the primary drivers of the contemporary extinction crisis as being unsustainable fishing, hunting and logging.

Of the 20 biodiversity goals laid out by the Aichi Biodiversity Targets in 2010, only 6 were "partially achieved" by the deadline of 2020. The report highlighted that if the status quo is not changed, biodiversity will continue to decline due to "currently unsustainable patterns of production and consumption, population growth and technological developments". The report also singled out Australia, Brazil and Cameroon and the Galapagos Islands (Ecuador) for having had one of its animals lost to extinction in the past 10 years. Following this, the leaders of 64 nations and the European Union pledged to halt environmental degradation and restore the natural world. Leaders from some of the world's biggest polluters, namely China, India, Russia, Brazil and the United States, were not among them. Some experts contend that the refusal of the United States to ratify the Convention on Biological Diversity is harming global efforts to halt the extinction crisis. Top scientists say that even if the 2010 targets had been met, it likely would not have resulted in any substantive reductions of current extinction rates. Others have raised concerns that the Convention on Biological Diversity does not go far enough, and argue the goal should be zero extinctions by 2050, along with cutting the impact of unsustainable food production on nature by half. That the targets are not legally binding has also been subject to criticism.

In 2020, with passing of the 2020 target date for the Aichi Biodiversity Targets, scientists proposed a measurable, near-term biodiversity target – comparable to the below 2 °C global warming target – of keeping described species extinctions to well below 20 per year over the next 100 years across all major groups (fungi, plants, invertebrates, and vertebrates) and across all ecosystem types (marine, freshwater, and terrestrial).

On December 19, 2022, every country on earth, except the United States and the Holy See, signed onto the Kunming-Montreal Global Biodiversity Framework, which includes protecting 30% of land and oceans by 2030 (30 by 30) and 22 other targets intended to reduce biodiversity loss. When the agreement was signed only 17% of land territory and 10% of ocean territory were protected. The agreement includes protecting the rights of Indigenous peoples and changing the current subsidy policy to one better for biodiversity protection. However, it makes a step backward in protecting species from extinction in comparison to the Aichi Targets. Some countries said the agreement does not go far enough to protect biodiversity, and that the process was rushed.

7.3 WATER SCARCITY: SOCIETAL AND ECONOMIC IMPACTS

Water scarcity (closely related to water stress or water crisis) is the lack of fresh water resources to meet the standard water demand. There are two types of water scarcity namely physical and economic water scarcity. Physical water scarcity is where there is not enough water to meet all demands, including that needed for ecosystems to function. Arid areas for example Central Asia, West Asia, and North Africa often experience physical water scarcity. Economic water scarcity on the other hand, is the result of lack of investment in infrastructure or technology to draw water from rivers, aquifers, or other water sources. It also results from weak human capacity to meet water demand. Much of Sub-Saharan Africa experiences economic water scarcity.

There is enough freshwater available globally and averaged over the year to meet demand. As such, water scarcity is caused by a mismatch between when and where people need water, and when and where it is available. The main drivers of the increase in global water demand are the increasing world population, rise in living conditions, changing diets (to more animal products), and expansion of irrigated agriculture. Climate change (including droughts or floods), deforestation, water pollution and wasteful use of water can also cause insufficient water supply. Scarcity varies over time as a result of natural variability in hydrology. These variations in scarcity may also be a function of prevailing economic policy and planning approaches.

Water scarcity assessments need to incorporate information on green water (soil moisture), water quality, environmental flow requirements, globalization, and virtual water trade. There is a need for collaboration between hydrological, water quality, aquatic ecosystem science and social science communities in water scarcity assessment. "Water stress" has been used as parameter to measure water scarcity, for example in the context of Sustainable Development Goal 6. Half a billion people live in areas with severe water scarcity throughout the year, and around four billion people face severe water scarcity at least one month per year. Half of the world's largest cities experience water scarcity. There are 2.3 billion people who reside in nations with water scarcities, which means that each individual receives less than 1 700 m3 of water annually. However, 380 billion m3 of municipal wastewater are produced globally each year.

Reducing water scarcity requires supply and demand side management, cooperation between countries and water conservation (including prevention of water pollution). It also requires expanding sources of usable water (through wastewater reuse or desalination) and virtual water trade.

7.3.1 Definitions

Water scarcity has been defined as the "volumetric abundance, or lack thereof, of freshwater resources" and it is thought to be "human-driven". This can also be called "physical water scarcity". There are two types of water scarcity: physical water scarcity and economic water scarcity. Environmental water requirements are sometimes included in water scarcity determinations but the approach to this varies from one organization to another.

Related Concepts

There are several definitions of "water scarcity", "water stress", and "water risk" provided in the literature, and therefore a harmonization has been proposed by the CEO Water Mandate in 2014. In their discussion paper they state that these three terms should not be used interchangeably.

Water Stress

Some organizations define "water stress" as a broader concept. Under that definition, it would include aspects of water availability, water quality and the accessibility of water. The latter is related to existing infrastructure and whether customers can afford to pay for the water. This is termed by others as "economic water scarcity". FAO defines water stress as the "symptoms of water scarcity or shortage". Such symptoms could be "growing conflict between users, and competition for water, declining standards of reliability and service, harvest failures and food insecurity". This is measured with a range of Water Stress Indices.

Another definition for water stress is as follows: "Water stress refers to the impact of high water use (either withdrawals or consumption) relative to water availability." Water stress is therefore viewed as a "demand-driven scarcity".

Types

Two types of water scarcity have been defined: physical and economic water scarcity. These terms were first defined in a 2007 study led by the International Water Management Institute o the use of water in agriculture over the previous 50 years of practitioners, researchers and policymakers. The study aimed to find out if the world had sufficient water resources to produce food for the growing population in the future.

Physical Water Scarcity

Physical water scarcity occurs when natural water resources are not enough to meet all demands, including that needed for ecosystems to function well. Dry regions often suffer from physical water scarcity. Human influence on climate has led to increased water scarcity in areas where water was previously hard to come by. It also occurs where water seems abundant but where resources are over-committed. For example, when there is overdevelopment of hydraulic infrastructure, often for irrigation or energy generation. Symptoms of physical water scarcity include "severe environmental degradation, declining groundwater and water allocations that favor some groups over others".

Another indicator, called ecological water scarcity has been proposed that considers water quantity, water quality, and environmental flow requirements.

Water is physically scarce in densely populated arid areas (for example Central and West Asia, and North Africa), with projected availabilities of less than 1000 cubic meters per capita per year. A study in 2007 found that more than 1.2 billion people live in areas of physical water scarcity. This water scarcity relates to water available for food production, rather than for drinking water which is a much smaller amount.

Some academics favour expanding the two types of water scarcity (i.e., the physical and economic) to a third type which would be called ecological water scarcity. This third type would focus on the water demand of ecosystems. It would refer to the minimum quantity and quality of water discharge that are required to maintain sustainable and functional ecosystems. However, other publications state that this aspect is simply part of the definition of physical water scarcity.

Economic Water Scarcity

Economic water scarcity is caused by a lack of investment in infrastructure or technology to draw water from rivers, aquifers, or other water sources, or insufficient human capacity to meet the demand for water. It causes people without reliable water access to travel long distances to fetch water, which is often unclean, for domestic and agricultural uses (irrigation).

According to the United Nations Development Programme, economic water scarcity is the most common cause of water scarcity. This is because most countries or regions have enough water to meet household, industrial, agricultural, and environmental needs, but lack the means to provide it in an accessible manner. Around a fifth of the world's population currently live in regions affected by physical water scarcity.

A quarter of the world's population is affected by economic water scarcity. Much of Sub-Saharan Africa is characterized by economic water scarcity. Developing water infrastructure there could therefore help to reduce poverty. Investing in water retention and irrigation infrastructure would also help to increase food production, especially in developing countries that largely rely on low-yield agriculture. Being able to provide a community with water that is adequate for consumption would also greatly benefit the people's health. Overcoming this type of scarcity, however, can require more than just new infrastructure. It requires socio-economic and socio-political interventions that address poverty and social inequality. Still, because there is a lack of funding hence much planning must come into play.

Although much emphasis is put on improving water sources for drinking and domestic purposes, we know that much more water is used for other uses such as bathing, laundry, livestock and cleaning than for drinking and cooking alone. This observation suggests that putting too much emphasis on drinking water needs, addresses a minor part of the problem and therefore limits the range of solutions available.

7.3.2 Water Security

The aim of water security is to make the most of water's benefits for humans and ecosystems. The second aim is to limit the risks of destructive impacts of water to an acceptable level. These risks include for example too much water (flood), too little water (drought and water scarcity) or poor quality (polluted) water. People who live with a high level of water security always have access to "an acceptable quantity and quality of water for health, livelihoods and production". For example, access to water, sanitation and hygiene services is one part of water security. Some organizations use the term water security more narrowly for water supply aspects only.

Fig. 7.6: A Typical Dry Lakebed is Seen in California

7.3.3 Water Risk

Water risk refers to the possibility of problems to do with water. Examples are water scarcity, water stress, flooding, infrastructure decay and drought. There exists an inverse relationship between water risk and water security. This means as water risk increases, water security decreases. Water risk is complex and multilayered. It includes risks flooding and drought. These can lead to infrastructure failure and worsen hunger. When these disasters take place, they result in water scarcity or other problems. The potential economic effects of water risk are important to note. Water risks threaten entire industries. Examples are the food and beverage sector, agriculture, oil and gas and utilities. Agriculture uses 69% of total freshwater in the world. So this industry is very vulnerable to water stress.

7.3.4 Assessments and Indicators

Simple indicators

Indicators include: the water use to availability ratio (or criticality ratio), physical and economic water scarcity—The IWMI Indicator, water poverty index.

"Water stress" has been used as a criterion to measure water scarcity, for example in the context of Sustainable Development Goal 6. In this context, a report by FAO in 2018 has defined water stress as: "the ratio between total freshwater withdrawn (TFWW) by all major sectors and total renewable freshwater resources (TRWR), after taking into account environmental flow requirements (EFR)". This means that the value for TFWW is divided by the difference between TRWR minus EFR. Environmental flows are water flows required to sustain freshwater and estuarine ecosystems. Previously, a definition used for Millennium Development Goal 7, target 7.A was simply the proportion of total water resources used, without taking into consideration EFR. With this definition, water stress is defined by the following categories: <10% is low stress; 10-20% is low-to-medium; 20-40% medium-to-high; 40-80% high; >80% very high.

Indicators are used to measure the extent of the water scarcity. One way to measure water scarcity is to calculate the amount of annual water resources available per person. For example, according to the "Falkenmark Water Stress Indicator" (developed by Malin Falkenmark), a country or region is said to experience "water stress" when annual water supplies drop below 1,700 cubic meters per person per year. At levels between 1,700 and 1,000 cubic meters per person per year, periodic or limited water shortages can be expected. When water supplies drop below 1,000 cubic meters per person per year, the country faces "water scarcity". However, the Falkenmark Water Stress Indicator does not help to explain the true nature of water scarcity.

Renewable Freshwater Resources

Renewable freshwater supply is a metric often used in conjunction when evaluating water scarcity. This metric is informative because it can describe the total available water resource each country contains. By knowing the total available water source, an idea can be gained about whether a country is prone to experiencing physical water scarcity. This metric has its faults in that it is an average; precipitation delivers water unevenly across the planet each year and annual renewable water resources vary from year to year. This metric also does not describe the ease of water access to individuals, households, industries, or the government. Lastly, as this metric is a description of a whole country, it does not accurately portray whether a country is experiencing water scarcity. For example, Canada and Brazil both have very high levels of available water supply but still face various water-related problems. Also, some tropical countries in Asia and Africa have low levels of freshwater resources.

More Sophisticated Indicators

Water scarcity assessments need to incorporate information on green water (soil moisture), water quality, environmental flow requirements, globalization, and virtual water trade. Since the beginning of the 2000s, water scarcity assessments have applied more complex models which are supported with spatial analysis tools. They include: Green-blue water scarcity, water footprint-based water scarcity assessment, cumulative abstraction to demand ratio—considering temporal variations, LCA-based water stress indicators (life cycle assessments), integrated water quantity–quality environment flow in the water scarcity assessment. Since the early 2010s assessments have combined both quantity- and quality-induced water scarcity.

Another indicator, called ecological water scarcity has been proposed that considers water quantity, water quality, and environmental flow requirements. For example, results from a modelling study in 2022 show that northern China suffered more severe ecological water scarcity than southern China. The main driving factor of ecological water scarcity in most provinces was water pollution rather than human water use.

Overall, there is a need for collaboration between hydrological, water quality, aquatic ecosystem science and social science communities in water scarcity assessment.

7.3.5 Available Water

The United Nations (UN) estimates that, of 1.4 billion cubic kilometers (1 quadrillion acre-feet) of water on Earth, just 200,000 cubic kilometers (162.1 billion acre-feet) represent freshwater available for human consumption. A mere 0.014% of all water on Earth is both fresh and easily accessible. Of the remaining water, 97% is saline, and a little less than 3% is difficult to access. The fresh water available to us on the planet is around 1% of the total water on earth. The total amount of easily accessible freshwater on Earth, in the form of surface water (rivers and lakes) or groundwater (in aquifers, for example), is 14,000 cubic kilometers (nearly 3359 cubic miles). Of this total amount, 'just' 5,000 cubic kilometers are being used and reused by humanity. Technically, there is a sufficient amount of freshwater on a global scale. Hence, in theory, there is more than enough freshwater available to meet the demands of the current world population of more than 7 billion people, and even support population growth to 9 billion or more. Due to the unequal geographical distribution and especially the unequal consumption of water, however, it is a scarce resource in some parts of the world and for some parts of the population.

Apart from the common surface water sources of freshwater such as rivers and lakes, other resources of freshwater such as groundwater and glaciers have become more developed sources of freshwater, becoming the main source of clean water. Groundwater is water that has pooled below the surface of the Earth and can provide a usable quantity of water through springs or wells. These areas where groundwater is collected are also known as aquifers. More and more of these sources are being drawn upon as conventional sources' usability decreases due to factors such as pollution or disappearance due to climate changes. Human population growth is a significant contributing factor in the increasing use of these types of water resources.

7.3.6 Scale

Current Estimates

Water scarcity was listed in 2019 by the World Economic Forum as one of the largest global risks in terms of potential impact over the next decade. It is manifested by partial or no satisfaction of expressed demand, economic competition for water quantity or quality, disputes between users, irreversible depletion of groundwater, and negative impacts on the environment.

About half of the world's population urrently experience severe water scarcity for at least some part of the year. Half a billion people in the world face severe water scarcity all year round. Half of the world's largest cities experience water scarcity. Almost two billion people do not currently have access to clean drinking water.

A study in 2016 calculated that globally, the population under water scarcity increased from 0.24 billion (14% of global population) in the 1900s to 3.8 billion (58%) in the 2000s. This study analyzed water scarcity using the fundamental concepts of shortage (impacts due to low availability per capita) and stress (impacts due to high consumption relative to availability).

Future Predictions

In the 20th century, water use has been growing at more than twice the rate of the population increase. Specifically, water withdrawals are predicted to rise by 50 percent by 2025 in developing countries, and 18 per cent in developed countries. One continent, for example, Africa, has been predicted to have 75 to 250 million inhabitants lacking access to fresh water. By 2025, 1.8 billion people will be living in countries or regions with absolute water scarcity, and two-thirds of the world population could be under stress conditions. By 2050, more than half of the world's population will live in water-stressed areas, and another billion may lack sufficient water, MIT researchers find.

With the increase in global temperatures and in an increase in waterdemand, six out of ten people are at risk of being water-stressed. The drying out of wetlands globally, at around 67%, was a direct cause of a large number of people at risk of water stress. As the global demand for water increases and as climate temperatures rise, it is estimated that two-thirds of the population, in 2025, will live under water stress.

According to a projection by the United Nations, by 2040, there can be about 4.5 billion people affected by a water crisis (or water scarcity). Additionally, with the increase in population, there will be a demand for food, for the food output to match the population growth, there would be an increased demand for water to irrigate crops. The World Economic Forum estimate that global water demand will surpass global supply 40% by 2030. Increasing the demand for water as well as increasing the population results in a water crisis where there is not enough water to share in healthy levels. The crises are not only due to quantity but quality also matters.

A study found that of ~39 million groundwater wells 6-20% are at high risk of running dry if local groundwater levels decline by a few meters, or – as with many areas and possibly more than half of major aquifers – continue to decline.

7.3.7 Impacts

There are several impacts and symptoms of water scarcity. These include serious restrictions on water use, "growing conflict between users and competition for water, declining standards of reliability and service, harvest failures and food insecurity".

Specific examples include:

1. Food insecurity in the Middle East and North Africa Region.
2. Inadequate access to safe drinking water for about 885 million people.
3. Groundwater overdrafting (excessive use) leading to diminished agricultural yields.
4. Overuse and pollution of water resources harming ecosystems and biodiversity.
5. Regional conflicts over scarce water resources sometimes resulting in warfare.

Water Supply Shortages

Water is the underlying tenuous balance of safe water supply, but controllable factors such as the management and distribution of the water supply itself contribute to further scarcity. A 2006 United Nations report focuses on issues of governance as the core of the water crisis.

The report noted that "There is enough water for everyone" and "Water insufficiency is often due to mismanagement, corruption, lack of appropriate institutions, bureaucratic inertia and a shortage of investment in both human capacity and physical infrastructure".

It has also been claimed, primarily by economists, that the water situation has occurred because of a lack of property rights, government regulations and subsidies in the water sector. These factors cause prices to be too low and consumption too high, making a point for water privatization.

The clean water crisis is an emerging global crisis that affects approximately 785 million people around the world. 1.1 billion people lack access to water and 2.7 billion experience water scarcity at least one month in a year. 2.4 billion people suffer from the contamination of water and poor sanitation. Contamination of water can lead to deadly diarrheal diseases such as cholera and typhoid fever, and other waterborne diseases causing 80% of illnesses around the world.

Environment

Fig. 7.7: Deforestation of the Madagascar Highland Plateau has led to Extensive Siltation and Unstable Fflows of Western Rivers

Abstraction of water for domestic, food and industrial uses has major impacts on ecosystems in many parts of the world. This can apply even to regions not considered "water scarce".

Water scarcity has many negative impacts on the environment, such as adverse effects on lakes, rivers, ponds, wetlands and other fresh water resources. The resulting water overuse that is related to water scarcity, often located in areas of irrigation agriculture, harms the environment in several ways.

This includes increased salinity, nutrient pollution, and the loss of floodplains and wetlands. Furthermore, water scarcity makes flow management in the rehabilitation of urban streams problematic.

Through the last hundred years, more than half of the Earth's wetlands have been destroyed and have disappeared. These wetlands are important not only because they are the habitats of numerous inhabitants such as mammals, birds, fish, amphibians, and invertebrates, but they support the growing of rice and other food crops as well as provide water filtration and protection from storms and flooding. Freshwater lakes such as the Aral Sea in central Asia have also suffered. Once the fourth largest freshwater lake, it has lost more than 58,000 square km of area and vastly increased in salt concentration over the span of three decades.

Fig. 7.8: An Abandoned Ship in the Former Aral Sea, near Aral, Kazakhstan

Subsidence, or the gradual sinking of landforms, is another result of water scarcity. The U.S. Geological Survey estimates that subsidence has affected more than 17,000 square miles in 45 U.S. states, 80 percent of it due to groundwater usage.

Vegetation and wildlife are fundamentally dependent upon adequate freshwater resources. Marshes, bogs and riparian zones are more clearly dependent upon sustainable water supply. Forests and other upland ecosystems are equally at risk of significant productivity changes as water availability is diminished. In the case of wetlands, considerable area has been simply taken from wildlife use to feed and house the expanding human population. But other areas have suffered reduced productivity from gradual diminishing of freshwater inflow, as upstream sources are diverted for human use.

7.3.8 Causes and Contributing Factors

Population Growth

Around fifty years ago, the common view was that water was an infinite resource. At that time, there were fewer than half the current number of people on the planet. People were not as wealthy as today, consumed fewer calories and ate less meat, so less water was needed to produce their food. They required a third of the volume of water we presently take from rivers. Today, the competition for water resources is much more intense. This is because there are now seven billion people on the planet and their consumption of water-thirsty meat is rising. Also, there is increasing competition for water from industry, urbanization, biofuel crops, and water reliant food items. In the future, even more water will be needed to produce food because the Earth's population is forecast to rise to 9 billion by 2050.

In 2000, the world population was 6.2 billion. The UN estimates that, by 2050, there will be an additional 3.5 billion people with most of the growth in developing countries that already suffer water stress. Thus, water demand will increase unless there are corresponding increases in water conservation and recycling of this vital resource. In building on the data presented here by the UN, the World Bank goes on to explain that access to water for producing food will be one of the main challenges in the decades to come. Access to water will need to be balanced with the importance of managing water itself in a sustainable way while taking into account the impact of climate change, and other environmental and social variables.

In 60% of European cities with more than 100,000 people, groundwater is being used at a faster rate than it can be replenished.

Over-exploitation of Groundwater

Owing to expanding human population, competition for water is growing such that many of the world's major aquifers are becoming depleted. This is due both to direct human consumption as well as agricultural irrigation by groundwater. Millions of pumps of all sizes are currently extracting groundwater throughout the world. Irrigation in dry areas such as northern China, Nepal and India is supplied by groundwater and is being extracted at an unsustainable rate. Cities that have experienced aquifer drops between 10 and 50 meters include Mexico City, Bangkok, Beijing, Madras and Shanghai.

Until recently, groundwater was not a highly used resource. In the 1960s, more and more groundwater aquifers developed. Changes in knowledge, technology and funding have allowed for focused development into abstracting water from groundwater resources away from surface water resources. These changes allowed for progress in society such as the "agricultural groundwater revolution", expanding the irrigation sector allowing for increased food production and development in rural areas. Groundwater supplies nearly half of all drinking water in the world. The large volumes of water stored underground in most aquifers have a considerable buffer capacity allowing for water to be withdrawn during periods of drought or little rainfall. This is crucial for people that live in regions that cannot depend on precipitation or surface water as a supply alone, instead providing reliable access to water all year round. As of 2010, the world's aggregated groundwater abstraction is estimated at 1,000 kmper year, with 67% used for irrigation, 22% used for domestic purposes and 11% used for industrial purposes. The top ten major consumers of abstracted water (India, China, United States of America, Pakistan, Iran, Bangladesh, Mexico, Saudi Arabia, Indonesia, and Italy) make up 72% of all abstracted water use worldwide.

Although groundwater sources are quite prevalent, one major area of concern is the renewal rate or recharge rate of some groundwater sources. Extracting from groundwater sources that are non-renewable could lead to exhaustion if not properly monitored and managed. Another concern of increased groundwater usage is the diminished water quality of the source over time. Reduction of natural outflows, decreasing stored volumes, declining water levels and water degradation are commonly observed in groundwater systems. Groundwater depletion may result in many negative effects such as increased cost of groundwater pumping, induced salinity and other water quality changes, land subsidence, degraded springs and reduced baseflows.

Expansion of Agricultural and Industrial Users

Scarcity as a result of consumption is caused primarily by the extensive use of water in agriculture/livestock breeding and industry. People in developed countries generally use about 10 times more water daily than those in developing countries. A large part of this is indirect use in water-intensive agricultural and industrial production processes of consumer goods, such as fruit, oilseed crops and cotton. Because many of these production chains have been globalized, a lot of water in developing countries is being used and polluted in order to produce goods destined for consumption in developed countries.

Many aquifers have been over-pumped and are not recharging quickly. Although the total fresh water supply is not used up, much has become polluted, salted, unsuitable or otherwise unavailable for drinking, industry and agriculture. To avoid a global water crisis, farmers will have to strive to increase productivity to meet growing demands for food, while industry and cities find ways to use water more efficiently.

Business activities such as tourism continue to expand quickly. This expansion requires increased water services including both supply and sanitation, which can lead to more pressure on water resources and natural ecosystem. The approximate 50% growth in world energy use by 2040 will also increase the need for efficient water use. It may shift some irrigation water sources towards industrial use, as thermal power generation uses water for steam generation and cooling.

Water Pollution

Water pollution (or aquatic pollution) is the contamination of water bodies, usually as a result of human activities, so that it negatively affects its uses. Water bodies include lakes, rivers, oceans, aquifers, reservoirs and groundwater. Water pollution results when contaminants mix with these water bodies. Contaminants can come from one of four main sources: sewage discharges, industrial activities, agricultural activities, and urban runoff including stormwater. Water pollution is either surface water pollution or groundwater pollution. This form of pollution can lead to many problems, such as the degradation of aquatic ecosystems or spreading water-borne diseases when people use polluted water for drinking or irrigation. Another problem is that water pollution reduces the ecosystem services (such as providing drinking water) that the water resource would otherwise provide.

Sources of water pollution are either point sources or non-point sources. Point sources have one identifiable cause, such as a storm drain, a wastewater treatment plant or an oil spill. Non-point sources are more diffuse, such as agricultural runoff. Pollution is the result of the cumulative effect over time. Pollution may take the form of toxic substances (e.g., oil, metals, plastics, pesticides, persistent organic pollutants, industrial waste products), stressful conditions (e.g., changes of pH, hypoxia or anoxia, increased temperatures, excessive turbidity, changes of salinity), or the introduction of pathogenic organisms. Contaminants may include organic and inorganic substances. A common cause of thermal pollution is the use of water as a coolant by power plants and industrial manufacturers.

Climate Change

Climate change could have significant impacts on water resources around the world because of the close connections between the climate and hydrological cycle. Rising temperatures will increase evaporation and lead to increases in precipitation, though there will be regional variations in rainfall. Both droughts and floods may become more frequent and more severe in different regions at different times, generally less snowfall and more rainfall under a warmer climate, and dramatic changes in snowfall and snow melt are expected in mountainous areas. Higher temperatures will also affect water quality in ways that are not well understood. Possible impacts include increased eutrophication. Climate change could also mean an increase in demand for farm irrigation, garden sprinklers, and perhaps even swimming pools. There is now ample evidence that increased hydrologic variability and change in climate has and will continue to have a profound impact on the water sector. These effects will be seen through the hydrologic cycle, water availability, water demand, and water allocation at the global, regional, basin, and local levels.

The United Nations' FAO states that by 2025, 1.9 billion people will live in countries or regions with absolute water scarcity, and two-thirds of the world population could be under stress conditions. The World Bank adds that climate change could profoundly alter future patterns of both water availability and use, thereby increasing levels of water stress and insecurity, both at the global scale and in sectors that depend on water.

Overall, the effects of changes in population on water scarcity were found to be about four times more important than changes in water availability as a result of long-term climate change.

Glaciers

About 2% of Earth's water is frozen freshwater found in glaciers. Glaciers provide freshwater in the form meltwater, or freshwater melted from snow or ice, that supply streams or springs as temperatures rise. This water is used by locals for a number of reasons like agriculture, livestock, and hydropower. This is beneficial in helping reduce water scarcity as more water is available to a select number of people. It has been projected that total glaciers worldwide will be 60% of what they are now, in the year 2100. The main reason for the melting of these glaciers is climate change. Glaciers reflect sunlight from the sun back into space providing a decrease in temperatures worldwide. This process is called albedo and without the glaciers reflecting sunlight, temperatures would slowly begin to rise. As temperatures rise, glaciers will melt quicker overall reducing the total amount of sunlight being reflected worldwide. Melting glaciers, over a long period of time, begin receding and will be difficult to recover once seasonal changes occur. Glacier's losing mass may decrease their annual run-off, coupled with receding glaciers, which will change the availability of water in many cold regions of the world. About a third of glaciers may experience a 10% run-off reduction in some seasons.

In the Himalayas, retreating glaciers could reduce summer water flows by up to two-thirds. In the Ganges area, this would cause a water shortage for 500 million people. Climate change impacts potable water in the Hindu Kush Himalaya (HKH) area, where around 1.4 billion people are dependent on the five main rivers of Himalaya mountains. Although the impact will vary from place to place, it is predicted that the amount of meltwater will initially increase due to retreating glaciers and then gradually decrease because of reducing in glacier mass. In those areas where the amount of available water decreases, climate change makes it difficult to improve access to safe drinkable water. HKH area faces rapid urbanization causing a severe shortage of water and pressure on water resources. Rural areas will also suffer because of a lack of effective water management infrastructure and limited access to drinking water. More people will migrate because of the scarcity of drinking water. This situation will increase inequality by leaving the poor behind leading to higher mortality and suicide rate, and increased urbanization.

7.3.9 Options for Improvements

Supply and Demand Side Management

A review in 2006 stated that "It is surprisingly difficult to determine whether water is truly scarce in the physical sense at a global scale (a supply problem) or whether it is available but should be used better (a demand problem)".

The International Resource Panel of the UN states that governments have tended to invest heavily in largely inefficient solutions: mega-projects like dams, canals, aqueducts, pipelines and water reservoirs, which are generally neither environmentally sustainable nor economically viable. The most cost-effective way of decoupling water use from economic growth, according to the scientific panel, is for governments to create holistic water management plans that take into account the entire water cycle: from source to distribution, economic use, treatment, recycling, reuse and return to the environment.

In general, there is enough water on an annual and global scale, but the issue is more of a temporal and spatial variation. Therefore, reservoirs and pipelines are needed to address the temporal and spatial variations. It is necessary to have a well-planned infrastructure with demand side management. Both supply-side and demand-side management have advantages and disadvantages.

Co-operation between Countries

Lack of cooperation may give rise to regional water conflicts in many parts of the world, specially in developing countries, largely because of the disputes regarding the availability, use and management of water. For example, the dispute between Egypt and Ethiopia over the Grand Ethiopian Renaissance Dam has escalated in 2020. Egypt sees the dam as an existential threat, fearing that the dam will reduce the amount of water it receives from the Nile.

Water Conservation

Water conservation includes all the policies, strategies and activities to sustainably manage the natural resource of fresh water, to protect the hydrosphere, and to meet the current and future human demand (thus avoiding water scarcity). Population, household size and growth and affluence all affect how much water is used. Factors such as climate change have increased pressures on natural water resources especially in manufacturing and agricultural irrigation. Many countries have already implemented policies aimed at water conservation, with much success. The key activities to conserve water are as follows: any beneficial reduction in water loss, use and waste of resources, avoiding any damage to water quality; and improving water management practices that reduce the use or enhance the beneficial use of water. Technology solutions exist for households, commercial and agricultural applications. Water conservation programs involved in social solutions are typically initiated at the local level, by either municipal water utilities or regional governments.

Expanding Sources of Usable Water

Artificial sources of fresh water can include treated wastewater (reclaimed water), atmospheric water generators, and desalinated seawater. However, the economic and environmental side effects of these technologies must also be taken into consideration.

Wastewater Treatment and Reclaimed Water

Water reclamation (also called wastewater reuse, water reuse or water recycling) is the process of converting municipal wastewater (sewage) or industrial wastewater into water that can be reused for a variety of purposes. Types of reuse include: urban reuse, agricultural reuse (irrigation), environmental reuse, industrial reuse, planned potable reuse, and de facto wastewater reuse (unplanned potable reuse). For example, reuse may include irrigation of gardens and agricultural fields or replenishing surface water and groundwater (i.e., groundwater recharge). Reused water may also be directed toward fulfilling certain needs in residences (e.g., toilet flushing), businesses, and industry, and could even be treated to reach drinking water standards. The injection of reclaimed water into the water supply distribution system is known as direct potable reuse. However, drinking reclaimed water is not a typical practice. Treated municipal wastewater reuse for irrigation is a long-established practice, especially in arid countries. Reusing wastewater as part of sustainable water management allows water to remain as an alternative water source for human activities. This can reduce scarcity and alleviate pressures on groundwater and other natural water bodies.

Wastewater treatment is a process which removes and eliminates contaminants from wastewater and converts this into an effluent that can be returned to the water cycle. Once returned to the water cycle, the effluent creates an acceptable impact on the environment or is reused for various purposes (called water reclamation). The treatment process takes place in a wastewater treatment plant. There are several kinds of wastewater which are treated at the appropriate type of wastewater treatment plant. For domestic wastewater (also called municipal wastewater or sewage), the treatment plant is called a Sewage Treatment. For industrial wastewater, treatment either takes place in a separate Industrial wastewater treatment, or in a sewage treatment plant (usually after some form of pre-treatment). Further types of wastewater treatment plants include Agricultural wastewater treatment and leachate treatment plants.

Processes commonly used in wastewater treatment include phase separation (such as sedimentation), biological and chemical processes (such as oxidation) or polishing. The main by-product from wastewater treatment plants is a type of sludge that is usually treated in the same or another wastewater treatment plant. Biogas can be another by-product if anaerobic treatment processes are used. Treated wastewater can be reused as reclaimed water. The main purpose of wastewater treatment is for the treated wastewater to be able to be disposed or reused safely. However, before it is treated, the options for disposal or reuse must be considered so the correct treatment process is used on the wastewater. Bangladesh has officially inaugurated the largest single sewage treatment plant (STP) in South Asia, located in the Khilgaon area of the city. With a capacity to treat five million sewage per day, the STP marks a significant step towards addressing the country's wastewater management challenges.

Desalination

Desalination is a process that takes away mineral components from saline water. More generally, desalination is the removal of salts and minerals from a target substance, as in soil desalination, which is anissue for agriculture. Saltwater (especially sea water) is desalinated to produce water suitable for human consumption or irrigation. The by-product of the desalination process is brine. Desalination is used on many seagoing ships and submarines. Most of the modern interest in desalination is focused on cost-effective provision of fresh water for human use. Along with recycled wastewater, it is one of the few rainfall-independent water resources.

Virtual Water Trade

The virtual water trade (also known as embedded or embodied water) is the hidden flow of water in food or other commodities that are traded from one place to another. The virtual water trade is the idea that when goods and services are exchanged, so is virtual water.

Virtual water trade allows a new, amplified perspective on water problems: In the framewond balancing different perspectives, basic conditions, and interests. Analytically, the concept enables one to distinguish between global, regional, and local levels and their linkages. However, the use of virtual water estimates may offer no guidance for policymakers seeking to ensure that environmental objectives are being met.

For example, cereal grains have been major carriers of virtual water in countries where water resources are scarce. Therefore, cereal imports can play a crucial role in compensating local water deficit. However, low-income countries may not be able to afford such imports in the future which could lead to food insecurity and starvation.

7.3.10 Regional Examples

Africa

Water scarcity in Africa is predicted to reach dangerously high levels by 2025 when it is estimated that about two-thirds of the world's population may suffer from fresh water shortage. The main causes of water scarcity in Africa are physical and economic scarcity, rapid population growth, and climate change. Water scarcity is the lack of fresh water resources to meet the standard water demand. Although Sub-Saharan Africa has a plentiful supply of rainwater, it is seasonal and unevenly distributed, leading to frequent floods and droughts. Additionally, prevalent economic development and poverty issues, compounded with rapid population growth and rural-urban migration have rendered Sub-Saharan Africa as the world's poorest and least developed region.

The 2012 Report by the Food and Agriculture Organization of the United Nations indicates that growing water scarcity is now one of the leading challenges for sustainable development. This is because an increasing number of the river basins have reached conditions of water scarcity through the combined demands of agriculture and other sectors. Impacts of water scarcity in Africa range from health (women and children are particularly affected) to education, agricultural productivity, sustainable development as well as the potential for more water conflicts.

West Africa and North Africa

Water scarcity in Yemen is a growing problem that has resulted from population growth, poor water management, climate change, shifts in rainfall, water infrastructure deterioration, poor governance, and other anthropogenic effects. As of 2011, it has been estimated that Yemen is experiencing water scarcity to a degree that affects its political, economic and social dimensions. As of 2015, Yemen is among the most water scarce countries in the world. The majority of Yemen's population experiences water scarcity for at least one month during the year. In Nigeria, some reports have suggested that increase in extreme heat, drought and the shrinking of Lake Chad is causing water shortage and environmental migration that is forcing thousands to migrate to neighboring Chad and towns.

7.3.11 Asia

According to a major report compiled in 2019 by more than 200 researchers, the Himalayan glaciers that are the sources of Asia's biggest rivers – Ganges, Indus, Brahmaputra, Yangtze, Mekong, Salween and Yellow – could lose 66 percent of their ice by 2100. Approximately 2.4 billion people live in the drainage basin of the Himalayan rivers. India, China, Pakistan, Bangladesh, Nepal and Myanmar could experience floods followed by droughts in coming decades. In India alone, the Ganges provides water for drinking and farming for more than 500 million people. Even with the overpumping of its aquifers, China is developing a grain deficit. When this happens, it will almost certainly drive grain prices upward. Most of the 3 billion people projected to be added worldwide by mid-century will be born in countries already experiencing water shortages. Unless population growth can be slowed quickly, it is feared that there may not be a practical non-violent or humane solution to the emerging world water shortage. It is highly likely that climate change in Turkey will cause its southern river basins to be water scarce before 2070, and increasing drought in Turkey.

7.3.12 Americas

Fig. 7.9: Folsom Lake Reservoir during the Drought in California in 2015

In the Rio Grande Valley, intensive agribusiness has exacerbated water scarcity issues and sparked jurisdictional disputes regarding water rights on both sides of the U.S.-Mexico border. Scholars, including Mexican political scientist Armand Peschard-Sverdrup, have argued that this tension has created the need for a re-developed strategic transnational water management. Some have likened the disputes to a "war" over diminishing natural resources.

The west coast of North America, which gets much of its water from glaciers in mountain ranges such as the Rocky Mountains and Sierra Nevada, also would be affected.

7.3.13 Australia

By far the largest part of Australia is desert or semi-arid lands commonly known as the outback. Water restrictions are in place in many regions and cities of Australia in response to chronic shortages resulting from drought. The Australian of the year 2007, environmentalist Tim Flannery, predicted that unless it made drastic changes, Perth in Western Australia could become the world's first ghost metropolis, an abandoned city with no more water to sustain its population. In 2010, Perth suffered its second-driest winter on record and the water corporation tightened water restrictions for spring.

Some countries have already proven that decoupling water use from economic growth is possible. For example, in Australia, water consumption declined by 40% between 2001 and 2009 while the economy grew by more than 30%.

7.4 CLIMATE CHANGE INTERACTIONS

7.4.1 Climate Change

Rising temperatures and changing precipitation patterns can exacerbate groundwater depletion by increasing evaporation and reducing recharge.

In common usage, climate change describes global warming—the ongoing increase in global average temperature—and its effects on Earth's climate system. Climate change in a broader sense also includes previous long-term changes to Earth's climate. The current rise in global average temperature is more rapid than previous changes, and is primarily caused by humans burning fossil fuels. Fossil fuel use, deforestation, and some agricultural and industrial practices add to greenhouse gases, notably carbon dioxide and methane. Greenhouse gases absorb some of the heat that the Earth radiates after it warms from sunlight. Larger amounts of these gases trap more heat in Earth's lower atmosphere, causing global warming.

Climate change has an increasing impact on the environment. Deserts are expanding, while heat waves and wildfires are becoming more common. Amplified warming in the Arctic has contributed to melting permafrost, glacial retreat and sea ice loss. Higher temperatures are also causing more intense storms, droughts, and other weather extremes. Rapid environmental change in mountains, coral reefs, and the Arctic is forcing many species to relocate or become extinct. Even if efforts to minimise future warming are successful, some effects will continue for centuries. These include ocean heating, ocean acidification and sea level rise.

Climate change threatens people with increased flooding, extreme heat, increased food and water scarcity, more disease, and economic loss. Human migration and conflict can also be a result. The World Health Organization (WHO) calls climate change the greatest threat to global health in the 21st century. Societies and ecosystems will experience more severe risks without action to limit warming. Adapting to climate change through efforts like flood control measures or drought-resistant crops partially reduces climate change risks, although some limits to adaptation have already been reached. Poorer communities are responsible for a small share of global emissions, yet have the least ability to adapt and are most vulnerable to climate change.

Many climate change impacts have been felt in recent years, with 2023 the warmest on record at +1.48 °C (2.66 °F). Additional warming will increase these impacts and can trigger tipping points, such as the melting of the Greenland ice sheet. Under the 2015 Paris Agreement, nations collectively agreed to keep warming "well under 2 °C". However, with pledges made under the Agreement, global warming would still reach about 2.7 °C (4.9 °F) by the end of the century. Limiting warming to 1.5 °C will require halving emissions by 2030 and achieving net-zero emissions by 2050.

Strategies to phase out fossil fuels involve conserving energy, generating electricity cleanly, and using electricity to power transportation, heat buildings, and operate industrial facilities. The electricity supply can be made cleaner and more plentiful by vastly increasing deployment of wind, and solar power, alongside other forms of renewable energy and nuclear power. Carbon can also be removed from the atmosphere, for instance by increasing forest cover and farming with methods that capture carbon in soil.

Global Temperature Rise

Temperature Records Prior to Global Warming

Human beings evolved over the last few million years in a climate that cycled through ice ages, with global average temperature ranging between current levels and 5–6 °C colder than today. The temperature record prior to human evolution includes hotter temperatures and occasional abrupt changes, such as the Paleocene–Eocene Thermal Maximum 55.5 million years ago.

Historical patterns of warming and cooling, like the Medieval Warm Period and the Little Ice Age, did not occur at the same time across different regions. Temperatures may have reached as high as those of the late 20th century in a limited set of regions. Climate information for that period comes from climate proxies, such as trees and ice cores.

Warming since the Industrial Revolution

Thermometer records began to provide global coverage around 1850. There was little net warming between the 18th century and 1970, as aerosols offset the warming impact of greenhouse gas emissions. Increasing accumulation of greenhouse gases coupled with aerosol polution controls caused temperatures to then begin markedly increasing.

Multiple independent instrumental datasets show that the climate system is warming in recent years. A so-called "global warming hiatus" from 1998 to 2013 when warming was relatively slow was likely caused by negative phases of the Pacific Decadal Oscillation (PDO) and Atlantic Multidecadal Oscillation (AMO). The 2013-2022 decade warmed to an average 1.15 °C [1.00–1.25 °] compared to the ≈ pre-industrial baseline (1850–1900). Surface temperatures are rising by about 0.2 °C per decade.

Evidence of warming from air temperature measurements is reinforced by a wide range of other observations. For example, changes to the natural water c~cle have been predicted and observed, such as an increase in the frequency and intensity of heavy precipitation, melting of snow and land ice, and increased atmospheric humidity. Flora and fauna are also behaving in a manner consistent with warming; for instance, plants are flowering earlier in spring. Another key indicator is the cooling of the upper atmosphere, which demonstrates that greenhouse gases are trapping heat near the Earth's surface and preventing it from radiating into space.

Differences by Region

Different regions of the world warm at different rates. The pattern is independent of where greenhouse gases are emitted, because the gases persist long enough to diffuse across the planet. Since the pre-industrial period, the average surface temperature over land regions has increased almost twice as fast as the global-average surface temperature. This is because of the larger heat capacity of oceans, and because oceans lose more heat by evaporation. The thermal energy in the global climate system has grown with only brief pauses since at least 1970, and over 90% of this extra energy has been stored in the ocean. The rest has heated the atmosphere, melted ice, and warmed the continents.

The Northern Hemisphere and the North Pole have warmed much faster than the South Pole and Southern Hemisphere. The Northern Hemisphere not only has much more land, but also more seasonal snow cover and sea ice. As these surfaces flip from reflecting a lot of light to being dark after the ice has melted, they start absorbing more heat. Local black carbon deposits on snow and ice also contribute to Arctic warming. Arctic temperatures are increasing at over twice the rate of the rest of the world. Melting of glaciers and ice sheets in the Arctic disrupts ocean circulation, including a weakened Gulf Stream, further changing the climate.

Future Global Temperatures

The IPCC Sixth Assessment Report uses a 20 year temperature average to define global warming temperature rise, and it expects that 1.5 °C limit to be exceeded in the early 2030s. As annual temperatures fluctuate, the World Meteorological Organization estimates that there is a 66% chance that global temperature will exceed a rise of 1.5 °C for at least one year between 2023 and 2027. By the end of the 21st century, the IPCC projects that global warming is very likely to reach 1.0 °C to 1.8 °C under a very low GHG emissions scenario, 2.1 °C to 3.5 °C under an intermediate emissions scenario, or 3.3 °C to 5.7 °C under a very high GHG emissions scenario.

The remaining carbon budget for staying beneath certain temperature increases is determined by modelling the carbon cycle and climate sensitivity to greenhouse gases. According to the IPCC, global warming can be kept below 1.5 °C with a two-thirds chance if emissions after 2018 do not exceed 420 or 570 gigatonnes of CO_2. This corresponds to 10 to 13 years of current emissions. There are high uncertainties about the budget. For instance, it may be 100 gigatonnes of CO_2 smaller due to methane release from permafrost and wetlands. However, it is clear that fossil fuel resources are too abundant for shortages to be relied on to limit carbon emissions in the 21st century.

Causes of Recent Global Temperature Rise

The climate system experiences various cycles on its own which can last for years, decades or even centuries. For example, El Niño events cause short-term spikes in surface temperature while La Niña events cause short term cooling. Their relative frequency can affect global temperature trends on a decadal timescale. Other changes are caused by an imbalance of energy from external forcings. Examples of these include changes in the concentrations of greenhouse gases, solar luminosity, volcanic eruptions, and variations in the Earth's orbit around the Sun.

To determine the human contribution to climate change, unique "fingerprints" for all potential causes are developed and compared with both observed patterns and known internal climate variability. For example, solar forcing—whose fingerprint involves warming the entire atmosphere—is ruled out because only the lower atmosphere has warmed. Atmospheric aerosols produce a smaller, cooling effect. Other drivers, such as changes in albedo, are less impactful.

Greenhouse Gases

Greenhouse gases are transparent to sunlight, and thus allow it to pass through the atmosphere to heat the Earth's surface. The Earth radiates it as heat, and greenhouse gases absorb a portion of it. This absorption slows the rate at which heat escapes into space, trapping heat near the Earth's surface and warming it over time.

While water vapour (50%) and clouds (25%) are the biggest contributors to the greenhouse effect, they primarily change as a function of temperature and are therefore mostly considered to be feedbacks that change climate sensitivity. On the other hand, concentrations of gases such as CO_2 (20%), tropospheric ozone, CFCs and nitrous oxide are added or removed independently from temperature, and are therefore considered to be external forcings that change global temperatures.

Before the Industrial Revolution, naturally-occurring amounts of greenhouse gases caused the air near the surface to be about 33 °C warmer than it would have been in their absence. Human activity since the Industrial Revolution, mainly extracting and burning fossil fuels (coal, oil, and natural gas), has increased the amount of greenhouse gases in the atmosphere, resulting in a radiative imbalance. In 2019, the concentrations of CO_2 and methane had increased by about 48% and 160%, respectively, since 1750. These CO_2 levels are higher than they have been at any time during the last 2 million years. Concentrations of methane are far higher than they were over the last 800,000 years.

Global anthropogenic greenhouse gas emissions in 2019 were equivalent to 59 billion tonnes of CO_2. Of these emissions, 75% was CO_2, 18% was methane, 4% was nitrous oxide, and 2% was fluorinated gases. CO_2 emissions primarily come from burning fossil fuels to provide energy for transport, manufacturing, heating, and electricity. Additional CO_2 emissions come from deforestation and industrial processes, which include the CO_2 released by the chemical reactions for making cement, steel, aluminum, and fertiliser. Methane emissions come from livestock, manure, rice cultivation, landfills, wastewater, and coal mining, as well as oil and gas extraction. Nitrous oxide emissions largely come from the microbial decomposition of fertiliser.

The Earth's surface absorbs CO_2 as part of the carbon cycle. Despite the contribution of deforestation to greenhouse gas emissions, the Earth's land surface, particularly its forests, remain a significant carbon sink for CO_2. Land-surface sink processes, such as carbon fixation in the soil and photosynthesis, remove about 29% of annual global CO_2 emissions. The ocean also serves as a significant carbon sink via a two-step process. First, CO_2 dissolves in the surface water. Afterwards, the ocean's overturning circulation distributes it deep into the ocean's interior, where it accumulates over time as part of the carbon cycle. Over the last two decades, the world's oceans have absorbed 20 to 30% of emitted CO_2.

Aerosols and Clouds

Air pollution, in the form of aerosols, affects the climate on a large scale. Aerosols scatter and absorb solar radiation. From 1961 to 1990, a gradual reduction in the amount of sunlight reaching the Earth's surface was observed. This phenomenon is popularly known as global dimming, and is primarily attributed to sulfate aerosols produced by the combustion of fossil fuels with heavy sulfur concentrations like coal and bunker fuel. Smaller contributions come from black carbon, organic carbon from combustion of fossil fuels and biofuels, and from anthropogenic dust. Globally, aerosols have been declining since 1990 due to pollution controls, meaning that they no longer mask greenhouse gas warming as much.

Aerosols also have indirect effects on the Earth's energy budget. Sulfate aerosols act as cloud condensation nuclei and lead to clouds that have more and smaller cloud droplets. These clouds reflect solar radiation more efficiently than clouds with fewer and larger droplets. They also reduce the growth of raindrops, which makes clouds more reflective to incoming sunlight. Indirect effects of aerosols are the largest uncertainty in radiative forcing.

While aerosols typically limit global warming by reflecting sunlight, black carbon in soot that falls on snow or ice can contribute to global warming. Not only does this increase the absorption of sunlight, it also increases melting and sea-level rise. Limiting new black carbon deposits in the Arctic could reduce global warming by 0.2 °C by 2050.

Land Surface Changes

Humans change the Earth's surface mainly to create more agricultural land. Today, agriculture takes up 34% of Earth's land area, while 26% is forests, and 30% is uninhabitable (glaciers, deserts, etc.). The amount of forested land continues to decrease, which is the main land use change that causes global warming. Deforestation releases CO_2 contained in trees when they are destroyed, plus it prevents those trees from absorbing more CO_2. The main causes of deforestation are: permanent land-use change from forest to agricultural land producing products such as beef and palm oil (27%), logging to produce forestry/forest products (26%), short term shifting cultivation (24%), and wildfires (23%).

The type of vegetation in a region affects the local temperature. It impacts how much of the sunlight gets reflected back into space (albedo), and how much heat is lost by evaporation. For instance, the change from a dark forest to grassland makes the surface lighter, causing it to reflect more sunlight. Deforestation can also affect temperatures by modifying the release of chemical compounds that influence clouds, and by changing wind patterns. In tropic and temperate areas the net effect is to produce significant warming, while at latitudes closer to the poles a gain of albedo (as forest is replaced by snow cover) leads to a cooling effect. Globally, these effects are estimated to have led to a slight cooling, dominated by an increase in surface albedo. According to FAO, forest degradation aggravates the impacts of climate change as it reduces the carbon sequestration abilities of forests. Indeed, among their many benefits, forests also have the potential to reduce the impact of high temperatures.

Solar and Volcanic Activity

As the Sun is the Earth's primary energy source, changes in incoming sunlight directly affect the climate system. Solar irradiance has been measured directly by satellites, and indirect measurements are available from the early 1600s onwards. Since 1880, there has been no upward trend in the amount of the Sun's energy reaching the Earth.

Explosive volcanic eruptions represent the largest natural forcing over the industrial era. Eruptions can produce gases, dust and ash that partially block sunlight and reduce temperatures, or they can produce water vapor, which adds to greenhouse gases and increases temperatures. Gases in the stratosphere can cause atmospheric changes that last a couple years, with the temperature signal lasting about twice as long. In the industrial era, volcanic activity has had negligible impacts on global temperature trends. Present-day volcanic CO_2 emissions are equivalent to less than 1% of current anthropogenic CO_2 emissions.

Physical climate models are unable to reproduce the rapid warming observed in recent decades when taking into account only variations in solar output and volcanic activity. Further evidence for greenhouse gases causing global warming comes from measurements that show a warming of the lower atmosphere (the troposphere), coupled with a cooling of the upper atmosphere (the stratosphere). If solar variations were responsible for the observed warming, the troposphere and stratosphere would both warm.

Climate Change Feedback

The response of the climate system to an initial forcing is modified by feedbacks: increased by "self-reinforcing" or "positive" feedbacks and reduced by "balancing" or "negative" feedbacks. The main reinforcing feedbacks are the water-vapour feedback, the ice–albedo feedback, and the net effect of clouds. The primary balancing mechanism is radiative cooling, as Earth's surface gives off more heat to space in response to rising temperature. In addition to temperature feedbacks, there are feedbacks in the carbon cycle, such as the fertilizing effect of CO_2 on plant growth.

Uncertainty over feedbacks, particularly cloud cover, is the major reason why different climate models project different magnitudes of warming for a given amount of emissions. As air warms, it can hold more moisture. Water vapour, as a potent greenhouse gas, holds heat in the atmosphere. If cloud cover increases, more sunlight will be reflected back into space, cooling the planet. If clouds become higher and thinner, they act as an insulator, reflecting heat from below back downwards and warming the planet.

Another major feedback is the reduction of snow cover and sea ice in the Arctic, which reduces the reflectivity of the Earth's surface. More of the Sun's energy is now absorbed in these regions, contributing to amplification of Arctic temperature changes. Arctic amplification is also melting permafrost, which releases methane and CO_2 into the atmosphere. Climate change can also cause methane releases from wetlands, marine systems, and freshwater systems. Overall, climate feedbacks are expected to become increasingly positive.

Around half of human-caused CO_2 emissions have been absorbed by land plants and by the oceans. Climate change increases droughts and heat waves that inhibit plant growth, which makes it uncertain whether this carbon sink will continue to grow. Soils contain large quantities of carbon and may release some when they heat up. As more CO_2 and heat are absorbed by the ocean, it acidifies, its circulation changes and phytoplankton takes up less carbon, decreasing the rate at which the ocean absorbs atmospheric carbon. Overall, at higher CO_2 concentrations the Earth will absorb a reduced fraction of our emissions.

Modelling

A climate model is a representation of the physical, chemical and biological processes that affect the climate system. Models include natural processes like changes in the Earth's orbit, historical changes in the Sun's activity, and volcanic forcing. Models are used to estimate the degree of warming future emissions will cause when accounting for the strength of climate feedbacks,. Models also predict the circulation of the oceans, the annual cycle of the seasons, and the flows of carbon between the land surface and the atmosphere. The physical realism of models is tested by examining their ability to simulate contemporary or past climates. Past models have underestimated the rate of Arctic shrinkage and underestimated the rate of precipitation increase. Sea level rise since 1990 was underestimated in older models, but more recent models agree well with observations. The 2017 United States-published National Climate Assessment notes that "climate models may still be underestimating or missing relevant feedback processes". Additionally, climate models may be unable to adequately predict short-term regional climatic shifts.

A subset of climate models add societal factors to a physical climate model. These models simulate how population, economic growth, and energy use affect—and interact with—the physical climate. With this information, these models can produce scenarios of future greenhouse gas emissions. This is then used as input for physical climate models and carbon cycle models to predict how atmospheric concentrations of greenhouse gases might change. Depending on the socioeconomic scenario and the mitigation scenario, models produce atmospheric CO_2 concentrations that range widely between 380 and 1400 ppm.

Impacts

Environmental Effects

The environmental effects of climate change are broad and far-reaching, affecting oceans, ice, and weather. Changes may occur gradually or rapidly. Evidence for these effects comes from studying climate change in the past, from modelling, and from modern observations. Since the 1950s, droughts and heat waves have appeared simultaneously with increasing frequency. Extremely wet or dry events within the monsoon period have increased in India and East Asia. Monsoonal precipitation over the Northern Hemisphere has increased since 1980. The rainfall rate and intensity of hurricanes and typhoons is likely increasing, and the geographic range likely expanding poleward in response to climate warming. Frequency of tropical cyclones has not increased as a result of climate change.

Global sea level is rising as a consequence of glacial melt, melt of the Greenland ice sheets and Antarctica, and thermal expansion. Between 1993 and 2020, the rise increased over time, averaging 3.3 ± 0.3 mm per year. Over the 21st century, the IPCC projects that in a very high emissions scenario the sea level could rise by 61–110 cm. Increased ocean warmth is undermining and threatening to unplug Antarctic glacier outlets, risking a large melt of the ice sheet and the possibility of a 2-meter sea level rise by 2100 under high emissions.

Climate change has led to decades of shrinking and thinning of the Arctic sea ice. While ice-free summers are expected to be rare at 1.5 °C degrees of warming, they are set to occur once every three to ten years at a warming level of 2 °C. Higher atmospheric CO_2 concentrations have led to changes in ocean chemistry. An increase in dissolved CO_2 is causing oceans to acidify. In addition, oxygen levels are decreasing as oxygen is less soluble in warmer water. Dead zones in the ocean, regions with very little oxygen, are expanding too.

Tipping Points and Long-term Impacts

Greater degrees of global warming increase the risk of passing through 'tipping points'—thresholds beyond which certain impacts can no longer be avoided even if temperatures are reduced. An example is the collapse of West Antarctic and Greenland ice sheets, where a temperature rise of 1.5 to 2 °C may commit the ice sheets to melt, although the time scale of melt is uncertain and depends on future warming. Some large-scale changes could occur over a short time period, such as a shutdown of certain ocean currents like the Atlantic meridional overturning circulation (AMOC). Tipping points can also include irreversible damage to ecosystems like the Amazon rainforest and coral reefs.

The long-term effects of climate change on oceans include further ice melt, ocean warming, sea level rise, and ocean acidification. The timescale of long term impacts are centuries to millennia due to CO_2's long atmospheric lifetime. When net emissions stabilize surface air temperatures will also stabilize, but oceans and ice caps will continue to absorb excess heat from the atmosphere. The result is an estimated total sea level rise of 2.3 metres per degree Celsius (4.2 ft/°F) after 2000 years. Oceanic CO_2 uptake is slow enough that ocean acidification will also continue for hundreds to thousands of years.

Nature and Wildlife

Recent warming has driven many terrestrial and freshwater species poleward and towards higher altitudes. Higher atmospheric CO_2 levels and an extended growing season have resulted in global greening. However, heatwaves and drought have reduced ecosystem productivity in some regions. The future balance of these opposing effects is unclear. Climate change has contributed to the expansion of drier climate zones, such as the expansion of deserts in the subtropics. The size and speed of global warming is making abrupt changes in ecosystems more likely. Overall, it is expected that climate change will result in the extinction of many species.

The oceans have heated more slowly than the land, but plants and animals in the ocean have migrated towards the colder poles faster than species on land. Just as on land, heat waves in the ocean occur more frequently due to climate change, harming a wide range of organisms such as corals, kelp, and seabirds. Ocean acidification makes it harder for marine calcifying organisms such as mussels, barnacles and corals to produce shells and skeletons; and heatwaves have bleached coral reefs. Harmful algal blooms enhanced by climate change and eutrophication lower oxygen levels, disrupt food webs and cause great loss of marine life. Coastal ecosystems are under particular stress. Almost half of global wetlands have disappeared due to climate change and other human impacts.

Humans

The effects of climate change are impacting humans everywhere in the world. Impacts can be observed on all continents and ocean regions, with low-latitude, less developed areas facing the greatest risk. Continued warming has potentially "severe, pervasive and irreversible impacts" for people and ecosystems. The risks are unevenly distributed, but are generally greater for disadvantaged people in developing and developed countries.

Food and Health

The WHO calls climate change the greatest threat to global health in the 21st century. Extreme weather leads to injury and loss of life, and crop failures to malnutrition. Various infectious diseases are more easily transmitted in a warmer climate, such as dengue fever and malaria. Young children are the most vulnerable to food shortages. Both children and older people are vulnerable to extreme heat. The World Health Organization (WHO) has estimated that between 2030 and 2050, climate change would cause around 250,000 additional deaths per year. They assessed deaths from heat exposure in elderly people, increases in diarrhea, malaria, dengue, coastal flooding, and childhood malnutrition. Over 500,000 more adult deaths are projected yearly by 2050 due to reductions in food availability and quality. By 2100, 50% to 75% of the global population may face climate conditions that are life-threatening due to combined effects of extreme heat and humidity.

Climate change is affecting food security. It has caused reduction in global yields of maize, wheat, and soybeans between 1981 and 2010. Future warming could further reduce global yields of major crops. Crop production will probably be negatively affected in low-latitude countries, while effects at northern latitudes may be positive or negative. Up to an additional 183 million people worldwide, particularly those with lower incomes, are at risk of hunger as a consequence of these impacts. Climate change also impacts fish populations. Globally, less will be available to be fished. Regions dependent on glacier water, regions that are already dry, and small islands have a higher risk of water stress due to climate change.

Inequality

Economic damages due to climate change may be severe and there is a chance of disastrous consequences. Climate change has likely already increased global economic inequality, and this trend is projected to continue. Severe impacts are expected in South-East Asia and sub-Saharan Africa, where most of the local inhabitants are dependent upon natural and agricultural resources. The World Bank estimates that climate change could drive over 120 million people into poverty by 2030.

Inequalities based on wealth and social status have worsened due to climate change. Major difficulties in mitigating, adapting, and recovering to climate shocks are faced by marginalized people who have less control over resources. Indigenous people, who are subsistent on their land and ecosystems, will face endangerment to their wellness and lifestyles due to climate change. An expert elicitation concluded that the role of climate change in armed conflict has been small compared to factors such as socio-economic inequality and state capabilities.

While women are not inherently more at risk from climate change and shocks, limits on women's resources and discriminatory gender norms constrain their adaptive capacity and resilience. For example, women's work burdens, including hours worked in agriculture, tend to decline less than men's during climate shocks such as heat stress.

Climate Migration

Low-lying islands and coastal communities are threatened by sea level rise, which makes urban flooding more common. Sometimes, land is permanently lost to the sea. This could lead to statelessness for people in island nations, such as the Maldives and Tuvalu. In some regions, the rise in temperature and humidity may be too severe for humans to adapt to. With worst-case climate change, models project that almost one-third of humanity might live in extremely hot and uninhabitable climates.

These factors can drive climate or environmental migration, within and between countries. More people are expected to be displaced because of sea level rise, extreme weather and conflict from increased competition over natural resources. Climate change may also increase vulnerability, leading to "trapped populations" who are not able to move due to a lack of resources.

Reducing and Recapturing Emissions

Climate change can be mitigated by reducing the rate at which greenhouse gases are emitted into the atmosphere, and by increasing the rate at which carbon dioxide is removed from the atmosphere. In order to limit global warming to less than 1.5 °C global greenhouse gas emissions needs to be net-zero by 2050, or by 2070 with a 2 °C target. This requires far-reaching, systemic changes on an unprecedented scale in energy, land, cities, transport, buildings, and industry.

The United Nations Environment Programme estimates that countries need to triple their pledges under the Paris Agreement within the next decade to limit global warming to 2 °C. An even greater level of reduction is required to meet the 1.5 °C goal. With pledges made under the Paris Agreement as of October 2021, global warming would still have a 66% chance of reaching about 2.7 °C (range: 2.2–3.2 °C) by the end of the century. Globally, limiting warming to 2 °C may result in higher economic benefits than economic costs.

Although there is no single pathway to limit global warming to 1.5 or 2 °C, most scenarios and strategies see a major increase in the use of renewable energy in combination with increased energy efficiency measures to generate the needed greenhouse gas reductions. To reduce pressures on ecosystems and enhance their carbon sequestration capabilities, changes would also be necessary in agriculture and forestry, such as preventing deforestation and restoring natural ecosystems by reforestation. Other approaches to mitigating climate change have a higher level of risk. Scenarios that limit global warming to 1.5 °C typically project the large-scale use of carbon dioxide removal methods over the 21st century. There are concerns, though, about over-reliance on these technologies, and environmental impacts. Solar radiation modification (SRM) is also a possible supplement to deep reductions in emissions. However, SRM raises significant ethical and legal concerns, and the risks are imperfectly understood.

Clean Energy

Renewable energy is key to limiting climate change. For decades, fossil fuels have accounted for roughly 80% of the world's energy use. The remaining share has been split between nuclear power and renewables (including hydropower, bioenergy, wind and solar power and geothermal energy). Fossil fuel use is expected to peak in absolute terms prior to 2030 and then to decline, with coal use experiencing the sharpest reductions. Renewables represented 75% of all new electricity generation installed in 2019, nearly all solar and wind. Other forms of clean energy, such as nuclear and hydropower, currently have a larger share of the energy supply. However, their future growth forecasts appear limited in comparison.

While solar panels and onshore wind are now among the cheapest forms of adding new power generation capacity in many locations, green energy policies are needed to achieve a rapid transition from fossil fuels to renewables. To achieve carbon neutrality by 2050, renewable energy would become the dominant form of electricity generation, rising to 85% or more by 2050 in some scenarios. Investment in coal would be eliminated and coal use nearly phased out by 2050.

Electricity generated from renewable sources would also need to become the main energy source for heating and transport. Transport can switch away from internal combustion engine vehicles and towards electric vehicles, public transit, and active transport (cycling and walking). For shipping and flying, low-carbon fuels would reduce emissions. Heating could be increasingly decarbonised with technologies like heat pumps.

There are obstacles to the continued rapid growth of clean energy, including renewables. For wind and solar, there are environmental and land use concerns for new projects. Wind and solar also produce energy intermittently and with seasonal variability. Traditionally, hydro dams with reservoirs and conventional power plants have been used when variable energy production is low. Going forward, battery storage can be expanded, energy demand and supply can be matched, and long-distance transmission can smooth variability of renewable outputs. Bioenergy is often not carbon-neutral and may have negative consequences for food security. The growth of nuclear power is constrained by controversy around radioactive waste, nuclear weapon proliferation, and accidents. Hydropower growth is limited by the fact that the best sites have been developed, and new projects are confronting increased social and environmental concerns.

Low-carbon energy improves human health by minimising climate change as well as reducing air pollution deaths, which were estimated at 7 million annually in 2016. Meeting the Paris Agreement goals that limit warming to a 2 °C increase could save about a million of those lives per year by 2050, whereas limiting global warming to 1.5 °C could save millions and simultaneously increase energy security and reduce poverty. Improving air quality also has economic benefits which may be larger than mitigation costs.

Energy Conservation

Reducing energy demand is another major aspect of reducing emissions. If less energy is needed, there is more flexibility for clean energy development. It also makes it easier to manage the electricity grid, and minimises carbon-intensive infrastructure development. Major increases in energy efficiency investment will be required to achieve climate goals, comparable to the level of investment in renewable energy. Several COVID-19 related changes in energy use patterns, energy efficiency investments, and funding have made forecasts for this decade more difficult and uncertain.

Strategies to reduce energy demand vary by sector. In the transport sector, passengers and freight can switch to more efficient travel modes, such as buses and trains, or use electric vehicles. Industrial strategies to reduce energy demand include improving heating systems and motors, designing less energy-intensive products, and increasing product lifetimes. In the building sector the focus is on better design of new buildings, and higher levels of energy efficiency in retrofitting. The use of technologies like heat pumps can also increase building energy efficiency.

Agriculture and Industry

Agriculture and forestry face a triple challenge of limiting greenhouse gas emissions, preventing the further conversion of forests to agricultural land, and meeting increases in world food demand. A set of actions could reduce agriculture and forestry-based emissions by two thirds from 2010 levels. These include reducing growth in demand for food and other agricultural products, increasing land productivity, protecting and restoring forests, and reducing greenhouse gas emissions from agricultural production.

On the demand side, a key component of reducing emissions is shifting people towards plant-based diets. Eliminating the production of livestock for meat and dairy would eliminate about 3/4ths of all emissions from agriculture and other land use. Livestock also occupy 37% of ice-free land area on Earth and consume feed from the 12% of land area used for crops, driving deforestation and land degradation.

Steel and cement production are responsible for about 13% of industrial CO_2 emissions. In these industries, carbon-intensive materials such as coke and lime play an integral role in the production, so that reducing CO_2 emissions requires research into alternative chemistries.

Carbon Sequestration

Natural carbon sinks can be enhanced to sequester significantly larger amounts of CO_2 beyond naturally occurring levels. Reforestation and afforestation (planting forests where there were none before) are among the most mature sequestration techniques, although the latter raises food security concerns. Farmers can promote sequestration of carbon in soils through practices such as use of winter cover crops, reducing the intensity and frequency of tillage, and using compost and manure as soil amendments. Forest and landscape restoration yields many benefits for the climate, including greenhouse gas emissions sequestration and reduction.

Restoration/recreation of coastal wetlands, prairie plots and seagrass meadows increases the uptake of carbon into organic matter. When carbon is sequestered in soils and in organic matter such as trees, there is a risk of the carbon being re-released into the atmosphere later through changes in land use, fire, or other changes in ecosystems.

Where energy production or CO_2-intensive heavy industries continue to produce waste CO_2, the gas can be captured and stored instead of released to the atmosphere. Although its current use is limited in scale and expensive, carbon capture and storage (CCS) may be able to play a significant role in limiting CO_2 emissions by mid-century. This technique, in combination with bioenergy (BECCS) can result in net negative emissions as CO_2 is drawn from the atmosphere. It remains highly uncertain whether carbon dioxide removal techniques will be able to play a large role in limiting warming to 1.5 °C. Policy decisions that rely on carbon dioxide removal increase the risk of global warming rising beyond international goals.

Adaptation

Adaptation is "the process of adjustment to current or expected changes in climate and its effects". Without additional mitigation, adaptation cannot avert the risk of "severe, widespread and irreversible" impacts. More severe climate change requires more transformative adaptation, which can be prohibitively expensive. The capacity and potential for humans to adapt is unevenly distributed across different regions and populations, and developing countries generally have less. The first two decades of the 21st century saw an increase in adaptive capacity in most low- and middle-income countries with improved access to basic sanitation and electricity, but progress is slow. Many countries have implemented adaptation policies. However, there is a considerable gap between necessary and available finance.

Adaptation to sea level rise consists of avoiding at-risk areas, learning to live with increased flooding, and building flood controls. If that fails, managed retreat may be needed. There are economic barriers for tackling dangerous heat impact. Avoiding strenuous work or having air conditioning is not possible for everybody. In agriculture, adaptation options include a switch to more sustainable diets, diversification, erosion control, and genetic improvements for increased tolerance to a changing climate. Insurance allows for risk-sharing, but is often difficult to get for people on lower incomes. Education, migration and early warning systems can reduce climate vulnerability. Planting mangroves or encouraging other coastal vegetation can buffer storms.

Ecosystems adapt to climate change, a process that can be supported by human intervention. By increasing connectivity between ecosystems, species can migrate to more favourable climate conditions. Species can also be introduced to areas acquiring a favorable climate. Protection and restoration of natural and semi-natural areas helps build resilience, making it easier for ecosystems to adapt. Many of the actions that promote adaptation in ecosystems, also help humans adapt via ecosystem-based adaptation. For instance, restoration of natural fire regimes makes catastrophic fires less likely, and reduces human exposure. Giving rivers more space allows for more water storage in the natural system, reducing flood risk. Restored forest acts as a carbon sink, but planting trees in unsuitable regions can exacerbate climate impacts.

There are synergies but also trade-offs between adaptation and mitigation. An example for synergy is increased food productivity, which has large benefits for both adaptation and mitigation. An example of a trade-off is that increased use of air conditioning allows people to better cope with heat, but increases energy demand. Another trade-off example is that more compact urban development may reduce emissions from transport and construction, but may also increase the urban heat island effect, exposing people to heat-related health risks.

Policies and Politics

Countries that are most vulnerable to climate change have typically been responsible for a small share of global emissions. This raises questions about justice and fairness. Limiting global warming makes it much easier to achieve the UN's Sustainable Development Goals, such as eradicating poverty and reducing inequalities. The connection is recognised in Sustainable Development Goal 13 which is to "take urgent action to combat climate change and its impacts". The goals on food, clean water and ecosystem protection have synergies with climate mitigation.

The geopolitics of climate change is complex. It has often been framed as a free-rider problem, in which all countries benefit from mitigation done by other countries, but individual countries would lose from switching to a low-carbon economy themselves. Sometimes mitigation also has localized benefits though. For instance, the benefits of a coal phase-out to public health and local environments exceed the costs in almost all regions. Furthermore, net importers of fossil fuels win economically from switching to clean energy, causing net exporters to face stranded assets: fossil fuels they cannot sell.

Policy Options

A wide range of policies, regulations, and laws are being used to reduce emissions. As of 2019, carbon pricing covers about 20% of global greenhouse gas emissions. Carbon can be priced with carbon taxes and emissions trading systems. Direct global fossil fuel subsidies reached $319 billion in 2017, and $5.2 trillion when indirect costs such as air pollution are priced in. Ending these can cause a 28% reduction in global carbon emissions and a 46% reduction in air pollution deaths. Money saved on fossil subsidies could be used to support the transition to clean energy instead. More direct methods to reduce greenhouse gases include vehicle efficiency standards, renewable fuel standards, and air pollution regulations on heavy industry. Several countries require utilities to increase the share of renewables in power production.

Climate Justice

Policy designed through the lens of climate justice tries to address human rights issues and social inequality. According to proponents of climate justice, the costs of climate adaptation should be paid by those most responsible for climate change, while the beneficiaries of payments should be those suffering impacts. One way this can be addressed in practice is to have wealthy nations pay poorer countries to adapt.

Oxfam found that in 2023 the wealthiest 10% of people were responsible for 50% of global emissions, while the bottom 50% were responsible for just 8%. Production is another way to look at responsibility, with a 2023 study published in One Earth estimating that the top 21 fossil fuel companies would owe cumulative climate reparations of $5.4 trillion over the period 2025–2050. To achieve a just transition, people working in the fossil fuel sector would also need other jobs, and their communities would need investments.

International Climate Agreements

Nearly all countries in the world are parties to the 1994 United Nations Framework Convention on Climate Change (UNFCCC). The goal of the UNFCCC is to prevent dangerous human interference with the climate system. As stated in the convention, this requires that greenhouse gas concentrations are stabilised in the atmosphere at a level where ecosystems can adapt naturally to climate change, food production is not threatened, and economic development can be sustained. The UNFCCC does not itself restrict emissions but rather provides a framework for protocols that do. Global emissions have risen since the UNFCCC was signed. Its yearly conferences are the stage of global negotiations.

The 1997 Kyoto Protocol extended the UNFCCC and included legally binding commitments for most developed countries to limit their emissions. During the negotiations, the G77 (representing developing countries) pushed for a mandate requiring developed countries to "[take] the lead" in reducing their emissions, since developed countries contributed most to the accumulation of greenhouse gases in the atmosphere. Per-capita emissions were also still relatively low in developing countries and developing countries would need to emit more to meet their development needs.

The 2009 Copenhagen Accord has been widely portrayed as disappointing because of its low goals, and was rejected by poorer nations including the G77. Associated parties aimed to limit the global temperature rise to below 2 °C. The Accord set the goal of sending $100 billion per year to developing countries for mitigation and adaptation by 2020, and proposed the founding of the Green Climate Fund. As of 2020, only 83.3 billion were delivered. Only in 2023 the target is expected to be achieved.

In 2015 all UN countries negotiated the Paris Agreement, which aims to keep global warming well below 2.0 °C and contains an aspirational goal of keeping warming under 1.5 °C. The agreement replaced the Kyoto Protocol. Unlike Kyoto, no binding emission targets were set in the Paris Agreement. Instead, a set of procedures was made binding. Countries have to regularly set ever more ambitious goals and reevaluate these goals every five years. The Paris Agreement restated that developing countries must be financially supported. As of October 2021, 194 states and the European Union have signed the treaty and 191 states and the EU have ratified or acceded to the agreement.

The 1987 Montreal Protocol, an international agreement to stop emitting ozone-depleting gases, may have been more effective at curbing greenhouse gas emissions than the Kyoto Protocol specifically designed to do so. The 2016 Kigali Amendment to the Montreal Protocol aims to reduce the emissions of hydrofluorocarbons, a group of powerful greenhouse gases which served as a replacement for banned ozone-depleting gases. This made the Montreal Protocol a stronger agreement against climate change.

National Responses

In 2019, the United Kingdom parliament became the first national government to declare a climate emergency. Other countries and jurisdictions followed suit. That same year, the European Parliament declared a "climate and environmental emergency". The European Commission presented its European Green Deal with the goal of making the EU carbon-neutral by 2050. In 2021, the European Commission released its "Fit for 55" legislation package, which contains guidelines for the car industry; all new cars on the European market must be zero-emission vehicles from 2035.

Major countries in Asia have made similar pledges: South Korea and Japan have committed to become carbon-neutral by 2050, and China by 2060. While India has strong incentives for renewables, it also plans a significant expansion of coal in the country. Vietnam is among very few coal-dependent fast developing countries that pledged to phase out unabated coal power by the 2040s or as soon as possible thereafter.

As of 2021, based on information from 48 national climate plans, which represent 40% of the parties to the Paris Agreement, estimated total greenhouse gas emissions will be 0.5% lower compared to 2010 levels, below the 45% or 25% reduction goals to limit global warming to 1.5 °C or 2 °C, respectively.

Feedback Loops

Depleting groundwater can contribute to climate change by altering land surface characteristics and impacting vegetation, further intensifying the feedback loop.

The feedback loops associated with depleting groundwater create a complex web of challenges that demand urgent and coordinated action. Sustainable water management practices, informed by scientific research, technological innovations, and community engagement, are essential components of breaking the vicious cycle of groundwater depletion. As the world grapples with increasing water demands and climate uncertainties, a holistic approach that addresses both the causes and consequences of groundwater depletion is crucial for ensuring the long-term viability of this precious resource. Through collaborative efforts and a commitment to sustainable practices, it is possible to navigate the delicate balance between meeting current water needs and safeguarding the integrity of groundwater systems for future generations.

7.4.2 Excessive Pumping and Lowering Water Tables

The primary driver of groundwater depletion is excessive pumping from wells for various purposes, including irrigation, industrial processes, and municipal water supply. As demand rises, particularly in regions heavily reliant on groundwater, the rate of extraction surpasses natural recharge rates. This results in a lowering of water tables, reducing the volume of available groundwater.

1. Consequences:
   a. Lowering water tables impact the accessibility of wells, requiring deeper drilling.
   b. Decreased water availability for agriculture, affecting crop yields and food security.
   c. The compounding effect on ecosystems, as natural discharge zones are unable to replenish surface water bodies.

7.4.3 Impact on Agricultural Practices

Agriculture is a major consumer of groundwater, with irrigation being a significant contributor to depletion. As water tables drop, farmers intensify pumping to meet crop water requirements, creating a feedback loop. The pursuit of higher yields leads to increased water use, exacerbating the depletion problem.

1. Consequences:
   a. Declining agricultural productivity due to insufficient water for crops.
   b. Escalating costs for farmers as they invest in deeper wells and more powerful pumps.
   c. Shifts in cropping patterns towards more water-efficient or drought-tolerant crops.

7.4.4 Subsidence and Infrastructure Damage

The excessive withdrawal of groundwater can lead to land subsidence, a phenomenon where the land surface sinks. As aquifers are depleted, the pore spaces in the subsurface collapse, causing the land above to settle. Subsidence poses a significant threat to infrastructure, including buildings, roads, and pipelines.

1. Consequences:
   a. Structural damage to buildings and infrastructure, impacting public safety.

b. Increased costs for repairing and maintaining subsidence-affected infrastructure.

c. Challenges in sustainable urban planning due to the risk of subsidence.

7.4.5 Saltwater Intrusion

Coastal areas are particularly vulnerable to the feedback loop of groundwater depletion due to the intrusion of saltwater. Over-pumping in coastal aquifers lowers the freshwater table, allowing saltwater from the ocean to migrate inland. This intrusion further reduces the availability of freshwater, creating a self-perpetuating cycle.

1. Consequences:

   a. Contaminated freshwater resources with increased salinity.

   b. Impact on coastal ecosystems and biodiversity.

   c. Threats to agriculture in coastal regions as soil salinity rises.

7.4.6 Climate Change Feedback

Climate change adds another layer to the feedback loops associated with groundwater depletion. Changes in precipitation patterns and temperature affect groundwater recharge rates. In regions experiencing reduced precipitation and increased evaporation, groundwater recharge diminishes, amplifying the depletion problem.

1. Consequences:

   a. Increased reliance on groundwater during drought periods.

   b. Compromised resilience of ecosystems dependent on groundwater.

   c. Challenges in water resource management due to climate-induced variability.

7.4.7 Ecosystem Degradation

Groundwater depletion directly affects ecosystems dependent on consistent water availability. Springs, wetlands, and rivers, sustained by groundwater discharge, face reduced flow or complete drying. This disrupts habitats, endangers species, and triggers a negative feedback loop as degraded ecosystems contribute less to groundwater recharge.

1. Consequences:

   a. Loss of biodiversity and disruption of ecological balance.

   b. Diminished ecosystem services such as water purification and flood control.

   c. Challenges in restoring ecosystems once they have reached a critical state.

7.4.8 Interconnected Surface Water and Groundwater Depletion

Surface water and groundwater systems are interconnected, and excessive pumping from aquifers can deplete not only groundwater resources but also interconnected surface water bodies. This feedback loop intensifies as reduced base flow in rivers and streams further limits natural recharge to aquifers.

1. Consequences:
    a. Impact on aquatic ecosystems dependent on consistent flow.
    b. Increased vulnerability to droughts and climate variability.
    c. Challenges in maintaining reliable water supplies for downstream users.

7.4.9 Breaking the Vicious Cycle

Addressing the feedback loops associated with depleting groundwater requires a multifaceted approach involving sustainable management practices, technological innovations, and policy interventions.

1. Adopting Sustainable Water Use Practices:
    a. Implementing efficient irrigation methods to reduce water consumption in agriculture.
    b. Encouraging the use of drought-resistant crop varieties.
    c. Promoting water conservation in industrial processes and domestic use.
2. Integrated Water Resource Management:
    a. Implementing aquifer management plans that consider both surface water and groundwater interactions.
    b. Incorporating climate change projections into water resource planning to account for variability in precipitation patterns and recharge rates.
    c. Establishing regulations on groundwater extraction rates to ensure sustainable use.
3. Investing in Water Recharge and Storage:
    a. Implementing artificial recharge methods, such as spreading basins and injection wells, to enhance groundwater recharge.
    b. Developing and maintaining green infrastructure to enhance natural recharge processes.
    c. Investing in the construction of water storage facilities to capture excess surface water during periods of abundance for later use.
4. Promoting Water-Efficient Technologies:
    a. Encouraging the adoption of water-efficient technologies in agriculture, such as precision irrigation and soil moisture monitoring.

   b. Investing in water recycling and reuse systems to reduce reliance on fresh groundwater for non-potable uses.

5. Monitoring and Research:
   a. Implementing comprehensive groundwater monitoring networks to track water levels, quality, and subsidence.
   b. Supporting research on innovative technologies for sustainable groundwater management.
   c. Promoting collaboration between scientists, policymakers, and local communities to share knowledge and insights.
6. Public Awareness and Community Engagement:
   a. Educating the public about the importance of sustainable groundwater use.
   b. Engaging local communities in water conservation efforts.
   c. Fostering a sense of responsibility and ownership in the community for sustainable water management.
7. Policy and Regulatory Frameworks:
   a. Enforcing regulations on groundwater extraction and implementing permits to control pumping rates.
   b. Integrating groundwater considerations into broader water management policies.
   c. Developing and enforcing land-use planning regulations to prevent over-extraction in vulnerable areas.

CHAPTER 8

GROUNDWATER AND GEOLOGICAL PROCESSES

8.1 GEOLOGICAL PROCESSES DRIVING FLUID FLOW

The Earth's subsurface is a dynamic realm where geological processes drive the movement of fluids, shaping landscapes, sustaining ecosystems, and influencing the distribution of natural resources. The intricate interplay of geological features, such as rock formations, faults, and fractures, plays a pivotal role in determining how fluids, including water, oil, and gases, flow through the subsurface. The Earth's subsurface is a dynamic and interconnected realm where geological processes act as the architects of fluid flow. From the dissolution of soluble rocks in karst landscapes to the migration of hydrocarbons through fault networks, these processes shape the distribution of resources, influence landscape features, and contribute to Earth's geological evolution. Understanding the mechanisms and impacts of geological processes driving fluid flow is essential for sustainable resource management, environmental conservation, and predicting natural hazards. As scientific advancements continue to unravel the complexities of subsurface dynamics, our appreciation for the intricate interplay between Earth's geological processes and fluid flow deepens, paving the way for informed decision-making and responsible stewardship of our planet's geological heritage.

8.1.1 Porosity and Permeability

Porosity and permeability are fundamental properties that govern fluid flow in geological formations. Porosity refers to the volume of void spaces or pores within a rock or sediment, while permeability is the ability of a material to transmit fluids. These properties vary depending on the type of rock and the geological processes that have influenced its formation.

1. Mechanisms:
    a. Sedimentation and Diagenesis: The accumulation of sediments, such as sand and clay, over time leads to the formation of sedimentary rocks. Diagenetic processes, including compaction and cementation, further influence porosity and permeability.

b. Faulting and Fracturing: Tectonic activity, including faulting and fracturing, creates pathways for fluid migration. Faults can act as conduits or barriers to fluid flow, depending on their orientation and sealing characteristics.

c. Weathering: Mechanical and chemical weathering alter rock structures, creating openings and fractures that enhance porosity. Water, over time, exploits these openings, facilitating fluid movement.

2. Impacts:

   a. Highly porous and permeable formations, such as sandstone, act as prolific reservoirs for groundwater, oil, and gas.

   b. Low porosity and permeability, characteristic of dense rocks like shale, can impede fluid flow and act as seals for hydrocarbon reservoirs.

8.1.2 Regional and Local Groundwater Flow

Groundwater, a vital component of Earth's subsurface fluid system, is influenced by regional and local geological processes. Regional flow occurs over large scales, often following topographic gradients, while local flow is confined to specific geological features.

1. Mechanisms:

   a. Topography: Regional groundwater flow is influenced by topographic gradients, with water flowing from high elevation recharge areas to lower elevation discharge areas.

   b. Aquifer Properties: The permeability and porosity of aquifers, influenced by geological processes, determine the speed and direction of groundwater flow. Well-developed fractures and conduits can enhance local flow within aquifers.

   c. Geologic Structures: Faults, folds, and other structural features can act as barriers or conduits for groundwater, influencing its regional and local movement.

2. Impacts:

   a. Understanding regional groundwater flow is crucial for managing water resources, predicting contamination pathways, and ensuring sustainable extraction.

8.1.3 Hydrothermal Systems

Hydrothermal systems represent dynamic subsurface environments where high-temperature fluids, often associated with volcanic activity, circulate through Earth's crust. These systems play a crucial role in mineral deposition, geothermal energy generation, and the redistribution of elements.

1. Mechanisms:
   a. Magmatic Heating: Magma intrusions release heat, causing surrounding groundwater to circulate and become hydrothermal fluids. This process is associated with the formation of mineral deposits.
   b. Fracturing and Faulting: Geological features such as fractures and faults provide pathways for hydrothermal fluids to rise from depth. These conduits enhance fluid flow and facilitate mineralization.
2. Impacts:
   a. Hydrothermal systems contribute to the formation of economically significant ore deposits, including gold, silver, and copper.
   b. Geothermal energy, harnessed from subsurface heat, is a sustainable resource derived from hydrothermal systems.

8.1.4 Karst Landscapes and Speleogenesis

Karst landscapes, characterized by unique surface and subsurface features such as sinkholes, caves, and underground rivers, result from the dissolution of soluble rocks like limestone and gypsum. The process of speleogenesis, or cave formation, is driven by chemical reactions between groundwater and the soluble rock.

1. Mechanisms:
   a. Carbonation: Rainwater, slightly acidic due to dissolved carbon dioxide, reacts with calcium carbonate in limestone, initiating a chemical process called carbonation. This process dissolves the rock and creates conduits for fluid flow.
   b. Sinkhole Formation: As dissolution continues, large cavities may form underground. When the roof of these cavities collapses, surface sinkholes are created.
2. Impacts:
   a. Karst aquifers, with their complex network of conduits, contribute to rapid groundwater flow.
   b. Speleogenesis results in the formation of extensive cave systems with unique ecosystems.

8.1.5 Subsurface Oil and Gas Migration

The migration of oil and gas through the subsurface is a complex geological process influenced by the source rock, migration pathways, and trapping mechanisms. Understanding these processes is crucial for the exploration and extraction of hydrocarbons.

1. Mechanisms:

   a. Source Rock Formation: Organic-rich source rocks, subjected to heat and pressure over time, generate hydrocarbons such as oil and gas.
   b. Migration Pathways: Faults, fractures, and permeable rock layers act as migration pathways for hydrocarbons to move from source rocks to reservoir rocks.
   c. Trapping Mechanisms: Geological structures, such as anticlines and fault traps, prevent hydrocarbons from migrating further, creating reservoirs.
2. Impacts:
   a. Successful exploration and extraction of oil and gas depend on understanding the geological processes that govern their migration and accumulation.
   b. Geological formations, such as anticlines, serve as targets for oil and gas exploration.

8.1.6 Subsidence and Compaction

Geological processes such as subsidence and compaction influence the redistribution of fluids and sediments within the subsurface. These processes can have profound effects on landscapes, infrastructure, and the movement of fluids.

1. Mechanisms:
   a. Sediment Compaction: The weight of overlying sediments causes compaction, reducing pore spaces and influencing fluid migration.
   b. Subsidence: Withdrawal of fluids, such as groundwater or hydrocarbons, can lead to subsidence as pore spaces collapse.
2. Impacts:
   a. Compaction influences fluid flow and reservoir properties in geological formations.
   b. Subsidence poses risks to infrastructure and alters landscape features.

8.1.7 Faults and Seismic Activity

Faults, fractures in the Earth's crust, and seismic activity contribute to the movement of fluids by creating conduits, altering permeability, and triggering fluid release during earthquakes.

1. Mechanisms:
   a. Fault Zones: Faults may act as barriers or conduits for fluid flow, depending on their sealing properties and orientation.
   b. Seismic Activity: Earthquakes can fracture rocks, creating new pathways for fluids and influencing groundwater flow.

2. Impacts:
   a. Fault zones play a critical role in compartmentalizing or connecting fluid reservoirs.
   b. Seismic activity can trigger changes in subsurface fluid pressures and release fluids, affecting groundwater and hydrocarbon systems.

8.2 TOPOGRAPHY-DRIVEN FLOW IN THE CONTEXT OF GEOLOGICAL PROCESSES

The intricate dance of geological processes unfolds beneath the Earth's surface, orchestrating the movement of fluids in a symphony of topography-driven flow. This dynamic interaction between topography and fluid flow is a fundamental aspect of Earth's geology, influencing landscapes, shaping subsurface features, and playing a pivotal role in the distribution of water, oil, and gases. As water navigates the labyrinthine channels beneath our feet and hydrocarbons traverse geological formations, topography emerges as a maestro, directing the movements of these essential fluids. In the grand tapestry of Earth's geology, topography emerges as a master conductor, directing the flow of fluids and shaping the landscapes we inhabit. From the highest mountain peaks to the deepest subsurface reservoirs, the influence of topography-driven flow is omnipresent. Understanding this intricate relationship provides valuable insights into the sustainable management of water resources, the exploration of hydrocarbons, and the preservation of landscapes. As we navigate the complexities of Earth's geological processes, the symphony of topography-driven flow continues to play a central role in the ever-evolving narrative of our planet's dynamic history.

8.2.1 The Mechanisms of Topography-Driven Flow

At the heart of topography-driven flow lies the gravitational force, orchestrating the movement of fluids from elevated regions to lower elevations. This mechanism is intricately linked to the creation of landscapes and the geological processes that shape them. Rainfall, the precursor to many subsurface journeys, descends upon elevated regions, initiating a cascade of fluid movement that transforms the topographical canvas.

8.2.2 Terrain Sculpting and Surface Runoff

Elevated terrains, sculpted by tectonic forces, weathering, and erosion, serve as starting points for topography-driven flow. The uneven distribution of elevations, from towering mountain ranges to expansive plateaus, sets the stage for diverse hydrological pathways. Rainwater, guided by gravity, meanders through valleys, forming rivulets that converge into streams, eventually swelling into rivers. This surface runoff is not only a testament to topography's guiding hand but also a sculptor of landscapes over geological time scales.

8.2.3 Influence on Groundwater Dynamics

Topography is equally influential in orchestrating the subsurface movement of groundwater. Elevated regions, often characterized by porous and permeable rock formations, become recharge zones where rainwater infiltrates the ground. Gravity pulls this infiltrated water downward, creating a subterranean journey through aquifers and geological formations. The fluid's journey may be prolonged or accelerated based on the heterogeneity of the subsurface materials, with topography acting as the invisible conductor guiding the groundwater flow.

8.2.4 Mountain Front Recharge and Discharge

Mountainous regions, with their towering peaks and intricate valleys, play a distinctive role in topography-driven flow. The phenomenon known as mountain front recharge occurs as precipitation infiltrates the base of mountains. This process replenishes aquifers, contributing to the sustenance of downstream rivers. Conversely, mountain fronts also serve as discharge zones, where groundwater emerges as springs, nourishing ecosystems and contributing to river flow. This interplay exemplifies how topography shapes the recharge-discharge dynamics of groundwater systems.

8.2.5 Influence on Hydrocarbon Migration

Topography-driven flow extends its influence beyond the aqueous realm to the migration of hydrocarbons within the subsurface. In sedimentary basins, where vast reservoirs of oil and gas reside, topography plays a crucial role in determining the migration pathways of these valuable resources. Faults, fractures, and stratigraphic variations guided by topographical features create conduits for hydrocarbon movement. The gravitational pull, once again, directs these fluids along pathways dictated by the subtle undulations of the Earth's surface.

8.2.6 Faulting and Tectonics

Topography-driven flow intertwines with tectonic forces, especially along fault zones where the Earth's crust is fractured. Faults, products of tectonic movements, become avenues for fluid migration. In some cases, topography accentuates faulting, creating preferred pathways for subsurface fluids. Conversely, fault zones may also influence topography by promoting or inhibiting erosion, showcasing the intricate feedback loop between geological structures and surface features.

8.2.7 Impacts on Landscape Evolution

The marriage of topography and fluid flow has profound implications for landscape evolution. Over geological time scales, the relentless action of flowing water, driven by topography, sculpts valleys, carves canyons, and shapes the very fabric of the Earth's surface. From the Grand Canyon, etched by the Colorado River, to the sinuous meanders of the Amazon, topography-driven flow leaves an indelible mark on the landscapes we see today.

8.2.8 Challenges and Hazards

While topography-driven flow is essential for the sustenance of ecosystems and the formation of valuable resources, it also presents challenges and hazards. Flash floods, landslides, and riverbank erosion are manifestations of the delicate balance between topography and fluid dynamics. Human activities, including urbanization and alterations to natural drainage patterns, can amplify these hazards, underscoring the importance of understanding and managing topography-driven flow in the context of geological processes.

8.2.9 Management and Sustainability

Recognizing the pivotal role of topography-driven flow in geological processes is crucial for sustainable resource management and environmental conservation. Watershed management, groundwater recharge strategies, and land-use planning should consider the intricate interplay between topography and fluid flow. Additionally, sustainable practices in the extraction of groundwater and hydrocarbons must account for the geological features that guide subsurface flow.

8.3 COMPACTION-DRIVEN FLUID FLOW

In the labyrinthine realm beneath the Earth's surface, an intricate dance unfolds—a symphony of geological processes shaping the movement of fluids. Among these processes, compaction-driven fluid flow emerges as a key player, steering the migration of water, hydrocarbons, and other substances through the porous and permeable subsurface. As sediments settle and geological formations undergo compaction, a dynamic interplay is set in motion, influencing fluid flow patterns and contributing to the evolution of Earth's subsurface architecture. In conclusion, compaction-driven fluid flow is an indispensable chapter in the geological narrative, shaping the subsurface landscape and influencing the distribution of fluids essential to life and industry. From the diagenetic evolution of sedimentary rocks to the migration of hydrocarbons within reservoirs, the fingerprints of compaction are omnipresent. As we navigate the complexities of Earth's subsurface, an appreciation for the symphony of compaction-driven fluid flow enhances our ability to harness resources sustainably, mitigate hazards, and unravel the geological mysteries hidden beneath our feet.

8.3.1 The Genesis of Compaction

Compaction is a geological process born from the weight of overlying sediments and the relentless forces of gravity. It begins with the accumulation of loose particles—ranging from clay and silt to sand and gravel—that form sedimentary deposits over time. As these sediments pile up, the pressure exerted by the overlying layers increases, initiating compaction. The grains within the sedimentary matrix adjust and rearrange, leading to a reduction in pore spaces and an increase in sediment density.

8.3.2 Porosity, Permeability, and Fluid Migration

The reduction in pore spaces during compaction has profound implications for the porosity and permeability of geological formations. Porosity, representing the volume of voids within a rock or sediment, diminishes as compaction progresses. Simultaneously, permeability—the ability of a material to transmit fluids—may either increase or decrease depending on the nature of the sediments and their arrangement.

Compaction-driven changes in porosity and permeability play a pivotal role in guiding fluid flow through the subsurface. Initially, during the early stages of compaction, sedimentary layers may exhibit high porosity and permeability, facilitating the movement of fluids such as water or hydrocarbons. However, as compaction intensifies, the reduction in pore spaces and the closure of interconnected pathways can impede fluid flow, creating distinct zones of low permeability.

8.3.3 Fluid Expulsion and Diagenesis

Compaction-induced pressure within sediments can lead to the expulsion of pore fluids. This expulsion process, known as fluid expulsion or dewatering, is a critical phase in the diagenetic evolution of sedimentary rocks. As compaction drives fluids out of the pore spaces, dissolved minerals may precipitate, resulting in the cementation of the sedimentary matrix. This diagenetic alteration further influences the porosity and permeability of the rocks, shaping their mechanical and hydrological properties.

8.3.4 Overpressure and Geomechanical Effects

In some cases, compaction-driven fluid expulsion can generate overpressure within geological formations. Overpressure occurs when the rate of fluid expulsion lags behind the rate of sediment compaction. The excess pressure can have significant geomechanical effects, leading to phenomena such as faulting, fracturing, and the initiation of fluid migration pathways. Overpressure zones may act as conduits for fluid flow, providing permeable channels through otherwise low-permeability formations.

8.3.5 Compaction-Driven Fluid Flow in Sedimentary Basins

Sedimentary basins, repositories of immense geological history, are arenas where compaction-driven fluid flow orchestrates intricate subsurface movements. As sediments accumulate over millions of years, compaction-induced changes in porosity and permeability create diverse fluid flow regimes. The expulsion of fluids, the migration of hydrocarbons, and the diagenetic evolution of sedimentary rocks collectively contribute to the geological tapestry of sedimentary basins.

In the context of petroleum geology, compaction-driven fluid flow is particularly relevant. The burial of organic-rich sediments in sedimentary basins sets the stage for the generation and migration of hydrocarbons. As these sediments undergo compaction, organic matter transforms into hydrocarbons under elevated temperature and pressure conditions. The expelled fluids, enriched with hydrocarbons, seek migration pathways through porous and permeable formations. Structural traps, faults, and stratigraphic variations guided by compaction-induced changes in permeability become reservoirs for economically significant oil and gas accumulations.

8.3.6 Challenges and Hazards

While compaction-driven fluid flow is a fundamental aspect of subsurface dynamics, it also presents challenges and hazards, particularly in the context of water resources and geotechnical engineering. Excessive fluid expulsion during compaction can lead to the formation of overpressure zones, influencing the stability of geological formations. Overpressured zones may pose challenges in drilling operations, as they are associated with increased risks of blowouts and drilling complications. In the realm of water resources, understanding compaction-driven fluid flow is crucial for managing groundwater systems. Overextraction of groundwater can induce compaction in aquifer systems, leading to land subsidence—a phenomenon where the land surface sinks. Subsidence poses risks to infrastructure, alters drainage patterns, and impacts ecosystems. Recognizing the interconnectedness between compaction, fluid flow, and subsurface stability is essential for mitigating these hazards.

8.3.7 Engineering Considerations and Reservoir Management

In geotechnical engineering, an understanding of compaction-driven fluid flow is paramount for infrastructure projects. The assessment of subsurface conditions, including the degree of compaction, fluid expulsion, and overpressure, guides decisions related to foundation design, tunneling, and slope stability. Moreover, in the management of subsurface reservoirs, such as those containing hydrocarbons or water, accounting for compaction-induced changes is crucial for sustainable extraction and reservoir optimization.

8.3.8 Future Perspectives

As our understanding of subsurface processes deepens and technological advancements in geophysics and geochemistry continue, new insights into compaction-driven fluid flow are likely to emerge. Numerical modeling and simulations enable scientists and engineers to explore the intricate dynamics of fluid migration in response to compaction, offering valuable tools for predicting subsurface behavior and optimizing resource management strategies.

8.4 VARIABLE-DENSITY DRIVEN FLUID FLOW

In the intricate ballet of subsurface processes, variable-density driven fluid flow takes center stage, orchestrating the movement of fluids beneath the Earth's crust. Unlike traditional fluid dynamics, where buoyancy is uniform, variable-density flow introduces a nuanced dimension. Variable-density driven fluid flow stands as a testament to the complexity and richness of Earth's subsurface processes. From the depths of hydrothermal systems to the intricacies of salt tectonics, the interplay between temperature, chemical composition, and fluid movement shapes the landscapes we see and the resources we extract. As we delve deeper into the mysteries of variable-density flow, scientific inquiry continues to illuminate the pathways and patterns that govern subsurface dynamics. The symphony of variable-density driven fluid flow echoes through the geological history of our planet, inviting exploration, understanding, and the promise of sustainable resource management.

8.4.1 The Drivers of Variable-Density Flow

At the core of variable-density driven fluid flow lie the intricate interactions between temperature, salinity, and chemical composition within subsurface fluids. These parameters impart distinct densities to the fluids, creating buoyancy differences that set the stage for dynamic fluid movement. As fluids respond to changes in temperature gradients or encounter variations in solute concentrations, they embark on journeys guided by the gravitational pull and buoyancy forces.

8.4.2 Thermal Density Gradients and Thermohaline Circulation

One of the primary drivers of variable-density flow is the variation in temperature within subsurface fluids. Thermal density gradients arise as fluids experience changes in temperature with depth. As fluids warm or cool, their density fluctuates, creating buoyancy differences that drive fluid flow.

Thermohaline circulation, a subset of variable-density flow, is particularly prominent in oceanic and deep groundwater systems. Here, variations in temperature and salinity contribute to the density contrasts that induce large-scale circulation patterns. In the ocean, for instance, cold, dense waters sink in polar regions, initiating a global conveyor belt that redistributes heat and nutrients across vast distances.

8.4.3 Chemical Density Gradients and Solute Transport

Chemical density gradients, arising from variations in solute concentration, contribute significantly to variable-density flow. This phenomenon is particularly relevant in subsurface environments where solutes such as salts, dissolved minerals, or contaminants influence fluid density. In groundwater systems, variable-density flow driven by solute transport plays a crucial role in shaping aquifer dynamics. As fluids with different solute concentrations encounter each other, density contrasts lead to fluid displacements. This process, known as density-driven flow or solute-driven flow, is fundamental to understanding the transport of contaminants in aquifers and the migration of dissolved substances through porous media.

8.4.4 Hydrothermal Systems and Variable-Density Flow

Hydrothermal systems, where fluids interact with Earth's internal heat, provide a fascinating arena for variable-density flow. In these systems, subsurface fluids are heated through contact with magma or geothermal heat sources, leading to changes in density and initiating buoyancy-driven circulation.

Hydrothermal convection, driven by variable-density flow, is a cornerstone in the formation of mineral deposits and the redistribution of elements within Earth's crust. As hot fluids rise due to buoyancy forces, they mobilize and transport minerals, contributing to the enrichment of ore deposits. Hydrothermal systems, with their intricate interplay of temperature, pressure, and fluid flow, exemplify the complex nature of variable-density-driven processes.

8.4.5 Salt Tectonics and Compaction

In sedimentary basins, the interplay between variable-density flow and salt tectonics adds another layer of complexity to subsurface dynamics. Evaporite deposits, rich in salt minerals, experience compaction and deformation over geological time scales. As salt undergoes compaction, variations in density create buoyancy differences, driving fluid flow and influencing the structural evolution of sedimentary basins.

Salt diapirs, or salt domes, are iconic features shaped by variable-density flow in sedimentary basins. As salt rises buoyantly through overlying sediments, it creates structural highs and lows, impacting the distribution of fluids and influencing the accumulation of hydrocarbons. The interaction between variable-density-driven processes and salt tectonics is a testament to the intricate dance occurring beneath the Earth's surface.

8.4.6 Biogeochemical Reactions and Variable-Density Flow

The subsurface is not only a realm of geological processes but also a stage for biogeochemical reactions that contribute to variable-density flow. Microbial activity, organic matter degradation, and the transformation of dissolved substances create changes in fluid composition, influencing density and buoyancy.

In groundwater systems, variable-density flow induced by biogeochemical reactions is pivotal for nutrient cycling and the fate of contaminants. The movement of electron acceptors and donors within subsurface environments, driven by variations in fluid density, shapes the redox conditions and microbial communities present. Understanding these biogeochemical interactions is essential for predicting the behavior of groundwater contaminants and their impact on ecosystems.

8.4.7 Challenges and Environmental Implications

While variable-density driven fluid flow unveils the complexities of subsurface dynamics, it also poses challenges and environmental implications. In the context of groundwater systems, understanding the transport of contaminants driven by density contrasts is crucial for assessing the risks of groundwater pollution. Variable-density flow can influence the migration of pollutants over long distances, complicating efforts to manage and remediate contaminated aquifers.

In hydrocarbon reservoirs, the interaction between variable-density flow and fluid displacement impacts the efficiency of enhanced oil recovery techniques. Understanding how density differences influence the distribution of injected fluids and the displacement of hydrocarbons is essential for optimizing reservoir management strategies.

8.4.8 Engineering Applications and Enhanced Resource Recovery

Variable-density driven fluid flow finds practical applications in engineering disciplines, from reservoir management to groundwater remediation. In hydrocarbon reservoir engineering, modeling variable-density flow is essential for predicting the behavior of injected fluids during enhanced oil recovery processes. The optimization of injection strategies, based on an understanding of density-driven displacement, enhances the recovery of hydrocarbons from subsurface formations.

In the realm of geothermal energy, harnessing variable-density flow in hydrothermal systems is a promising avenue. Understanding the dynamics of fluid circulation driven by density contrasts is crucial for designing efficient geothermal energy extraction systems. Enhanced resource recovery through geothermal exploitation exemplifies how variable-density flow can be leveraged for sustainable energy solutions.

8.4.9 Future Directions and Scientific Advances

As our understanding of variable-density driven fluid flow evolves, scientific research continues to push the boundaries of knowledge. Advanced modeling techniques, incorporating complex interactions between temperature, chemical composition, and fluid flow, provide insights into the intricacies of subsurface dynamics. High-resolution imaging technologies, such as seismic and borehole logging, enable scientists to visualize and quantify variable-density flow in unprecedented detail.

The integration of interdisciplinary approaches, including geochemistry, microbiology, and fluid dynamics, promises a more holistic understanding of variable-density-driven processes. Unraveling the intricacies of biogeochemical reactions, the role of microbial communities, and the impact of variable-density flow on subsurface ecosystems remains a frontier for future research.

8.5 SALINITY GRADIENTS LEADING TO VARIABLE-DENSITY FLOW

In the hidden tapestry of subsurface processes, the interplay of salinity gradients emerges as a defining force, orchestrating the movement of fluids beneath the Earth's crust. Salinity, the concentration of dissolved salts in water, introduces a nuanced dimension to fluid dynamics, creating density contrasts that drive variable-density flow. Salinity-driven variable-density flow weaves a complex narrative within the subsurface realm, influencing aquifer dynamics, solute transport, and ecosystem health. From coastal estuaries to geothermal systems, the interplay of salinity gradients with buoyancy forces shapes the landscapes we see and the resources we depend on. As we navigate the intricate tapestry of subsurface processes, an appreciation for the role of salinity-driven variable-density flow enhances our ability to manage water resources sustainably, protect coastal ecosystems, and unlock the geological mysteries hidden beneath our feet.

8.5.1 The Salinity Conundrum

Salinity gradients, resulting from variations in the concentration of dissolved salts, are a ubiquitous feature of subsurface environments. These gradients can arise from diverse geological and hydrological processes, including the dissolution of minerals, evaporation of water, and interactions with geological formations. Understanding the salinity conundrum is crucial for unraveling the intricacies of variable-density flow and its far-reaching implications.

8.5.2 Buoyancy Forces and Density Gradients

At the heart of salinity-driven variable-density flow lies the principle of buoyancy. As water acquires dissolved salts, its density increases, creating buoyancy differences compared to less saline water. These density gradients become the guiding force for fluid movement, setting in motion the intricate dance of variable-density flow.

In coastal environments, where freshwater from rivers meets saline seawater, sharp salinity gradients create distinct layers with different densities. This buoyancy-driven stratification influences the circulation patterns of estuaries and coastal seas, impacting nutrient transport, ecosystem dynamics, and the distribution of marine organisms.

8.5.3 Aquifer Dynamics and Groundwater Salinity

In groundwater systems, the influence of salinity gradients on variable-density flow is particularly pronounced. Aquifers, subsurface formations that store and transmit groundwater, exhibit intricate salinity profiles influenced by geological and hydrological factors.

The intrusion of saline water into freshwater aquifers, a phenomenon known as saltwater intrusion, is a critical manifestation of salinity-driven variable-density flow. Over-extraction of freshwater, sea-level rise, and geological factors can contribute to the movement of saline water into coastal aquifers. This intrusion poses challenges to coastal communities that depend on freshwater resources, impacting both the quantity and quality of groundwater.

8.5.4 Density-Driven Solute Transport

Salinity gradients not only influence the movement of water but also play a crucial role in the transport of solutes through variable-density flow. The density contrasts between saline and freshwater drive the migration of dissolved substances, shaping the fate of nutrients, contaminants, and geochemical constituents in subsurface environments.

In the context of groundwater contamination, salinity-driven variable-density flow can facilitate the migration of pollutants over long distances. Understanding how density gradients influence solute transport is essential for predicting the spread of contaminants, designing effective remediation strategies, and safeguarding water quality in aquifers.

8.5.5 Coastal Upwelling and Downwelling

Coastal regions, where land meets the sea, exemplify the intric≈te dynamics of salinity-driven variable-density flow. The interaction between freshwater discharge from rivers and saline seawater creates buoyancy-driven circulation patterns known as coastal upwelling and downwelling.

In coastal upwelling, less dense freshwater from rivers forms a surface layer that moves seaward. This process can enhance nutrient supply to surface waters, supporting vibrant ecosystems and fisheries. Conversely, coastal downwelling involves the sinking of denser saline water, transporting nutrients to deeper layers and influencing the vertical distribution of marine life. The delicate balance between these processes is intricately linked to salinity gradients, temperature variations, and the geomorphology of coastal environments.

8.5.6 Estuarine Dynamics and Ecosystems

Estuaries, where freshwater rivers meet the salty sea, epitomize the synergy of salinity-driven variable-density flow and ecosystem dynamics. The unique characteristics of estuaries, including dynamic salinity gradients, tidal influence, and nutrient transport, create habitats that support diverse and productive ecosystems.

Salinity-driven stratification in estuaries influences the vertical distribution of nutrients and organisms. Tidal mixing, driven by the rise and fall of tides, further complicates the salinity structure, shaping the ecological niches within estuarine environments. Understanding these dynamics is crucial for the conservation and management of estuarine ecosystems, which serve as vital nurseries for marine life and provide essential ecological services.

8.5.7 Geothermal Systems and Salinity-Driven Convection

Beyond coastal and freshwater environments, salinity-driven variable-density flow plays a role in geothermal systems, where heat from Earth's interior influences subsurface fluid dynamics. In these systems, variations in salinity create buoyancy-driven convection, influencing the movement of hydrothermal fluids and contributing to the formation of mineral deposits.

As hydrothermal fluids circulate through subsurface fractures and conduits, salinity gradients induced by interactions with rocks and minerals become a driving force for fluid flow. This process, known as salinity-driven convection, is a key mechanism in the redistribution of elements and the formation of economically significant ore deposits in geothermal systems.

8.5.8 Challenges and Mitigation Strategies

While the role of salinity-driven variable-density flow is central to understanding subsurface processes, it also poses challenges and necessitates careful management. In coastal aquifers, saltwater intrusion due to over-pumping of freshwater resources requires mitigation strategies to preserve the quality and sustainability of groundwater supplies. Implementing sustainable groundwater management practices, including the reduction of excessive withdrawals and the use of alternative water sources, is crucial for addressing these challenges.

In estuarine and coastal ecosystems, anthropogenic activities such as urbanization, agriculture, and industrial discharges can alter salinity gradients, affecting the health of marine life. Integrated coastal zone management, conservation of freshwater inflows, and the restoration of estuarine habitats are essential components of mitigating the impacts of human activities on salinity-driven variable-density flow in these critical ecosystems.

8.5.9 Future Directions and Scientific Frontiers

Advancements in scientific research, supported by cutting-edge technologies, continue to deepen our understanding of salinity-driven variable-density flow. High-resolution monitoring, isotopic tracing, and numerical modeling enable scientists to unravel the complexities of subsurface processes with unprecedented precision. Interdisciplinary approaches, combining hydrogeology, geochemistry, and environmental science, provide holistic insights into the dynamics of salinity-driven flow.

The exploration of extraterrestrial environments, such as Mars, introduces new frontiers in understanding variable-density flow. Salinity gradients, though challenging to detect remotely, may play a role in subsurface processes on other planets. Studying analogous environments on Earth, such as hypersaline lakes and deep-sea brine pools, contributes to our broader understanding of the role of salinity in shaping subsurface dynamics.

8.6 MOUNTAIN FORMATION

Mountain formation refers to the geological processes that underlie the formation of mountains. These processes are associated with large-scale movements of the Earth's crust (tectonic plates). Folding, faulting, volcanic activity, igneous intrusion and metamorphism can all be parts of the orogenic process of mountain building. The formation of mountains is not necessarily related to the geological structures found on it.

The understanding of specific landscape features in terms of the underlying tectonic processes is called tectonic geomorphology, and the study of geologically young or ongoing processes is called neotectonics.

From the late 18th century until its replacement by plate tectonics in the 1960s, geosyncline theory was used to explain much mountain-building.

8.6.1 Types of Mountains

There are five main types of mountains: volcanic, fold, plateau, fault-block and dome. A more detailed classification useful on a local scale predates plate tectonics and adds to these categories.

Volcanic Mountains

Movements of tectonic plates create volcanoes along the plate boundaries, which erupt and form mountains. A volcanic arc system is a series of volcanoes that form near a subduction zone where the crust of a sinking oceanic plate melts and drags water down with the subducting crust.

Fig. 8.1: The Dome of Vitosha Mountain Next to Sofia

Most volcanoes occur in a band encircling the Pacific Ocean (the Pacific Ring of Fire), and in another that extends from the Mediterranean across Asia to join the Pacific band in the Indonesian Archipelago. The most important types of volcanic mountain are composite cones or stratovolcanoes (Vesuvius, Kilimanjaro and Mount Fuji are examples) and shield volcanoes (such as Mauna Loa on Hawaii, a hotspot volcano).

A shield volcano has a gently sloping cone due to the low viscosity of the emitted material, primarily basalt. Mauna Loa is the classic example, with a slope of 4°-6°. (The relation between slope and viscosity falls under the topic of angle of repose.) The composite volcano or stratovolcano has a more steeply rising cone (33°-40°), due to the higher viscosity of the emitted material, ad eruptions are more violent and less frequent than for shield volcanoes. Besides the examples already mentioned re Mount Shasta, Mount Hood and Mount Rainier. Vitosha - the domed mountain next to Sofia, capital of Bulgaria, is also formed by volcanic activity.

Fold Mountains

When plates collide or udergo subduction (that is – ride one over another), the plates tend to buckle and fold, forming mountains. Most of the major continental mountain ranges are associated with thrusting and folding or orogenesis. Examples are the Balkan Mountains, the Jura and the Zagros mountains.

Block Mountains

When a fault block is raised or tilted, block mountains can result. Higher blocks are called horsts and troughs are called grabens. A spreading apart of the surface causes tensional forces. When the tensional forces are strong enough to cause a plate to split apart, it does so such that a center block drops down relative to its flanking blocks.

An example of this is the Sierra Nevada Range, where delamination created a block 650 km long and 80 km wide that consists of many individual portions tipped gently west, with east facing slips rising abruptly to produce the highest mountain front in the continental United States.

Another good example is the Rila - Rhodope mountain Massif in Bulgaria, Southeast Europe, including the well dfined horsts of Belasitsa (linear horst), Rila mountain (vaulted domed shaped horst) and Pirin mountain - a horst forming a massive anticline situated between the complex graben valleys of Struma and that of Mesta.

Uplifted Passive Margins

Unlike orogenic mountains there is no widely accepted geophysical model that explains elevated passive continental margins such as the Scandinavian Mountains, Eastern Greenland, the Brazilian Highlands or Australia's Great Dividing Range. Different elevated passive continental margins most likely share the same mechanism of uplift. This mechanism is possibly related to far-field stresses in Earth's lithosphere. According to this view elevated passived margins can be likened to giant anticlinal lithospheric folds, where folding is caused by horizontal compression acting on a thin to thick crust transition zone (as are all passive margins).

8.6.2 Models

Hotspot Volcanoes

Hotspots are supplied by a magma source in the Earth's mantle called a mantle plume. Although originally attributed to a melting of subducted oceanic crust, recent evidence belies this connection. The mechanism for plume formation remains a research topic.

Fault Blocks

Several movements of the Earth's crust that lead to mountains are associated with faults. These movements actually are amenable to analysis that can predict, for example, the height of a raised block and the width of an intervening rift between blocks using the rheology of the layers and the forces of isostasy. Early bent plate models predicting fractures and fault movements have evolved into today's kinematic and flexural models.

8.7 EROSION

Erosion is the action of surface processes (such as water flow or wind) that removes soil, rock, or dissolved material from one location on the Earth's crust and then transports it to another location where it is deposited. Erosion is distinct from weathering which involves no movement. Removal of rock or soil as clastic sediment is referred to as physical or mechanical erosion; this contrasts with chemical erosion, where soil or rock material is removed from an area by dissolution. Eroded sediment or solutes may be transported just a few millimetres, or for thousands of kilometres.

Agents of erosion include rainfall; bedrock wear in rivers; coastal erosion by the sea and waves; glacial plucking, abrasion, and scour; areal flooding; wind abrasion; groundwater processes; and mass movement processes in steep landscapes like landslides and debris flows. The rates at which such processes act control how fast a surface is eroded. Typically, physical erosion proceeds the fastest on steeply sloping surfaces, and rates may also be sensitive to some climatically controlled properties including amounts of water supplied (e.g., by rain), storminess, wind speed, wave fetch, or atmospheric temperature (especially for some ice-related processes). Feedbacks are also possible between rates of erosion and the amount of eroded material that is already carried by, for example, a river or glacier. The transport of eroded materials from their original location is followed by deposition, which is arrival and emplacement of material at a new location.

While erosion is a natural process, human activities have increased by 10 40 times the rate at which soil erosion is occurring globally. At agriculture sites in the Appalachian Mountains, intensive farming practices have caused erosion at up to 100 times the natural rate of erosion in the region. Excessive (or accelerated) erosion causes both "on-site" and "off-site" problems. On-site impacts include decreases in agricultural productivity and (on natural landscapes) ecological collapse, both because of loss of the nutrient-rich upper soil layers. In some cases, this leads to desertification. Off-site effects include sedimentation of waterways and eutrophication of water bodies, as well as sediment-related damage to roads and houses. Water and wind erosion are the two primary causes of land degradation; combined, they are responsible for about 84% of the global extent of degraded land, making excessive erosion one of the most significant environmental problems worldwide.

Intensive agriculture, deforestation, roads, anthropogenic climate change and urban sprawl are amongst the most significant human activities in regard to their effect on stimulating erosion. However, there are many prevention and remediation practices that can curtail or limit erosion of vulnerable soils.

Fig. 8.2: A Natural Arch Produced by the Wind Erosion of Differentially Weathered Rock in Jebel Kharaz, Jordan

Fig. 8.3: A Wave-like Sea Cliff Produced by Coastal Erosion, in Jinshitan Coastal National Geopark, Dalian, Liaoning Province, China

8.7.1 Physical Processes

Rainfall and Surface Runoff

Fig. 8.4: Soil and Water being Splashed by the Impact of a Single Raindrop

Rainfall, and the surface runoff which may result from rainfall, produces four main types of soil erosion: splash erosion, sheet erosion, rill erosion, and gully erosion. Splash erosion is generally seen as the first and least severe stage in the soil erosion process, which is followed by sheet erosion, then rill erosion and finally gully erosion (the most severe of the four).

In splash erosion, the impact of a falling raindrop creates a small crater in the soil, ejecting soil particles. The distance these soil particles travel can be as much as 0.6 m (2.0 ft) vertically and 1.5 m (4.9 ft) horizontally on level ground.

If the soil is saturated, or if the rainfall rate is greater than the rate at which water can infiltrate into the soil, surface runoff occurs. If the runoff has sufficient flow energy, it will transport loosened soil particles (sediment) down the slope. Sheet erosion is the transport of loosened soil particles by overland flow.

Fig. 8.5: A Spoil Tip Covered in Rills and Gullies due to Erosion Processes Caused by Rainfall: Rummu, Estonia

Rill erosion refers to the development of small, ephemeral concentrated flow paths which function as both sediment source and sediment delivery systems for erosion on hillslopes. Generally, where water erosion rates on disturbed upland areas are greatest, rills are active. Flow depths in rills are typically of the order of a few centimetres (about an inch) or less and along-channel slopes may be quite steep. This means that rills exhibit hydraulic physics very different from water flowing through the deeper, wider channels of streams and rivers.

Gully erosion occurs when runoff water accumulates and rapidly flows in narrow channels during or immediately after heavy rains or melting snow, removing soil to a considerable depth. A gully is distinguished from a rill based on a critical cross-sectional area of at least one square foot, i.e., the size of a channel that can no longer be erased via normal tillage operations.

Extreme gully erosion can progress to formation of badlands. These form under conditions of high relief on easily eroded bedrock in climates favorable to erosion. Conditions or disturbances that limit the growth of protective vegetation (rhexistasy) are a key element of badland formation.

Rivers and Streams

Valley or stream erosion occurs with continued water flow along a linear feature. The erosion is both downward, deepening the valley, and headward, extending the valley into the hillside, creating head cuts and steep banks. In the earliest stage of stream erosion, the erosive activity is dominantly vertical, the valleys have a typical V-shaped cross-section and the stream gradient is relatively steep. When some base level is reached, the erosive activity switches to lateral erosion, which widens the valley floor and creates a narrow floodplain. The stream gradient becomes nearly flat, and lateral deposition of sediments becomes important as the stream meanders across the valley floor. In all stages of stream erosion, by far the most erosion occurs during times of flood when more and faster-moving water is available to carry a larger sediment load. In such processes, it is not the water alone that erodes: suspended abrasive particles, pebbles, and boulders can also act erosively as they traverse a surface, in a process known as traction.

Bank erosion is the wearing away of the banks of a stream or river. This is distinguished from changes on the bed of the watercourse, which is referred to as scour. Erosion and changes in the form of river banks may be measured by inserting metal rods into the bank and marking the position of the bank surface along the rods at different times.

Thermal erosion is the result of melting and weakening permafrost due to moving water. It can occur both along rivers and at the coast. Rapid river channel migration observed in the Lena River of Siberia is due to thermal erosion, as these portions of the banks are composed of permafrost-cemented non-cohesive materials. Much of this erosion occurs as the weakened banks fail in large slumps. Thermal erosion also affects the Arctic coast, where wave action and near-shore temperatures combine to undercut permafrost bluffs along the shoreline and cause them to fail. Annual erosion rates along a 100-kilometre (62-mile) segment of the Beaufort Sea shoreline averaged 5.6 metres (18 feet) per year from 1955 to 2002.

Most river erosion happens nearer to the mouth of a river. On a river bend, the longest least sharp side has slower moving water. Here deposits build up. On the narrowest sharpest side of the bend, there is faster moving water so this side tends to erode away mostly.

Rapid erosion by a large river can remove enough sediments to produce a river anticline, as isostatic rebound raises rock beds unburdened by erosion of overlying beds.

Coastal Erosion

Shoreline erosion, which occurs on both exposed and sheltered coasts, primarily occurs through the action of currents and waves but sea level (tidal) change can also play a role.

Hydraulic action takes place when the air in a joint is suddenly compressed by a wave closing the entrance of the joint. This then cracks it. Wave pounding is when the sheer energy of the wave hitting the cliff or rock breaks pieces off. Abrasion or corrasion is caused by waves launching sea load at the cliff. It is the most effective and rapid form of shoreline erosion (not to be confused with corrosion). Corrosion is the dissolving of rock by carbonic acid in sea water. Limestone cliffs are particularly vulnerable to this kind of erosion. Attrition is where particles/sea load carried by the waves are worn down as they hit each other and the cliffs. This then makes the material easier to wash away. The material ends up as shingle and sand. Another significant source of erosion, paricularly on carbonate coastlines, is boring, scraping and grinding of organisms, a process termed bioerosion.

Sediment is transported along the coast in the direction of the prevailing current (longshore drift). Whe the upcurrent supply of sediment is less than the amount being carried away, erosion occurs. When the upcurrent amount of sediment is greater, sand or gravel banks will tend to form as a result of deposition. These banks may slowly migrate along the coast in the direction of the longshore drift, alternately protecting and exposing parts of the coastline.

Where there is a bend in the coastline, quite often a buildup of eroded material occurs forming a long narrow bank (a spit). Armoured beaches and submerged offshore sandbanks may also protect parts of a coastline from erosion. Over the years, as the shoals gradually shift, the erosion may be redirected to attack different parts of the shore.

Erosion of a coastal surface, followed by a fall in sea level, can produce a distinctive landform called a raised beach.

Chemical Erosion

Chemical erosion is the loss of matter in a landscape in the form of solutes. Chemical erosion is usually calculated from the solutes found in streams. Anders Rapp pioneered the study of chemical erosion in his work about Kärkevagge published in 1960. Formation of sinkholes and other features of karst topography is an example of extreme chemical erosion.

Glaciers

Glaciers erode predominantly by three different processes: abrasion/scouring, plucking, and ice thrusting. In an abrasion process, debris in the basal ice scrapes along the bed, polishing and gouging te underlying rocks, similar to sandpaper on wood. Scientists have shown that, in addition to the role of temperature played in valley-deepening, other glaciological processes, such as erosion also control cross-valley variations.

In a homogeneous bedrock erosion pattern, curved channel cross-section beneath the ice is created. Though the glacier continues to incise vertically, the shape of the channel beneath the ice eventually remain constant, reaching a U-shaped parabolic steady-state shape as we now see in glaciated valleys.

Scientists also provide a numerical estimate of the time required for the ultimate formation of a steady-shaped U-shaped valley—approximately 100,000 years. In a weak bedrock (containing material more erodible than the surrounding rocks) erosion pattern, on the contrary, the amount of over deepening is limited because ice velocities and erosion rates are reduced. Glaciers can also cause pieces of bedrock to crack off in the process of plucking. In ice thrusting, the glacier freezes to its bed, then as it surges forward, it moves large sheets of frozen sediment at the base along with the glacier. This method produced some of the many thousands of lake basins that dot the edge of the Canadian Shield. Differences in the height of mountain ranges are not only being the result tectonic forces, such as rock uplift, but also local climate variations. Scientists use global analysis of topography to show that glacial erosion controls the maximum height of mountains, as the relief between mountain peaks and the snow line are generally confined to altitudes less than 1500 m. The erosion caused by glaciers worldwide erodes mountains so effectively that the term glacial buzzsaw has become widely used, which describes the limiting effect of glaciers on the height of mountain ranges.

As mountains grow higher, they generally allow for more glacial activity (especially in the accumulation zone above the glacial equilibrium line altitude), which causes increased rates of erosion of the mountain, decreasing mass faster than isostatic rebound can add to the mountain. This provides a good example of a negative feedback loop. Ongoing research is showing that while glaciers tend to decrease mountain size, in some areas, glaciers can actually reduce the rate of erosion, acting as a glacial armor. Ice can not only erode mountains but also protect them from erosion. Depending on glacier regime, even steep alpine lands can be preserved through time with the help of ice. Scientists have proved this theory by sampling eight summits of northwestern Svalbard using Be10 and Al26, showing that northwestern Svalbard transformed from a glacier-erosion state under relatively mild glacial maxima temperature, to a glacier-armor state occupied by cold-based, protective ice during much colder glacial maxima temperatures as the Quaternary ice age progressed.

These processes, combined with erosion and transport by the water network beneath the glacier, leave behind glacial landforms such as moraines, drumlins, ground moraine (till), glaciokarst, kames, kame deltas, moulins, and glacial erratics in their wake, typically at the terminus or during glacier retreat.

The best-developed glacial valley morphology appears to be restricted to landscapes with low rock uplift rates (less than or equal to 2mm per year) and high relief, leading to long-turnover times.

Where rock uplift rates exceed 2mm per year, glacial valley morphology has generally been significantly modified in postglacial time. Interplay of glacial erosion and tectonic forcing governs the morphologic impact of glaciations on active orogens, by both influencing their height, and by altering the patterns of erosion during subsequent glacial periods via a link between rock uplift and valley cross-sectional shape.

Floods

At extremely high flows, kolks, or vortices are formed by large volumes of rapidly rushing water. Kolks cause extreme local erosion, plucking bedrock and creating pothole-type geographical features called rock-cut basins. Examples can be seen in the flood regions result from glacial Lake Missoula, which created the channeled scablands in the Columbia Basin region of eastern Washington.

Fig. 8.6: The Mouth of the River Seaton in Cornwall after Heavy Rainfall Caused Flooding in the Area and Cause a Significant Amount of the Beach to Erode; Leaving Behind a Tall Sand Bank in its Place

Wind Erosion

Fig. 8.7: Árbol de Piedra, a Rock Formation in the Altiplano, Bolivia Sculpted by Wind Erosion

Wind erosion is a major geomorphological force, especially in arid and semi-arid regions. It is also a major source of land degradation, evaporation, desertification, harmful airborne dust, and crop damage—especially after being increased far above natural rates by human activities such as deforestation, urbanization, and agriculture.

Wind erosion is of two primary varieties: deflation, where the wind picks up and carries away loose particles; and abrasion, where surfaces are worn down as they are struck by airborne particles carried by wind. Deflation is divided into three categories: (1) surface creep, where larger, heavier particles slide or roll along the ground; (2) saltation, where particles are lifted a short height into the air, and bounce and saltate across the surface of the soil; and (3) suspension, where very small and light particles are lifted into the air by the wind, and are often carried for long distances. Saltation is responsible for the majority (50–70%) of wind erosion, followed by suspension (30–40%), and then surface creep (5–25%). Wind erosion is much more severe in arid areas and during times of drought. For example, in the Great Plains, it is estimated that soil loss due to wind erosion can be as much as 6100 times greater in drought years than in wet years.

Mass Wasting

Fig. 8.8: A Wadi in Makhtesh Ramon, Israel, Showing Gravity Collapse Erosion on its Banks

Mass wasting or mass movement is the downward and outward movement of rock and sediments on a sloped surface, mainly due to the force of gravity.

Mass wasting is an important part of the erosional process and is often the first stage in the breakdown and transport of weathered materials in mountainous areas. It moves material from higher elevations to lower elevations where other eroding agents such as streams and glaciers can then pick up the material and move it to even lower elevations. Mass-wasting processes are always occurring continuously on all slopes; some mass-wasting processes act very slowly; others occur very suddenly, often with disastrous results. Any perceptible down-slope movement of rock or sediment is often referred to in general terms as a landslide. However, landslides can be classified in a much more detailed way that reflects the mechanisms responsible for the movement and the velocity at which the movement occurs. One of the visible topographical manifestations of a very slow form of such activity is a scree slope.

Slumping happens on steep hillsides, occurring along distinct fracture zones, often within materials like clay that, once released, may move quite rapidly downhill. They will often show a spoon-shaped isostatic depression, in which the material has begun to slide downhill. In some cases, the slump is caused by water beneath the slope weakening it. In many cases it is simply the result of poor engineering along highways where it is a regular occurrence.

Surface creep is the slow movement of soil and rock debris by gravity which is usually not perceptible except through extended observation. However, the term can also describe the rolling of dislodged soil particles 0.5 to 1.0 mm (0.02 to 0.04 in) in diameter by wind along the soil surface.

Submarine Sediment Gravity Flows

On the continental slope, erosion of the ocean floor to create channels and submarine canyons can result from the rapid downslope flow of sediment gravity flows, bodies of sediment-laden water that move rapidly downslope as turbidity currents. Where erosion by turbidity currents creates oversteepened slopes it can also trigger underwater landslides and debris flows. Turbidity currents can erode channels and canyons into substrates ranging from recently deposited unconsolidated sediments to hard crystalline bedrock. Almost all continental slopes and deep ocean basins display such channels and canyons resulting from sediment gravity flows and submarine canyons act as conduits for the transfer of sediment from the continents and shallow marine environments to the deep sea. Turbidites, which are the sedimentary deposits resulting from turbidity currents, comprise some of the thickest and largest sedimentary sequences on Earth, indicating that the associated erosional processes must also have played a prominent role in Earth's history.

8.7.2 Factors Affecting Erosion Rates

Climate

The amount and intensity of precipitation is the main climatic factor governing soil erosion by water. The relationship is particularly strong if heavy rainfall occurs at times when, or in locations where, the soil's surface is not well protected by vegetation. This might be during periods when agricultural activities leave the soil bare, or in semi-arid regions where vegetation is naturally sparse. Wind erosion requires strong winds, particularly during times of drought when vegetation is sparse and soil is dry (and so is more erodible). Other climatic factors such as average temperature and temperature range may also affect erosion, via their effects on vegetation and soil properties. In general, given similar vegetation and ecosystems, areas with more precipitation (especially high-intensity rainfall), more wind, or more storms are expected to have more erosion.

In some areas of the world (e.g., the mid-western US), rainfall intensity is the primary determinant of erosivity (for a definition of erosivity check,) with higher intensity rainfall generally resulting in more soil erosion by water. The size and velocity of rain drops is also an important factor. Larger and higher-velocity rain drops have greater kinetic energy, and thus their impact will displace soil particles by larger distances than smaller, slower-moving rain drops.

In other regions of the world (e.g., western Europe), runoff and erosion result from relatively low intensities of stratiform rainfall falling onto the previously saturated soil. In such situations, rainfall amount rather than intensity is the main factor determining the severity of soil erosion by water. According to the climate change projections, erosivity will increase significantly in Europe and soil erosion may increase by 13–22.5% by 2050

In Taiwan, where typhoon frequency increased significantly in the 21st century, a strong link has been drawn between the increase in storm frequency with an increase in sediment load in rivers and reservoirs, highlighting the impacts climate change can have on erosion.

Vegetative Cover

Vegetation acts as an interface between the atmosphere and the soil. It increases the permeability of the soil to rainwater, thus decreasing runoff. It shelters the soil from winds, which results in decreased wind erosion, as well as advantageous changes in microclimate. The roots of the plants bind the soil together, and interweave with other roots, forming a more solid mass that is less susceptible to both water and wind erosion. The removal of vegetation increases the rate of surface erosion.

Topography

The topography of the land determines the velocity at which surface runoff will flow, which in turn determines the erosivity of the runoff. Longer, steeper slopes (especially those without adequate vegetative cover) are more susceptible to very high rates of erosion during heavy rains than shorter, less steep slopes. Steeper terrain is also more prone to mudslides, landslides, and other forms of gravitational erosion processes.

Tectonics

Tectonic processes control rates and distributions of erosion at the Earth's surface. If the tectonic action causes part of the Earth's surface (e.g., a mountain range) to be raised or lowered relative to surrounding areas, this must necessarily change the gradient of the land surface. Because erosion rates are almost always sensitive to the local slope, this will change the rates of erosion in the uplifted area. Active tectonics also brings fresh, unweathered rock towards the surface, where it is exposed to the action of erosion.

However, erosion can also affect tectonic processes. The removal by erosion of large amounts of rock from a particular region, and its deposition elsewhere, can result in a lightening of the load on the lower crust and mantle. Because tectonic processes are driven by gradients in the stress field developed in the crust, this unloading can in turn cause tectonic or isostatic uplift in the region. In some cases, it has been hypothesised that these twin feedbacks can act to localize and enhance zones of very rapid exhumation of deep crustal rocks beneath places on the Earth's surface with extremely high erosion rates, for example, beneath the extremely steep terrain of Nanga Parbat in the western Himalayas. Such a place has been called a "tectonic aneurysm".

Development

Human land development, in forms including agricultural and urban development, is considered a significant factor in erosion and sediment transport, which aggravate food insecurity. In Taiwan, increases in sediment load in the northern, central, and southern regions of the island can be tracked with the timeline of development for each region throughout the 20th century. The intentional removal of soil and rock by humans is a form of erosion that has been named lisasion.

8.7.3 Erosion at Various Scales

Mountain Ranges

Mountain ranges are known to take many millions of years to erode to the degree they effectively cease to exist. Scholars Pitman and Golovchenko estimate that it takes probably more than 450 million years to erode a mountain mass similar to the Himalaya into an almost-flat peneplain if there are no major sea-level changes. Erosion of mountains massifs can create a pattern of equally high summits called summit accordance. It has been argued that extension during post-orogenic collapse is a more effective mechanism of lowering the height of orogenic mountains than erosion. Examples of heavily eroded mountain ranges include the Timanides of Northern Russia. Erosion of this orogen has produced sediments that are now found in the East European Platform, including the Cambrian Sablya Formation near Lake Ladoga. Studies of these sediments indicate that it is likely that the erosion of the orogen began in the Cambrian and then intensified in the Ordovician.

Soils

If the rate of erosion is higher than the rate of soil formation the soils are being destroyed by erosion. Where soil is not destroyed by erosion, erosion can in some cases prevent the formation of soil features that form slowly. Inceptisols are common soils that form in areas of fast erosion.

While erosion of soils is a natural process, human activities have increased by 10-40 times the rate at which erosion is occurring globally. Excessive (or accelerated) erosion causes both "on-site" and "off-site" problems. On-site impacts include decreases in agricultural productivity and (on natural landscapes) ecological collapse, both because of loss of the nutrient-rich upper soil layers. In some cases, the eventual result is desertification. Off-site effects include sedimentation of waterways and eutrophication of water bodies, as well as sediment-related damage to roads and houses. Water and wind erosion are the two primary causes of land degradation; combined, they are responsible for about 84% of the global extent of degraded land, making excessive erosion one of the most significant environmental problems.

In the United States, farmers cultivating highly erodible land must comply with a conservation plan to be eligible for certain forms of agricultural assistance.

CHAPTER 9

GPS, GIS, AND REMOTE SENSING IN HYDROGEOLOGY

Hydrogeology, the study of water beneath the Earth's surface, plays a crucial role in understanding groundwater resources, contamination, and sustainable water management.

As technology advances, tools like GPS (Global Positioning System), GIS (Geographic Information System), and remote sensing have become indispensable in hydrogeological research and applications.

9.1 UNDERSTANDING GPS (GLOBAL POSITIONING SYSTEM)

Hydrogeology, the study of groundwater and its interaction with geological formations, is crucial for water resource management, environmental protection, and sustainable development.

Global Positioning System (GPS) technology has become an indispensable tool in hydrogeological research, offering precise positioning capabilities that aid in various aspects of groundwater exploration, monitoring, and management.

9.1.1 GPS Technology

GPS is a satellite-based navigation system developed by the United States Department of Defense. It consists of a constellation of satellites orbiting the Earth, transmitting signals that are picked up by GPS receivers on the ground.

These receivers calculate their position by triangulating signals from multiple satellites, providing accurate three-dimensional positioning anywhere on Earth.

The GPS system operates 24/7, under all weather conditions, and is freely accessible to users worldwide. It has revolutionized navigation, surveying, mapping, and various other fields, including hydrogeology.

9.1.2 Applications of GPS in Hydrogeology

Site Characterization and Exploration

One of the primary applications of GPS in hydrogeology is site characterization and exploration. GPS enables hydrogeologists to accurately locate and map groundwater wells, monitoring stations, and other hydrological features. By precisely determining the coordinates of these points, researchers can create detailed maps of aquifer geometries, flow paths, and hydrostratigraphic units.

GPS also facilitates the integration of geological, hydrological, and topographical data, providing a comprehensive understanding of subsurface hydrological processes. This information is vital for groundwater resource assessment, contaminant transport modeling, and site remediation planning.

Groundwater Monitoring and Management

GPS technology plays a crucial role in groundwater monitoring and management programs.By installing GPS-enabled dataloggers in monitoring wells, hydrogeologists can continuously record groundwater levels with precise spatial information. This real-time data allows for the monitoring of groundwater fluctuations, seasonal variations, and long-term trends.

Furthermore, GPS helps in designing and optimizing groundwater monitoring networks by identifying suitable locations for monitoring wells based on hydrogeological parameters such as aquifer properties, groundwater flow directions, and potential contamination sources. This ensures efficient data collection and informed decision-making in groundwater management.

Tracer Studies and Flow Path Analysis

Tracer studies are commonly used in hydrogeology to track the movement of groundwater and delineate flow paths within aquifer systems. GPS technology enhances tracer studies by providing accurate positioning data for tracer injection points, monitoring wells, and sampling locations.

By precisely tracking the movement of tracers through the subsurface, hydrogeologists can determine groundwater flow velocities, directionality, and transit times. This information is essential for understanding aquifer heterogeneity, contaminant transport mechanisms, and the effectiveness of remediation strategies.

Groundwater Modeling and Simulation

GPS data serves as valuable input for groundwater modeling and simulation studies. By incorporating accurate spatial information into hydrogeological models, researchers can improve the accuracy and reliability of model predictions. GPS-derived data on well locations, topographic features, and hydrological boundaries enhance the spatial representation of aquifer systems, leading to more robust simulation results.

Groundwater models calibrated with GPS data can simulate various scenarios, such as groundwater recharge, extraction, and contaminant migration, enabling stakeholders to evaluate different management strategies and assess their potential impacts on groundwater resources.

9.1.3 Advantages and Limitations of GPS in Hydrogeology

Advantages

- High Accuracy: GPS provides precise positioning with centimeter-level accuracy, enabling detailed mapping and analysis of hydrogeological features.
- Real-time Monitoring: GPS-enabled dataloggers allow for continuous monitoring of groundwater levels and environmental parameters, facilitating timely decision-making.
- Cost-effectiveness: GPS technology has become more affordable and accessible, making it a cost-effective solution for hydrogeological applications.
- Integration with GIS: GPS data can be seamlessly integrated with Geographic Information Systems (GIS), enhancing spatial analysis and visualization capabilities.

Limitations

- Signal Interference: GPS signals may be obstructed or weakened by natural and man-made obstacles such as trees, buildings, and atmospheric conditions, affecting signal reception and accuracy.
- Limited Vertical Accuracy: GPS typically provides better horizontal than vertical accuracy, which may pose challenges in applications requiring precise elevation measurements.
- Satellite Availability: In remote or densely forested areas, satellite availability may be limited, leading to reduced GPS accuracy and coverage.

9.2 THE ROLE OF GIS (GEOGRAPHIC INFORMATION SYSTEM) IN HYDROGEOLOGY

Geographic Information Systems (GIS) have become indispensable tools in the field of hydrogeology, revolutionizing the way researchers study, manage, and visualize spatial data related to groundwater resources. GIS technology allows hydrogeologists to integrate various datasets, analyze spatial relationships, and generate valuable insights for groundwater exploration, monitoring, and management.

9.2.1 GIS in Hydrogeology

GIS is a computer-based system designed to capture, store, manipulate, analyze, manage, and present spatial or geographic data. In hydrogeology, GIS serves as a powerful platform for organizing and synthesizing diverse datasets, including geological formations, hydrological features, land use patterns, and groundwater monitoring data.

The primary components of GIS include data input, data storage, data analysis, and data output. GIS software allows users to overlay, query, and analyze spatial data layers, enabling comprehensive spatial analysis and decision-making.

9.2.2 Applications of GIS in Hydrogeology

Groundwater Resource Assessment

GIS facilitates groundwater resource assessment by integrating geological, hydrological, and environmental data layers to delineate aquifer extents, recharge areas, and groundwater flow paths. By analyzing factors such as topography, geology, soil properties, and land use, hydrogeologists can identify potential groundwater recharge zones and assess the sustainability of groundwater extraction rates.

GIS-based groundwater resource assessment also involves the estimation of aquifer properties, such as hydraulic conductivity and porosity, through spatial interpolation techniques. These estimates help in characterizing aquifer properties across large spatial extents and optimizing groundwater management strategies.

Hydrological Modeling and Simulation

GIS plays a critical role in hydrological modeling and simulation studies by providing a spatial framework for representing hydrological processes. Hydrological models simulate groundwater flow, surface water interactions, and contaminant transport within a spatially distributed framework, enabling researchers to assess the impacts of land use changes, climate variability, and groundwater pumping on water resources.

GIS-based hydrological models integrate spatial data on topography, land cover, soil properties, and climate variables to simulate hydrological processes at various spatial and temporal scales. These models help in predicting groundwater recharge rates, evaluating water balance components, and identifying vulnerable areas to groundwater contamination.

Groundwater Contamination Assessment

GIS facilitates groundwater contamination assessment by integrating spatial data on potential contamination sources, hydrogeological parameters, and groundwater quality indicators. By overlaying contaminant source maps with groundwater vulnerability maps, hydrogeologists can identify areas at high risk of groundwater pollution and prioritize monitoring and remediation efforts.

GIS-based groundwater contamination models simulate contaminant transport processes in aquifers, considering factors such as hydraulic conductivity, groundwater flow velocities, and attenuation mechanisms. These models help in predicting the spread of contaminants, estimating time to travel to receptors, and designing effective remediation strategies.

Decision Support Systems for Water Management

GIS-based decision support systems (DSS) provide stakeholders with tools for informed decision-making in water resources management. These systems integrate hydrogeological data, simulation models, and visualization techniques to analyze complex spatial relationships and evaluate alternative management scenarios.

GIS-based DSS allow users to explore spatial data interactively, perform scenario analysis, and visualize the potential impacts of management actions on groundwater resources. By incorporating stakeholder preferences and objectives, GIS-based DSS facilitate collaborative decision-making processes and promote sustainable water management practices.

9.2.3 Advantages and Challenges of GIS in Hydrogeology

Advantages

- Comprehensive Data Integration: GIS allows for the integration of diverse datasets from multiple sources, enabling comprehensive spatial analysis and visualization.
- Spatial Analysis Capabilities: GIS software provides powerful tools for spatial analysis, including overlay analysis, proximity analysis, and spatial interpolation, enhancing the understanding of hydrogeological processes.
- Decision Support: GIS-based decision support systems facilitate informed decision-making by integrating hydrogeological data, simulation models, and visualization tools into user-friendly interfaces.

Challenges

- Data Quality and Availability: Challenges related to data quality, accuracy, and availability may arise when integrating spatial datasets from different sources, requiring careful data validation and preprocessing.
- Spatial Uncertainty: Spatial analysis in GIS involves inherent uncertainties related to data interpolation, modeling assumptions, and parameter estimation, which may affect the reliability of analysis results.
- Technical Expertise: Effective use of GIS in hydrogeology requires technical expertise in spatial data management, GIS software operation, and spatial analysis techniques, necessitating training and skill development.

9.3 APPLICATIONS OF REMOTE SENSING IN HYDROGEOLOGY

Remote sensing, the acquisition of information about the Earth's surface from a distance, has become a valuable tool in hydrogeological research and applications. By utilizing sensors mounted on satellites, aircraft, drones, and other platforms, remote sensing technology provides valuable data for studying groundwater resources, monitoring hydrological processes, and assessing environmental changes.

9.3.1 Remote Sensing in Hydrogeology

Remote sensing technology captures electromagnetic radiation emitted or reflected by the Earth's surface, allowing for the generation of images and datasets depicting various land surface features. In hydrogeology, remote sensing data provides valuable information on surface water bodies, vegetation distribution, land use patterns, and geological formations, which are essential for understanding groundwater dynamics and interactions.

Remote sensing techniques include passive and active sensing, with passive sensors measuring naturally occurring radiation (e.g., sunlight) and active sensors emitting their own radiation (e.g., radar) and measuring the reflected or backscattered signals. These techniques offer different capabilities for mapping and monitoring hydrogeological features and processes.

9.3.2 Applications of Remote Sensing in Hydrogeology

Groundwater Exploration and Mapping

Remote sensing data is widely used for groundwater exploration and mapping, particularly in arid and semi-arid regions where surface water resources are limited. Satellite imagery and aerial photographs provide valuable information on surface water features such as rivers, lakes, and wetlands, which may indicate potential groundwater recharge areas.

Additionally, remote sensing techniques such as thermal infrared imaging and hyperspectral analysis can detect subsurface geological features associated with groundwater accumulation, such as faults, fractures, and karst formations. By identifying surface indicators of groundwater presence, hydrogeologists can target areas for further investigation and drilling.

Groundwater Recharge Assessment

Remote sensing data plays a crucial role in assessing groundwater recharge rates and mechanisms. Vegetation indices derived from satellite imagery, such as the Normalized Difference Vegetation Index (NDVI), provide information on vegetation health and biomass, which is correlated with groundwater recharge rates.

Changes in vegetation patterns observed through remote sensing can indicate variations in groundwater availability and recharge dynamics. By monitoring vegetation dynamics over time, hydrogeologists can assess the impacts of climate variability, land use changes, and water management practices on groundwater recharge processes.

Groundwater Contamination Detection

Remote sensing technology is valuable for detecting and monitoring groundwater contamination from anthropogenic sources such as industrial facilities, landfills, and agricultural activities. Hyperspectral imaging, for example, can identify spectral signatures associated with contaminant plumes, allowing for the mapping and monitoring of contaminated sites.

Furthermore, thermal infrared imaging can detect thermal anomalies caused by subsurface contamination, such as thermal plumes resulting from groundwater discharge of heated industrial effluents. By combining remote sensing data with ground-based monitoring techniques, hydrogeologists can assess the extent of contamination, track contaminant migration pathways, and prioritize remediation efforts.

Hydrological Modeling and Prediction

Remote sensing data serves as valuable input for hydrological modeling and prediction studies, facilitating the simulation of groundwater flow, surface water interactions, and hydrological processes at various spatial and temporal scales. Satellite-derived datasets such as precipitation, evapotranspiration, and soil moisture provide essential inputs for hydrological models.

Remote sensing techniques such as synthetic aperture radar (SAR) can also provide information on soil moisture content, which is a critical parameter for hydrological modeling and forecasting. By integrating remote sensing data with hydrological models, hydrogeologists can improve the accuracy and reliability of predictions related to groundwater availability, flood risk, and water resource management.

9.3.3 Advantages and Challenges of Remote Sensing in Hydrogeology

Advantages

- Wide Spatial Coverage: Remote sensing technology provides wide spatial coverage, allowing for regional-scale assessment of groundwater resources and hydrological processes.
- Multispectral and Multitemporal Data: Remote sensing datasets are available in multiple spectral bands and temporal resolutions, enabling the detection of subtle changes in land surface features over time.
- Non-invasive Data Collection: Remote sensing allows for non-invasive data

collection, reducing the need for costly and time-consuming field surveys and sample collection.

- Integration with Geographic Information Systems (GIS): Remote sensing data can be seamlessly integrated with GIS platforms, enhancing spatial analysis and visualization capabilities for hydrogeological applications.

Challenges

- Spatial and Temporal Resolution: Remote sensing data may have limitations in spatial and temporal resolution, which can affect the accuracy and detail of hydrogeological analysis and interpretation.
- Atmospheric Interference: Atmospheric conditions such as clouds, haze, and aerosols can interfere with remote sensing data acquisition, leading to reduced data quality and accuracy.
- Data Interpretation and Validation: Interpretation of remote sensing data requires expertise in image processing and spectral analysis, and validation with ground-based measurements is essential to ensure the accuracy and reliability of results.
- Cost and Accessibility: Access to high-resolution remote sensing data and specialized processing tools may be costly and limited, particularly for researchers in developing countries or with limited funding resources.

9.4 UTILIZING GPS FOR HYDROGEOLOGICAL SURVEYS

9.4.1 Groundwater Monitoring Network Design

Groundwater monitoring networks are essential for assessing the quantity and quality of groundwater resources, understanding hydrological processes, and managing water supplies sustainably. The design of these networks is a critical aspect of effective groundwater management, ensuring that monitoring wells are strategically located to provide representative data for decision-making. Global Positioning System (GPS) technology has revolutionized the design and implementation of groundwater monitoring networks, offering precise positioning capabilities that enhance the efficiency and effectiveness of network design.

Groundwater Monitoring Networks

Groundwater monitoring networks consist of a network of wells strategically distributed across an area of interest to monitor groundwater levels, quality, and hydrological parameters. These networks serve as valuable tools for assessing groundwater resources, identifying trends and changes over time, and guiding groundwater management decisions.

The design of groundwater monitoring networks involves selecting appropriate monitoring well locations based on hydrogeological characteristics, land use patterns, regulatory requirements, and monitoring objectives. Effective network design ensures adequate spatial coverage, representative data collection, and efficient resource allocation.

Importance of GPS in Groundwater Monitoring Network Design

Global Positioning System (GPS) technology has transformed the process of designing groundwater monitoring networks, offering precise positioning capabilities that enhance the accuracy and reliability of network design. GPS allows for the accurate determination of the geographic coordinates (latitude, longitude, and elevation) of monitoring well locations, enabling hydrogeologists to:

Site Selection and Characterization

GPS facilitates site selection and characterization by providing precise location data for potential monitoring well sites. Hydrogeologists can use GPS receivers to survey candidate locations, assess accessibility, and evaluate site suitability based on hydrogeological criteria such as aquifer properties, groundwater flow directions, and proximity to potential contamination sources.

By accurately determining the coordinates of monitoring well locations, GPS ensures that wells are positioned in representative hydrogeological settings, maximizing the effectiveness of data collection and interpretation.

Spatial Analysis and Optimization

GPS enables spatial analysis and optimization techniques to be applied in the design of groundwater monitoring networks. GIS (Geographic Information System) software, integrated with GPS data, allows hydrogeologists to overlay various spatial datasets, such as geological maps, land use maps, and hydrological models, to identify optimal locations for monitoring wells.

Spatial analysis techniques, such as proximity analysis, buffer analysis, and spatial interpolation, can be used to identify areas with high groundwater vulnerability, prioritize monitoring efforts, and optimize the spatial distribution of monitoring wells.

Network Connectivity and Redundancy

GPS technology helps ensure network connectivity and redundancy by accurately positioning monitoring wells in a well-connected network. Hydrogeologists can use GPS data to design monitoring networks that capture spatial variations in groundwater levels, flow directions, and contaminant plumes, ensuring comprehensive coverage of the study area.

By strategically locating monitoring wells based on hydrogeological heterogeneity and connectivity, GPS enhances the reliability of data collection and interpretation, reducing uncertainty in groundwater assessments and management decisions.

Challenges and Considerations

While GPS technology offers numerous benefits for designing groundwater monitoring networks, several challenges and considerations should be addressed:

- Data Accuracy and Precision: GPS data accuracy and precision may vary depending on factors such as satellite geometry, atmospheric conditions, and receiver quality. Careful attention should be paid to data collection protocols and quality control measures to ensure accurate positioning of monitoring wells.
- Satellite Availability and Signal Obstruction: Satellite availability and signal obstruction can affect GPS data acquisition, particularly in urban areas or densely vegetated regions. Adequate satellite coverage and signal reception are essential for obtaining reliable location data for monitoring well sites.
- Integration with Other Technologies: GPS technology should be integrated with other technologies, such as GIS, remote sensing, and geophysical methods, to enhance the effectiveness of groundwater monitoring network design. Combining multiple data sources and techniques allows for comprehensive site characterization and network optimization.

9.4.2 Groundwater Level Monitoring

GPS-enabled dataloggers attached to monitoring wells provide real-time groundwater level data with precise location information. This data is essential for assessing groundwater fluctuations, aquifer recharge rates, and seasonal variations in groundwater storage.

9.4.3 Tracer Tests and Hydrogeological Mapping

Tracer tests and hydrogeological mapping are two powerful techniques used in the field of hydrogeology to investigate groundwater flow dynamics, aquifer properties, and contaminant transport processes. Tracer tests involve the injection of a known quantity of tracer substance into a groundwater system, followed by the monitoring of tracer concentrations to track the movement of groundwater and delineate flow paths. Hydrogeological mapping, on the other hand, involves the creation of detailed maps and models depicting geological, hydrological, and hydrogeological features to understand subsurface water flow and storage characteristics.

Tracer Tests: Principles and Methods

Tracer tests are conducted to study groundwater flow velocity, directionality, pathways, and aquifer properties by introducing a tracer substance into a groundwater system and tracking its movement over time. Tracers can be natural (e.g., isotopes) or artificial (e.g., fluorescent dyes, chemical compounds), and their selection depends on factors such as hydrogeological conditions, sensitivity of detection methods, and environmental considerations.

The basic steps involved in conducting a tracer test include:

1. Preparation: Selection of appropriate tracer substance, site characterization, determination of injection and monitoring well locations, and establishment of baseline hydrogeological conditions.
2. Injection: Injection of a known quantity of tracer substance into the groundwater system through a monitoring well or injection borehole at a predetermined rate and duration.
3. Monitoring: Continuous or periodic monitoring of tracer concentrations at monitoring wells or observation points downstream to track tracer movement and dispersion.
4. Data Analysis: Analysis of tracer breakthrough curves, travel times, breakthrough velocities, and other parameters to interpret groundwater flow characteristics and aquifer properties.

Various techniques are used to analyze tracer test data, including graphical methods, mathematical modeling, and statistical analysis. These techniques help hydrogeologists quantify groundwater flow parameters such as hydraulic conductivity, porosity, dispersivity, and residence times, providing valuable insights into aquifer behavior and contaminant transport mechanisms.

Applications of Tracer Tests

Tracer tests have diverse applications in hydrogeology, including:

- Groundwater Flow Characterization: Tracer tests help characterize groundwater flow patterns, velocities, and directions within aquifers, facilitating the delineation of flow paths and recharge zones.
- Aquifer Property Estimation: Tracer tests provide data for estimating aquifer properties such as hydraulic conductivity, porosity, and dispersivity, which are essential for groundwater modeling and resource assessment.
- Contaminant Transport Studies: Tracer tests are used to simulate and study the transport of contaminants in groundwater systems, assess the effectiveness of remediation strategies, and predict contaminant fate and transport pathways.
- Wellhead Protection Zone Delineation: Tracer tests aid in delineating wellhead protection zones around drinking water wells, identifying areas vulnerable to contamination and guiding land use planning and regulatory measures.

Tracer tests are valuable tools for understanding groundwater flow dynamics and aquifer behavior, providing essential data for groundwater management, remediation, and protection efforts.

Hydrogeological Mapping: Principles and Methods

Hydrogeological mapping involves the creation of detailed maps and models depicting geological, hydrological, and hydrogeological features of an area to understand groundwater flow and storage characteristics. Mapping techniques include geological mapping, hydrological mapping, geophysical mapping, and GIS-based mapping, each providing unique insights into subsurface water resources and processes.

The key steps involved in hydrogeological mapping include:

1. Data Collection: Gathering geological, hydrological, hydrogeological, and geospatial data through field surveys, borehole logs, well tests, geophysical surveys, remote sensing, and GIS databases.
2. Data Analysis: Integration and analysis of collected data to identify geological formations, aquifer properties, groundwater flow paths, recharge areas, discharge zones, and potential contaminant sources.
3. Map Creation: Creation of thematic maps depicting various hydrogeological features, including geological formations, aquifer extents, groundwater levels, water quality parameters, and hydrological boundaries.
4. Model Development: Development of conceptual and numerical models based on hydrogeological mapping data to simulate groundwater flow, predict aquifer behavior, and assess the impacts of human activities and environmental changes.

Hydrogeological mapping techniques can be applied at different scales, from local groundwater studies to regional groundwater resource assessments, providing valuable information for water resources management, land use planning, environmental impact assessment, and policy formulation.

Applications of Hydrogeological Mapping

Hydrogeological mapping has diverse applications in hydrogeology, including:

- Groundwater Resource Assessment: Hydrogeological mapping provides valuable information for assessing groundwater quantity, quality, and sustainability, guiding groundwater management and development strategies.
- Aquifer Characterization: Hydrogeological mapping helps characterize aquifer properties such as lithology, stratigraphy, hydraulic conductivity, and storativity, facilitating the understanding of groundwater flow and storage processes.

- Contaminant Plume Delineation: Hydrogeological mapping aids in delineating contaminant plumes, identifying potential contaminant sources, and assessing the extent and migration pathways of groundwater contamination.
- Well Siting and Design: Hydrogeological mapping assists in siting and designing groundwater wells by identifying optimal locations based on hydrogeological conditions, minimizing drilling costs, and maximizing well performance and yield.

Hydrogeological mapping plays a crucial role in groundwater exploration, management, and protection efforts, providing essential data and information for informed decision-making and sustainable water resource management.

Significance of Tracer Tests and Hydrogeological Mapping

Tracer tests and hydrogeological mapping are complementary techniques that provide valuable insights into groundwater flow dynamics, aquifer properties, and contaminant transport processes. Tracer tests offer direct measurements of groundwater flow parameters and contaminant transport behavior, while hydrogeological mapping provides a spatial framework for understanding the geological and hydrological controls on groundwater systems.

By combining tracer tests and hydrogeological mapping, hydrogeologists can develop comprehensive conceptual models of groundwater flow and contaminant transport, optimize groundwater monitoring and management strategies, and mitigate risks associated with groundwater contamination and depletion.

9.5 INTEGRATING GIS IN HYDROGEOLOGICAL ANALYSIS

9.5.1 Groundwater Vulnerability Mapping

GIS-based groundwater vulnerability mapping combines various spatial datasets, including geology, land use, soil properties, and hydrological parameters, to assess the susceptibility of aquifers to contamination. By analyzing factors such as permeability, distance to potential pollution sources, and land use practices, researchers can identify areas at high risk of groundwater pollution.

9.5.2 Hydrological Modeling and Simulation

GIS platforms support hydrological modeling and simulation by integrating spatial data with hydrological models. These models simulate groundwater flow, recharge, and contaminant transport processes, allowing researchers to evaluate different management scenarios and assess the potential impacts of land use changes, climate variability, and groundwater extraction.

9.5.3 Decision Support Systems for Water Management

GIS-based decision support systems (DSS) provide stakeholders with tools for informed decision-making in water resources management. These systems integrate hydrogeological data, simulation models, and visualization techniques to analyze complex spatial relationships and identify sustainable water management strategies.

9.6 REMOTE SENSING APPLICATIONS IN HYDROGEOLOGY

9.6.1 Monitoring Groundwater Resources

Remote sensing data, such as satellite imagery and aerial photographs, can be used to monitor changes in land cover and vegetation density, which are indicators of groundwater availability. Vegetation indices derived from remote sensing data provide valuable information on groundwater recharge rates and aquifer conditions.

9.6.2 Detecting Groundwater Contamination

Thermal infrared imaging and hyperspectral remote sensing techniques can detect thermal anomalies and chemical signatures associated with groundwater contamination. By analyzing these data, researchers can identify contaminated sites, assess the extent of contamination, and prioritize remediation efforts.

9.6.3 Mapping Hydrological Processes

Remote sensing data can be used to map surface water bodies, wetlands, and riparian zones, which are critical components of the hydrological cycle. By monitoring changes in surface water extent and dynamics, researchers can assess groundwater-surface water interactions, evaluate ecosystem health, and predict flood risk.

CHAPTER 10

CASE STUDIES IN SUCCESSFUL AQUIFER RECHARGE PROJECTS

10.1 INTRODUCTION

Aquifer recharge, the intentional augmentation of groundwater through various methods, has become a vital strategy for managing water resources in the face of growing water demands, climate variability, and over-extraction. Successful aquifer recharge projects serve as beacons of innovation, demonstrating how careful planning, effective technologies, and community involvement can contribute to sustainable water management.

10.2 THE ORANGE COUNTY GROUNDWATER REPLENISHMENT SYSTEM (GWRS), USA

In the arid landscape of Southern California, where water scarcity poses a perennial challenge, the Orange County Groundwater Replenishment System (GWRS) stands as a beacon of innovation and sustainability. Facing the dual threats of depleting groundwater reserves and the intrusion of seawater into coastal aquifers, Orange County embarked on a pioneering journey to secure a reliable water supply through advanced water purification and aquifer recharge.

10.2.1 Addressing Water Scarcity

The backdrop of the GWRS's inception is the ever-growing demand for water in Orange County, exacerbated by a Mediterranean climate characterized by limited rainfall. The region's reliance on groundwater as a primary water source made it vulnerable to over-extraction and the encroachment of seawater, jeopardizing the quality and quantity of available freshwater. In the face of these challenges, the visionary leaders in Orange County sought a sustainable solution that could bolster water resources for the community.

10.2.2 Indirect Potable Reuse and Advanced Treatment

The GWRS represents a paradigm shift in water management through its embrace of indirect potable reuse. Unlike traditional water treatment plants that discharge purified water into rivers or oceans, the GWRS takes a bold step by treating wastewater to produce high-quality purified water suitable for drinking. The advanced treatment processes employed at the GWRS are designed to remove contaminants to levels that meet or exceed stringent drinking water standards.

The treatment train at GWRS involves multiple steps, including microfiltration, reverse osmosis, and advanced oxidation with ultraviolet light and hydrogen peroxide. This comprehensive treatment ensures the removal of pathogens, trace contaminants, and impurities, resulting in water that surpasses the quality of many traditional water sources. The commitment to excellence in water treatment is underscored by rigorous monitoring protocols that guarantee the purity and safety of the reclaimed water.

10.2.3 Aquifer Recharge as a Sustainable Solution

Central to the success of the GWRS is the integration of aquifer recharge into the water supply strategy. Instead of relying solely on direct discharge into surface waters, the purified water from the GWRS is judiciously injected into the local aquifer. This intentional recharge of the groundwater basin not only mitigates seawater intrusion but also augments the natural replenishment of underground water reserves.

The aquifer recharge component of the GWRS is a strategic move to store treated water underground, creating a resilient buffer against drought conditions. By replenishing the aquifer, Orange County ensures a sustainable and drought-resistant water supply that can be tapped into during periods of water scarcity. This approach aligns with the broader goals of water sustainability, minimizing reliance on imported water and enhancing local resilience in the face of climate variability.

10.2.4 Robust Monitoring and Quality Assurance

At the heart of the GWRS's success is a robust system of monitoring and quality assurance that guarantees the safety and reliability of the reclaimed water. The treated water undergoes extensive testing to verify its compliance with not only state and federal drinking water standards but also the stringent regulations set by the California Department of Public Health. This commitment to water quality assurance builds trust within the community and ensures that the reclaimed water from the GWRS is a dependable and safe source of drinking water. The monitoring process includes regular sampling and analysis at various stages of the treatment train, as well as ongoing assessments of the aquifer's water quality. The transparency in reporting and adherence to the highest standards of water quality contribute to the success of the GWRS as a model for indirect potable reuse projects globally.

10.2.5 Sustainable Water Supply for the Community

Perhaps the most significant impact of the GWRS is the transformation of water scarcity into water security for the community of Orange County. By embracing cutting-edge technologies and innovative approaches to water management, the GWRS has not only replenished the local groundwater basin but has also provided a sustainable and drought-resistant water supply.

The success of the GWRS has far-reaching implications for the broader water management landscape. It serves as a testament to the feasibility and effectiveness of indirect potable reuse, challenging preconceptions about the safety and acceptability of reclaimed water for drinking purposes. The Orange County community now enjoys a diversified and resilient water portfolio that reduces dependence on imported water sources, enhancing the region's ability to weather the uncertainties of climate change and population growth.

10.2.6 Global Relevance and Lessons Learned

The Orange County GWRS is not just a local success story; it resonates globally as a model for addressing water scarcity through advanced water purification and aquifer recharge. Other regions facing similar challenges can draw inspiration from Orange County's journey and adapt the principles of indirect potable reuse to their specific contexts. Key lessons from the GWRS include the importance of visionary leadership, robust regulatory frameworks, public outreach and education, and the integration of advanced treatment technologies. The success of the GWRS underscores the potential for transforming wastewater into a valuable resource, emphasizing the need for a paradigm shift in water management practices worldwide.

10.3 MANAGED AQUIFER RECHARGE IN THE NETHERLANDS

The Netherlands, renowned for its expertise in water management, has implemented successful managed aquifer recharge (MAR) projects to combat land subsidence and secure freshwater resources. In regions experiencing excessive groundwater extraction, MAR techniques involve diverting excess surface water during wet periods to infiltrate and recharge aquifers. The Midden-Delfland Polder, a low-lying area prone to subsidence, utilizes MAR to replenish groundwater levels. This method not only mitigates land subsidence but also helps maintain a sustainable balance between surface water and groundwater resources.

The Netherlands, renowned for its intricate network of waterways and polders, has emerged as a global leader in water management. Facing challenges such as land subsidence, saltwater intrusion, and the need to secure freshwater resources, the country has implemented a series of innovative strategies, with managed aquifer recharge (MAR) playing a pivotal role. The Dutch approach to MAR serves as a blueprint for sustainable water management, demonstrating how a combination of ancient water expertise and modern technology can address complex hydrological issues.

10.3.1 Addressing Land Subsidence

Land subsidence, a gradual sinking of the land surface, is a critical issue in parts of the Netherlands, particularly in peat and clay-rich areas. Excessive groundwater extraction for various purposes, including agriculture and drinking water supply, has historically contributed to this phenomenon. To counteract land subsidence and its associated impacts, the Netherlands has embraced MAR as a strategic tool for replenishing aquifers and restoring groundwater levels.

MAR techniques involve deliberately introducing water into the subsurface, either through infiltration basins, injection wells, or other methods, to augment natural recharge processes. In the Netherlands, the emphasis is on spreading basins and infiltration trenches strategically located in areas susceptible to land subsidence. These structures allow rainwater or surface water from rivers and canals to percolate into the underlying aquifers, replenishing groundwater levels and mitigating land subsidence.

10.3.2 Combating Saltwater Intrusion

With a significant portion of the Netherlands lying below sea level, saltwater intrusion poses a constant threat to freshwater resources. This is particularly challenging in coastal regions, where over-extraction of freshwater can lead to the encroachment of saline seawater into aquifers. MAR serves as a key tool in combating saltwater intrusion by maintaining a balance between freshwater and saltwater.

The Dutch approach involves creating artificial freshwater lenses within aquifers through MAR. By carefully managing the recharge of freshwater from surface water bodies, the Netherlands ensures the maintenance of a stable interface between freshwater and saltwater. This strategic control over groundwater levels helps to prevent saline intrusion into coastal aquifers, preserving the quality of freshwater resources vital for agriculture, industry, and domestic use.

10.3.3 Integrated MAR Strategies

The success of managed aquifer recharge in the Netherlands lies in its integration into a broader water management strategy. MAR is not viewed in isolation but rather as part of a holistic approach that includes surface water management, water treatment, and sustainable extraction practices. The Netherlands' intricate system of water management infrastructure, including dikes, sluices, and canals, works in tandem with MAR to maintain a delicate balance between water supply and demand.

One notable example is the Groene Hart (Green Heart) region, an area characterized by peat subsidence and vulnerability to saltwater intrusion. Here, MAR is integrated into a comprehensive water management plan that includes controlled flooding of agricultural land during wet periods. This deliberate inundation allows rainwater to recharge underlying aquifers, combating land subsidence and mitigating the risk of saltwater intrusion.

10.3.4 Collaboration and Community Involvement

Central to the success of MAR in the Netherlands is a collaborative approach that involves various stakeholders, including government agencies, water boards, research institutions, and local communities. Water boards, decentralized entities responsible for water management in specific regions, play a crucial role in implementing MAR projects. These boards collaborate with municipalities, farmers, and other stakeholders to ensure the sustainable use of water resources.

Community involvement is particularly evident in projects where MAR is used to enhance natural water retention in rural areas. In the Veenweide Innovatiecentrum (Peat Meadow Innovation Center), for instance, local farmers, water boards, and researchers collaborate to implement MAR techniques. This collaborative effort aims to address land subsidence issues in peat-rich areas by promoting water retention in the soil and preventing excessive groundwater extraction.

10.3.5 Technological Innovations in MAR

The Netherlands leverages technological innovations to enhance the efficiency and effectiveness of MAR projects. Advanced modeling tools and hydrological simulations are employed to optimize the placement of infiltration structures and predict the impact of MAR on groundwater levels. Real-time monitoring systems provide valuable data on water quality, groundwater levels, and subsurface conditions, enabling adaptive management of MAR initiatives.

In some instances, the Netherlands utilizes soil sensors and geophysical techniques to assess the impact of MAR on soil structure and subsurface dynamics. This information is crucial for fine-tuning MAR practices, ensuring that the introduced water is distributed effectively and does not lead to unintended consequences such as soil compaction.

10.3.6 Sustainability and Climate Adaptation

The Dutch approach to MAR aligns closely with broader sustainability goals and climate adaptation strategies. By replenishing aquifers and promoting sustainable water use, MAR contributes to the resilience of ecosystems and supports biodiversity. Additionally, MAR helps the Netherlands adapt to the challenges posed by climate change, including altered precipitation patterns and more frequent extreme weather events.

In response to changing climate conditions, the Netherlands continues to explore and implement innovative MAR techniques. This includes investigating the potential of agricultural drainage water as a source for MAR, particularly in regions where excess water needs to be managed efficiently. The adaptability and ongoing evolution of the Dutch MAR strategy exemplify a proactive approach to climate resilience and sustainable water management.

10.3.7 Challenges and Future Directions

While the Netherlands has achieved significant success with MAR, challenges persist, and ongoing efforts focus on addressing emerging issues. Balancing the competing demands for water resources, especially in densely populated and agriculturally productive regions, requires continuous adaptation and optimization of MAR practices. Climate uncertainties further underscore the need for flexible and dynamic water management strategies. Future directions for MAR in the Netherlands include exploring nature-based solutions, such as the restoration of wetlands and the creation of multifunctional landscapes that enhance water retention. Additionally, integrating MAR with circular economy principles, where water is viewed as a valuable resource to be reused and recycled, holds promise for the sustainable management of water resources in the years to come.

10.4 THE TEXAS MANAGED AQUIFER RECHARGE (TXMAR) PROGRAM, USA

In the expansive landscapes of Texas, where the demand for water is soaring, and aquifers face the threat of depletion, the Texas Managed Aquifer Recharge (TXMAR) Program emerges as a pioneering initiative to safeguard this precious resource. As groundwater becomes increasingly vital for agriculture, industries, and communities, the TXMAR Program embodies the state's commitment to proactive water management. By employing a suite of innovative techniques, including spreading basins, injection wells, and riverbank filtration, Texas aims to augment aquifer recharge and ensure the resilience of its major aquifers.

10.4.1 Facing Groundwater Depletion and Rising Demand

Texas, with its vast and varied terrain, is no stranger to the challenges of groundwater depletion. As urban centers expand, agricultural needs evolve, and climatic patterns shift, the demand for groundwater has surged, placing stress on aquifers that are critical for sustaining both rural and urban water supplies. Recognizing the urgency of addressing groundwater depletion, Texas has turned to the TXMAR Program as a strategic response to ensure the long-term viability of its aquifer systems.

10.4.2 The TXMAR Approach

The TXMAR Program is characterized by a multifaceted approach that combines various recharge methods tailored to the diverse hydrogeological conditions prevalent across the state. The utilization of spreading basins, injection wells, and riverbank filtration showcases a commitment to flexibility and adaptability in enhancing aquifer recharge.

Spreading Basins

Spreading basins form a fundamental component of the TXMAR strategy, particularly in regions with favorable surface conditions for infiltration. These basins are designed to capture excess surface water during periods of abundance, such as heavy rainfall or storm events. The captured water is then deliberately spread over designated areas to percolate into the underlying aquifer, recharging groundwater levels.

One noteworthy example of spreading basin implementation is in the Texas Hill Country, where the program leverages natural depressions and topographical features to create basins that facilitate efficient water infiltration. This method not only replenishes aquifers but also minimizes surface water runoff, contributing to flood mitigation and enhancing overall water management resilience.

Injection Wells

In regions where surface conditions may not be conducive to spreading basins, the TXMAR Program turns to injection wells as an alternative method of introducing water into the subsurface. These wells enable the controlled injection of treated or untreated water directly into the aquifer, bypassing surface pathways. This approach is particularly valuable in areas with less permeable surface soils or where geological conditions favor the use of injection wells.

The injection well strategy is deployed strategically to target specific aquifer layers, optimizing the recharge process. The TXMAR Program underscores the importance of scientific assessments and hydrogeological studies to determine the most effective locations for well installation, ensuring that injected water effectively replenishes aquifer reserves.

Riverbank Filtration

Riverbank filtration is another innovative technique incorporated into the TXMAR Program, especially in regions where rivers intersect with aquifers. This method involves the intentional recharge of aquifers by allowing surface water from rivers to naturally infiltrate through the riverbanks into the underlying groundwater system. The infiltrated water undergoes natural filtration processes, improving its quality before becoming part of the aquifer.

In the Lower Rio Grande Valley, the TXMAR Program harnesses the potential of riverbank filtration along the Rio Grande. This approach not only contributes to aquifer recharge but also enhances the availability of high-quality groundwater for agricultural and municipal use. Riverbank filtration serves as an example of how natural processes can be harnessed to augment aquifer recharge sustainably.

10.4.3 Collaborative Efforts

The success of the TXMAR Program is intricately linked to collaborative efforts that extend beyond traditional bureaucratic boundaries. Water authorities, researchers, and local communities unite under the common goal of securing water resources for the future. The collaboration is not only horizontal, involving different water management agencies and institutions, but also vertical, encompassing partnerships with local communities and stakeholders who intimately understand regional water dynamics.

Local communities play a vital role in the success of TXMAR projects. Their involvement ranges from providing insights into local hydrogeological conditions to participating in water conservation and education initiatives. This collaborative model fosters a sense of ownership and shared responsibility for water sustainability, aligning with the program's overarching objectives.

10.4.4 Importance of Diverse Recharge Techniques

The TXMAR Program serves as a testament to the importance of embracing diverse recharge techniques tailored to the specific hydrogeological conditions of each region. Texas encompasses a vast and geologically varied landscape, and a one-size-fits-all approach to aquifer recharge is impractical. By strategically deploying spreading basins, injection wells, and riverbank filtration based on regional characteristics, the program maximizes the efficacy of each technique.

Diverse recharge methods not only address the geological heterogeneity but also enhance the resilience of the aquifer recharge system. A multifaceted approach ensures that Texas is better equipped to adapt to changing climatic conditions, varying surface water availability, and evolving water demands from different sectors.

10.4.5 Adaptive Management and Scientific Innovation

The success of the TXMAR Program hinges on adaptive management informed by scientific innovation. Regular monitoring, data collection, and hydrogeological assessments allow for the refinement of recharge strategies based on real-time information. Advances in remote sensing technologies, groundwater modeling, and data analytics contribute to the program's ability to dynamically respond to changing conditions.

Scientific innovation also extends to water treatment technologies, ensuring that the water introduced into aquifers meets or exceeds stringent quality standards. The integration of advanced treatment processes not only safeguards the integrity of aquifers but also addresses concerns related to the potential mobilization of contaminants during recharge activities.

10.4.6 Challenges and Future Directions

Despite its successes, the TXMAR Program faces challenges that necessitate ongoing refinement and adaptation. Sustainable funding, regulatory frameworks, and community engagement are perennial issues that demand attention. Balancing the competing needs of agriculture, urban centers, and ecosystems requires careful navigation and continuous collaboration.

Future directions for the TXMAR Program involve a deeper integration of nature-based solutions, such as the restoration of riparian zones and the promotion of green infrastructure. Exploring the potential of stormwater harvesting, enhancing water use efficiency, and leveraging artificial intelligence for predictive modeling are avenues that align with the program's commitment to innovation.

10.5 THE ARTIFICIAL RECHARGE OF GROUNDWATER IN ISRAEL

Israel, a country renowned for navigating water scarcity in its arid and semi-arid regions, has emerged as a global leader in the implementation of innovative approaches to artificial groundwater recharge. At the heart of Israel's success lies a pioneering model that embraces the controlled reuse of treated wastewater for agricultural irrigation, creating a closed-loop system that not only sustains agriculture but significantly enhances aquifer recharge. Central to this model is the Shafdan Wastewater Treatment Plant, one of the world's largest and most advanced facilities, which plays a pivotal role in treating wastewater to the highest standards before distributing it for agricultural use. This intricate system exemplifies the potential of sustainable water reuse in arid environments, offering insights and lessons for regions worldwide grappling with water scarcity.

10.5.1 Water Scarcity in Arid Environments

Israel's commitment to innovative water management is rooted in the country's geographical reality. Characterized by arid and semi-arid climates, Israel faces perennial water scarcity challenges exacerbated by low annual precipitation and high evaporation rates. The limited availability of freshwater resources necessitates creative and sustainable solutions to meet the diverse water needs of its population, support agriculture, and ensure environmental sustainability.

10.5.2 The Shafdan Wastewater Treatment Plant

At the forefront of Israel's pioneering efforts is the Shafdan Wastewater Treatment Plant, located in the Dan Region near Tel Aviv. Serving as a critical component in the country's water management infrastructure, Shafdan is not merely a wastewater treatment facility; it is a linchpin in a holistic strategy that prioritizes water reuse and aquifer recharge.

The Shafdan plant is renowned for its size, sophistication, and efficiency. Treating wastewater from residential, industrial, and commercial sources, Shafdan employs a multi-stage treatment process that includes biological treatment, tertiary filtration, and advanced disinfection. The result is high-quality treated wastewater that meets stringent standards for agricultural use.

10.5.3 Closed-Loop System

The hallmark of Israel's approach to artificial groundwater recharge is the establishment of a closed-loop system, wherein treated wastewater becomes a valuable resource for agricultural irrigation. The Shafdan model exemplifies the seamless integration of water treatment, reuse, and recharge. Treated wastewater, once deemed a waste product, is transformed into a strategic asset that not only sustains agriculture but also addresses the critical challenge of aquifer replenishment.

Controlled application of treated wastewater in agriculture ensures a judicious balance between water availability and demand. Rather than allowing treated wastewater to be discharged into rivers or the sea, Israel leverages its nutrient-rich content to enhance soil fertility and promote sustainable agricultural practices. This approach mitigates the need for additional freshwater resources for irrigation, conserving precious freshwater reserves while simultaneously replenishing aquifers.

10.5.4 Regulatory Framework and Public Perception

The success of Israel's closed-loop system is bolstered by a robust regulatory framework that governs the treatment and reuse of wastewater. Stringent water quality standards, backed by comprehensive monitoring and enforcement mechanisms, ensure that treated wastewater meets or exceeds safety standards for agricultural use. This regulatory diligence instills confidence in the public and agricultural stakeholders, fostering a positive perception of treated wastewater as a valuable and safe resource.

Public awareness and acceptance are crucial components of the success of the Shafdan model. Israel has invested in extensive outreach and education programs to inform the public about the benefits and safety of treated wastewater reuse. Through these efforts, treated wastewater has transitioned from being stigmatized as "dirty water" to being recognized as a vital and secure component of the water supply.

10.5.5 Enhancing Aquifer Recharge

The controlled application of treated wastewater in agriculture goes beyond addressing water scarcity; it significantly contributes to enhancing aquifer recharge. The infiltrated water percolates through the soil, gradually replenishing underlying aquifers. This deliberate recharge mechanism is especially valuable in arid regions where natural recharge from rainfall is limited.

The success of aquifer recharge through treated wastewater application is evident in the increased levels of groundwater observed in monitoring wells. Israel's meticulous monitoring and data collection efforts allow authorities to track the impact of treated wastewater on aquifer levels, ensuring that recharge practices are optimized for long-term sustainability.

10.5.6 Sustainable Agriculture and Food Security

The intersection of water reuse, agriculture, and aquifer recharge in Israel's closed-loop system has far-reaching implications for sustainable agriculture and food security. By providing a reliable source of irrigation water, the Shafdan model supports agricultural productivity in arid regions, where water availability is a limiting factor. Sustainable farming practices, enabled by the reuse of treated wastewater, contribute to food security and economic resilience.

The nutrient-rich composition of treated wastewater further enhances soil fertility, reducing the dependence on synthetic fertilizers. This holistic approach aligns with global efforts to promote sustainable agricultural practices that minimize environmental impact while ensuring food production in the face of climate variability.

10.5.7 Replicable Model for Arid Regions

Israel's success in implementing a closed-loop system for wastewater treatment, reuse, and aquifer recharge offers valuable lessons for other arid regions facing similar water challenges. The adaptability of the Shafdan model to different climatic and geological conditions underscores its potential as a replicable and scalable solution.

Key factors contributing to the model's replicability include the integration of advanced treatment technologies, robust regulatory frameworks, public engagement strategies, and a commitment to ongoing monitoring and optimization. The Shafdan model serves as an inspiration for arid regions globally, demonstrating that sustainable water management is not only feasible but also essential for ensuring water security and environmental resilience.

10.5.8 Challenges and Future Directions

While Israel's closed-loop system has proven remarkably successful, it is not without challenges. As the demand for water continues to rise, balancing the needs of agriculture, industry, and domestic consumption requires ongoing adaptation and optimization. Striking a harmonious balance between water availability and demand remains a dynamic challenge that demands continuous innovation.

Future directions for Israel's water management include the exploration of additional water sources, such as desalination and stormwater harvesting, to complement the closed-loop system. Integrating smart water technologies, improving water use efficiency, and enhancing the resilience of water infrastructure are crucial components of the country's evolving water strategy.

10.6 THE SIDI SAAD MANAGED AQUIFER RECHARGE PROJECT, TUNISIA

In the face of mounting water scarcity intensified by the impacts of climate change, Tunisia has emerged as a pioneer in innovative water management through the successful implementation of the Sidi Saad Managed Aquifer Recharge Project. This transformative initiative showcases the adaptability of Managed Aquifer Recharge (MAR) techniques to the diverse climatic and hydrogeological conditions of the region. By constructing infiltration basins and percolation ponds, the project captures and stores surface water, allowing it to gradually infiltrate into the aquifer, thereby replenishing groundwater supplies. The Sidi Saad project not only addresses immediate water challenges but also serves as a model for sustainable water management in arid and changing environments.

10.6.1 Water Scarcity and Climate Change in Tunisia

Tunisia, situated in North Africa, faces a myriad of water challenges, including scarcity aggravated by the impacts of climate change. The country's semi-arid to arid climate, coupled with irregular rainfall patterns and increasing temperatures, has led to a heightened strain on water resources. In this context, innovative solutions are imperative to secure a sustainable water future for Tunisia's growing population and diverse water needs.

10.6.2 The Sidi Saad Managed Aquifer Recharge Project

The Sidi Saad Managed Aquifer Recharge Project stands out as a testament to Tunisia's commitment to addressing water scarcity through forward-thinking strategies. Located in the Sidi Saad region, the project utilizes MAR techniques to harness surface water for replenishing groundwater resources. The core components of the initiative involve the construction of infiltration basins and percolation ponds, creating a system that captures, stores, and gradually releases water into the underlying aquifer.

Infiltration Basins and Percolation Ponds

At the heart of the Sidi Saad project are infiltration basins and percolation ponds strategically designed to maximize the recharge of the aquifer. Infiltration basins are excavated depressions in the ground that capture runoff from rainfall or surface water. These basins serve as temporary storage, allowing water to slowly infiltrate into the underlying soil and, ultimately, the aquifer.

Percolation ponds complement this approach by providing larger storage areas where water can be held for extended periods. These ponds are carefully designed to facilitate controlled percolation, preventing rapid runoff and ensuring that water effectively recharges the aquifer. The combination of infiltration basins and percolation ponds creates a balanced and adaptable system capable of accommodating varying water availability and climatic conditions.

Adaptability to Diverse Conditions

One of the standout features of the Sidi Saad project is its adaptability to diverse climatic and hydrogeological conditions. Tunisia, characterized by regional variations in climate and geology, requires flexible solutions that can be tailored to specific contexts. The use of infiltration basins and percolation ponds allows the project to be implemented in areas with different soil types, hydraulic conductivity, and water availability. The adaptability of MAR techniques is particularly crucial in arid regions where traditional water sources are limited. By capturing and storing surface water, the Sidi Saad project provides a reliable and sustainable source of groundwater replenishment, mitigating the impact of water scarcity on communities, agriculture, and ecosystems.

10.6.3 The Recharge Process

The recharge process employed by the Sidi Saad project follows a carefully orchestrated sequence. Surface water, whether from rainfall or other sources, is channeled into the infiltration basins and percolation ponds. The controlled release of water into these storage structures allows for gradual infiltration into the underlying soil. The permeable nature of the soil facilitates the percolation of water, eventually reaching the aquifer.

This deliberate recharge process not only replenishes groundwater supplies but also promotes the natural filtration of water. As water percolates through the soil, impurities and contaminants are filtered out, resulting in improved water quality in the aquifer. This dual benefit of recharge and water purification contributes to the overall sustainability and resilience of the water supply.

10.6.4 Environmental and Socioeconomic Impacts

The Sidi Saad project extends its positive impacts beyond aquifer recharge, influencing both the environment and socioeconomic aspects of the region. Ecologically, the project supports the health of local ecosystems by maintaining natural hydrological processes. Groundwater-dependent ecosystems, including wetlands and vegetation, receive vital support through the sustained availability of water. Socioeconomically, the project has implications for agricultural practices and community resilience. The reliable recharge of aquifers provides a steady source of water for agricultural irrigation, reducing dependency on unsustainable groundwater extraction. This, in turn, enhances the resilience of agriculture to changing climatic conditions and promotes food security within the region.

10.6.5 Challenges and Lessons Learned

While the Sidi Saad Managed Aquifer Recharge Project has achieved notable success, it has not been without challenges. Adapting MAR techniques to diverse geological conditions requires meticulous planning and ongoing monitoring. Understanding the local hydrogeology, soil characteristics, and the potential for water contamination is critical to optimizing recharge efforts.

Community engagement and awareness also play a crucial role in the success of MAR projects. Local communities need to be informed about the benefits of aquifer recharge, the safety of the process, and the role they play in its success. Building trust and cooperation among stakeholders ensures the long-term sustainability of such initiatives.

10.6.6 Future Directions and Replicability

The success of the Sidi Saad project paves the way for future MAR initiatives in Tunisia and beyond. As climate change continues to pose challenges to water availability, the adaptability and effectiveness of infiltration basins and percolation ponds make them valuable tools in sustainable water management.

Replicating the Sidi Saad model in other regions facing similar water challenges requires careful consideration of local conditions. Hydrogeological assessments, feasibility studies, and community involvement are essential components of successful MAR projects. Tunisia's experience with the Sidi Saad project serves as a blueprint for other countries seeking innovative solutions to water scarcity in arid and changing climates.

10.7 THE RIEGO ATIRANTADO SYSTEM, SPAIN

In the sun-drenched landscapes of Spain, where a Mediterranean climate brings irregular rainfall and challenges agricultural water management, the Riego Atirantado System stands as a testament to innovative solutions for aquifer recharge. Spain's commitment to sustainable water use is exemplified by this system, which utilizes check dams and small-scale irrigation techniques to optimize water utilization in farmlands. By capturing rainwater runoff and allowing it to percolate into the soil, the Riego Atirantado System not only supports sustainable agriculture but also significantly contributes to aquifer recharge.

10.7.1 Challenges of Water Management in Spain

Spain's distinctive Mediterranean climate poses inherent challenges to water availability, particularly in the agricultural sector. Irregular rainfall patterns, coupled with high evaporation rates, create a scenario where water scarcity is a recurrent concern. Sustainable water management strategies are essential to ensure the resilience of agricultural practices, support local communities, and preserve aquifer health.

10.7.2 Introduction to the Riego Atirantado System

The Riego Atirantado System, rooted in the Spanish language where "Riego" translates to irrigation, represents a tailored approach to optimize water use in agricultural areas. At its core, the system revolves around the strategic implementation of check dams and small-scale irrigation techniques. These elements work in harmony to capture rainwater runoff and guide it into the soil, promoting both efficient irrigation practices and aquifer recharge.

Check Dams and Rainwater Capture

Central to the Riego Atirantado System are check dams strategically placed along natural watercourses or in agricultural fields. Check dams, typically constructed with locally available materials, serve a dual purpose. Firstly, they slow down the flow of rainwater runoff, preventing rapid erosion and soil loss. Secondly, the dams create temporary storage for captured runoff, allowing the water to percolate into the soil and recharge aquifers.

The design and placement of check dams are critical factors in the system's effectiveness. Dams are strategically located to maximize the capture of rainfall runoff while minimizing the risk of sedimentation. This careful planning ensures that the system optimally utilizes available water resources while mitigating potential negative impacts on downstream ecosystems.

Small-Scale Irrigation Techniques

Complementing the use of check dams, the Riego Atirantado System incorporates small-scale irrigation techniques that align with the principles of water efficiency. Drip irrigation and other precision irrigation methods are employed to deliver water directly to the root zones of crops, minimizing water wastage and maximizing the effectiveness of each drop. The integration of small-scale irrigation technologies not only conserves water but also enhances the adaptability of agriculture to variable climate conditions. Farmers can tailor irrigation schedules based on crop water requirements, fostering a more sustainable and resilient agricultural system.

10.7.3 Promoting Sustainable Agriculture

The Riego Atirantado System goes beyond aquifer recharge; it actively promotes sustainable agriculture in regions grappling with water scarcity. By optimizing water use through check dams and small-scale irrigation, the system contributes to the long-term viability of farming practices. This is particularly crucial in the Mediterranean regions of Spain, where agriculture plays a pivotal role in the economy and livelihoods of local communities. Efficient water use in agriculture is achieved through the strategic capture and distribution of rainwater. The check dams slow down runoff, allowing water to percolate into the soil and replenish aquifers. This not only conserves groundwater but also improves soil moisture levels, creating more favorable conditions for crop growth. In essence, the Riego Atirantado System provides a sustainable water management model that supports the dual goals of aquifer recharge and agricultural productivity.

10.7.4 Ecosystem Benefits and Biodiversity

Beyond its immediate impacts on agriculture, the Riego Atirantado System brings ecosystem benefits and fosters biodiversity. By preventing rapid erosion and sedimentation, the check dams contribute to the preservation of water quality downstream. This, in turn, supports aquatic ecosystems and the diverse flora and fauna that depend on reliable water sources.

The controlled release of captured rainwater into the soil enhances the resilience of surrounding ecosystems. Native vegetation, often adapted to sporadic rainfall patterns, benefits from the supplemental moisture, promoting biodiversity and ecological stability. The system thus underscores the interconnectedness of water management, agriculture, and the broader environment.

10.7.5 Community Involvement and Adaptation

Successful implementation of the Riego Atirantado System requires active community involvement and adaptation to local conditions. Local knowledge, often passed down through generations, plays a crucial role in identifying optimal locations for check dams and tailoring small-scale irrigation practices to the specific needs of crops and soils. Communities that actively participate in the planning and maintenance of the system are more likely to reap its benefits sustainably. This collaborative approach not only ensures the effectiveness of the Riego Atirantado System but also fosters a sense of ownership and responsibility among local stakeholders for the stewardship of water resources.

10.7.6 Challenges and Lessons Learned

While the Riego Atirantado System has proven effective in mitigating water scarcity and enhancing aquifer recharge, it is not without challenges. The system's success is contingent on careful planning, continuous monitoring, and adaptation to evolving climatic conditions. Balancing the water needs of agriculture with the conservation of natural ecosystems requires ongoing collaboration and flexibility in the face of change. Lessons learned from the implementation of the Riego Atirantado System highlight the importance of comprehensive hydrological assessments and community engagement. A nuanced understanding of local hydrogeology, soil characteristics, and climate variability is essential for optimizing the system's impact on aquifer recharge.

10.7.7 Future Directions and Replicability

The success of the Riego Atirantado System positions it as a replicable model for other regions facing similar water challenges. The principles of capturing rainwater runoff, utilizing check dams, and implementing small-scale irrigation techniques can be adapted to diverse climates and agricultural practices. As climate change intensifies, systems that enhance aquifer recharge while supporting sustainable agriculture become increasingly relevant globally.

Future directions for the Riego Atirantado System involve leveraging technological advancements and data-driven approaches for improved water management. Integrating real-time monitoring systems, remote sensing technologies, and precision agriculture techniques can enhance the efficiency and adaptability of the system to changing conditions.

10.8 THE SHEIKH ZAYED CANAL, UNITED ARAB EMIRATES (UAE)

In the relentless embrace of aridity, the United Arab Emirates (UAE) has embarked on a monumental endeavor to counter water scarcity through the Sheikh Zayed Canal. Spanning over 360 kilometers, this ambitious project symbolizes the UAE's commitment to enhancing aquifer recharge in the face of challenges posed by over-extraction and salinity intrusion in coastal aquifers. The canal, a conduit for desalinated seawater, epitomizes a strategic solution to sustainably manage water resources in the arid landscape.

10.8.1 Water Scarcity in the UAE

The UAE, characterized by its arid climate and limited freshwater resources, faces an ever-growing demand for water driven by population growth, urbanization, and industrial expansion. In this context, innovative solutions are imperative to secure a sustainable water future for the nation. The Sheikh Zayed Canal emerges as a testament to the UAE's proactive stance in addressing water scarcity through visionary water management strategies.

10.8.2 Introduction to the Sheikh Zayed Canal[h]

Named in honor of the UAE's founding father, Sheikh Zayed bin Sultan Al Nahyan, the Sheikh Zayed Canal represents a pioneering effort to augment aquifer recharge through the controlled transport of desalinated seawater. This extensive canal system serves as a lifeline, not only for sustaining agriculture and urban development but also for combating the detrimental effects of over-extraction and salinity intrusion in coastal aquifers.

Desalinated Seawater for Aquifer Recharge

The key innovation of the Sheikh Zayed Canal lies in its utilization of desalinated seawater for aquifer recharge. Desalination, a prominent technology in the UAE due to its limited freshwater sources, enables the conversion of seawater into a valuable resource. The canal leverages this desalinated water to recharge underlying aquifers, strategically addressing the dual challenges of declining groundwater levels and salinity intrusion. The desalinated seawater is carefully discharged into the canal, and its controlled release ensures optimal infiltration into the surrounding soil. This deliberate process mimics natural recharge mechanisms while minimizing the risk of soil salinization. By introducing high-quality desalinated water into the aquifer, the project not only replenishes groundwater reserves but also mitigates the intrusion of saline water, a common issue in coastal areas.

Addressing Over-Extraction and Salinity Intrusion

Over-extraction of groundwater is a pervasive challenge in the UAE, driven by the high demand for water in agriculture, industry, and domestic use. This unbridled extraction not only depletes aquifer reserves but also exacerbates the risk of salinity intrusion. Coastal aquifers, in particular, are vulnerable to the encroachment of seawater as groundwater levels decline.

The Sheikh Zayed Canal strategically intervenes in this scenario by replenishing aquifers with desalinated seawater. The controlled discharge and infiltration prevent excessive extraction of native groundwater, allowing aquifers to recover and maintain a healthier balance. Moreover, the introduction of low-salinity water helps create a hydraulic barrier, limiting the intrusion of seawater into coastal aquifers.

10.8.3 Challenges and Innovations

While the Sheikh Zayed Canal represents a pioneering solution to water challenges in the UAE, it is not without its complexities and challenges. Managing the delicate balance between canal discharge and infiltration requires sophisticated monitoring and control systems. Advanced technologies, including real-time sensors and data analytics, play a crucial role in optimizing the canal's operation and ensuring the efficient recharge of aquifers. One of the notable innovations is the integration of smart infrastructure to monitor groundwater quality and levels continuously. This data-driven approach allows authorities to adaptively manage the canal's discharge, responding to changing environmental conditions and maintaining the integrity of both the canal and the aquifers it seeks to recharge.

10.8.4 Economic and Environmental Impacts

The economic and environmental impacts of the Sheikh Zayed Canal extend beyond aquifer recharge. The canal serves as a catalyst for economic diversification, particularly in agriculture. The availability of a reliable water source enhances the feasibility of agricultural activities, promoting food security and reducing dependency on imported produce.

Furthermore, the environmental benefits of the canal are significant. The project contributes to the preservation of natural ecosystems by alleviating pressure on native groundwater sources. Coastal ecosystems, in particular, benefit from the prevention of saline intrusion, safeguarding biodiversity and supporting the health of critical habitats.

10.8.5 Community Engagement and Water Conservation

The success of the Sheikh Zayed Canal is intertwined with community engagement and a broader commitment to water conservation. Local communities play a pivotal role in the responsible use of water resources, and awareness programs educate residents about the importance of sustainable water practices. Encouraging water conservation at the individual and community levels is essential for the long-term success and resilience of the canal project.

10.8.6 Future Directions and Global Relevance

The Sheikh Zayed Canal sets a precedent for water management strategies in arid regions worldwide, particularly those facing similar challenges of water scarcity and salinity intrusion. The innovative use of desalinated seawater for aquifer recharge offers insights and inspiration for regions grappling with depleting groundwater reserves and the impacts of climate change.

Future directions for the Sheikh Zayed Canal project involve continuous refinement and adaptation. Embracing technological advancements, exploring alternative water sources, and expanding the canal's reach are considerations for enhancing the project's impact. As climate change intensifies, proactive and adaptive water management solutions like the Sheikh Zayed Canal become crucial for building resilience and ensuring sustainable water futures.

10.9 THE VOLKERAK-ZOOMMEER PROJECT, NETHERLANDS

In the low-lying expanse of the Netherlands, where water management is not just a necessity but a defining aspect of national identity, the Volkerak-Zoommeer Project stands as a testament to the country's ingenuity in harnessing water resources. Renowned for its expertise in aquatic engineering, the Netherlands has implemented this ambitious project to address water quality concerns and bolster aquifer recharge. The initiative involves the strategic transformation of the Volkerak-Zoommeer estuary into a freshwater reservoir, effectively isolating it from the encroachment of saline seawater. This innovative solution not only ensures the availability of freshwater resources but also prevents salinity intrusion, contributing to sustainable aquifer recharge in the region.

10.9.1 Water Management Expertise in the Netherlands

The Netherlands, with its intricate network of canals, dikes, and polders, is a global leader in water management. The nation's low-lying topography and proximity to the sea demand unparalleled expertise in safeguarding against flooding and optimizing water usage. The Volkerak-Zoommeer Project emerges as a pinnacle of this expertise, showcasing the country's commitment to innovative solutions for addressing water quality issues and sustaining aquifers.

10.9.2 Introduction to the Volkerak-Zoommeer Project

Named after the estuaries it transforms, the Volkerak-Zoommeer Project is a multifaceted endeavor designed to reconcile water quality and aquifer recharge concerns. The project's focal point is the Volkerak-Zoommeer estuary, an area prone to salinity intrusion due to its proximity to the North Sea. By enclosing and isolating this estuary, the initiative aims to create a freshwater reservoir, contributing to regional aquifer recharge while ensuring the quality of the available water resources.

Enclosing the Estuary

The heart of the Volkerak-Zoommeer Project lies in the strategic enclosure of the estuary, effectively creating a barrier between the saline waters of the North Sea and the freshwater resources within. Dikes and barriers are meticulously engineered to isolate the Volkerak-Zoommeer estuary from seawater, preventing the intrusion of salinity into the estuarine environment. This process transforms the area into a controlled freshwater reservoir, setting the stage for sustainable aquifer recharge.

Water Quality Improvement

One of the primary objectives of the project is to address water quality concerns within the Volkerak-Zoommeer estuary. Historically, the estuary has faced challenges related to nutrient loading and algal blooms, impacting both the ecological balance and the quality of water resources. By isolating the estuary from saline seawater, the project aims to mitigate these issues and improve the overall water quality. The reduced salinity levels create a more conducive environment for freshwater species, promoting biodiversity and ecological health. Additionally, the controlled conditions within the enclosed reservoir allow for targeted management of nutrient levels, minimizing the risk of eutrophication and enhancing the overall ecological resilience of the area.

10.9.3 Balancing Freshwater Availability

The creation of a freshwater reservoir within the Volkerak-Zoommeer Project addresses a critical need in the Netherlands - the balance between freshwater availability and the risk of salinity intrusion. In a country where a significant portion of the land lies below sea level, preserving freshwater resources is essential for agriculture, industry, and domestic use. The project ensures a reliable and sustainable source of freshwater for both immediate consumption and strategic aquifer recharge.

Contributing to Aquifer Recharge

The freshwater reservoir formed as part of the Volkerak-Zoommeer Project serves as a strategic contributor to regional aquifer recharge. Groundwater replenishment is a vital aspect of maintaining water balance in the Netherlands, where extensive land use and extraction of groundwater for various purposes can lead to aquifer depletion. The controlled release of freshwater from the reservoir facilitates gradual infiltration into the underlying soil, contributing to the replenishment of aquifers in the region. This intentional aquifer recharge aligns with the Netherlands' broader goals of ensuring water security, preserving natural ecosystems, and sustaining agricultural productivity. The availability of a reliable source of freshwater for recharge purposes helps maintain optimal groundwater levels, reducing the risk of subsidence and ensuring the long-term sustainability of water resources.

10.9.4 Adaptive Management and Continuous Monitoring

The success of the Volkerak-Zoommeer Project hinges on adaptive management and continuous monitoring. The engineered barriers and infrastructure must be regularly assessed to ensure their effectiveness in preventing salinity intrusion. Additionally, water quality parameters within the freshwater reservoir are closely monitored to track changes in nutrient levels, algal blooms, and overall ecological health.

Adaptive management allows authorities to fine-tune the operation of the reservoir based on evolving environmental conditions and emerging challenges. The integration of real-time monitoring systems and data analytics ensures a proactive approach to addressing issues, contributing to the long-term success of the project.

10.9.5 Environmental and Ecological Benefits

Beyond its immediate contributions to water quality improvement and aquifer recharge, the Volkerak-Zoommeer Project yields broader environmental and ecological benefits. The creation of a controlled freshwater reservoir supports diverse aquatic ecosystems adapted to lower salinity levels. Native flora and fauna thrive in the altered conditions, fostering biodiversity and contributing to the overall health of the estuarine environment.

The project's positive environmental impacts extend to the surrounding terrestrial ecosystems as well. The availability of freshwater resources can influence the health of wetlands, marshes, and other habitats dependent on reliable water sources. This interconnectedness highlights the project's role in preserving the delicate balance between aquatic and terrestrial ecosystems.

10.9.6 Community Engagement and Stakeholder Collaboration

The success of the Volkerak-Zoommeer Project is contingent on active community engagement and collaboration with various stakeholders. Local communities, including farmers, residents, and businesses, play a crucial role in the responsible use and management of water resources. Outreach programs and educational initiatives inform stakeholders about the project's objectives, benefits, and the importance of water conservation.

Collaboration with environmental organizations, research institutions, and governmental agencies ensures a holistic approach to water management. Stakeholder input and feedback contribute to the adaptive nature of the project, fostering a sense of shared responsibility for the sustainable use of water resources.

10.9.7 Future Directions and Global Relevance

The Volkerak-Zoommeer Project sets a precedent for sustainable water management strategies, not only within the Netherlands but globally. As climate change intensifies and water scarcity becomes a pervasive challenge, similar initiatives that balance freshwater availability and prevent salinity intrusion gain global relevance.

Future directions for the project involve leveraging technological advancements, such as artificial intelligence and remote sensing, to enhance monitoring and adaptive management. Additionally, the lessons learned from the Volkerak-Zoommeer Project can inspire and inform similar endeavors in regions facing analogous water management challenges.

10.10 CONCLUSION

The success stories of aquifer recharge projects around the world underscore the importance of innovative approaches, adaptive strategies, and community engagement in securing sustainable water resources. These case studies demonstrate the versatility of aquifer recharge methods, ranging from advanced purification systems to community-driven initiatives and large-scale infrastructure projects. As global water challenges persist, the lessons gleaned from these success stories offer valuable insights for designing effective and context-specific aquifer recharge strategies that promote groundwater sustainability.

BIBLIOGRAPHY

- Applied Hydrogeology by C.W. Fetter (2001)
- Applied Hydrogeology of Fractured Rocks by B.B.S. Singhal and R.P. Gupta (2010)
- Applied Hydrogeophysics by Harry Vereecken, Andrew Binley, and Shiguo Wu (2015)
- Environmental Hydrogeology by Andy D. Ward and Paul L. Williams (1998)
- Environmental Isotopes in Hydrogeology by Ian D. Clark and Peter Fritz (1997)
- Field Hydrogeology by Rick Brassington (2016)
- Field Methods for Geologists and Hydrogeologists by F. P. Haeni and H. R. Krouse (2009)
- Field Techniques for Estimating Water Fluxes between Surface Water and Ground Water by Allen M. Shapiro (2003)
- Geostatistics for Environmental Scientists by Richard Webster and Margaret A. Oliver (2007)
- Geothermal Reservoir Engineering by Malcolm Grant and Marcia Grant (2011)
- Groundwater and Seepage by Milton E. Harr (1962)
- Groundwater and Subsurface Environments: Human Impacts in Asian Coastal Cities by Mukand S. Babel and Shah Alam Khan (2011)
- Groundwater by R. Allan Freeze and John A. Cherry (1979)
- Groundwater Contamination: Sources, Control, and Prevention by David W. Mercer and J. T. Bennett (1990)
- Groundwater Contamination: Use of Models in Decision-Making by N.S. Robins (1995)

- Groundwater Geochemistry: A Practical Guide to Modeling of Natural and Contaminated Aquatic Systems by Broder J. Merkel and Britta Planer-Friedrich (2014)
- Groundwater Hydrology by B.M. Das (2018)
- Groundwater Hydrology by David Keith Todd and Larry W. Mays (2005)
- Groundwater Hydrology of Springs: Engineering, Theory, Management and Sustainability by Neven Kresic (2010)
- Groundwater Lowering in Construction: A Practical Guide to Dewatering by Martin Preene (2001)
- Groundwater Monitoring Handbook for Coal and Oil Shale Development by Charles H. Voss (1983)
- Groundwater Remediation and Treatment Technologies by Vikram J. Sahai and Krishna R. Reddy (2011)
- Groundwater Resources: Sustainability, Management, and Restoration by Kresic and William (2010)
- Groundwater Science and Policy: An International Overview by Jenny Pickerill and Pauline Smedley (2010)
- Groundwater Science by Charles R. Fitts (2012)
- Handbook of Groundwater Remediation using Permeable Reactive Barriers: Applications to Radionuclides, Trace Metals, and Nutrients by Peter S. C. Rao and J. Scott (2002)
- Hydrogeological Conceptual Site Models: Data Analysis and Visualization by Neven Kresic and Alex Mikszewski (2014)
- Hydrogeological Processes in Karst Terrains by Neven Kresic (2010)
- Hydrogeology and Groundwater Modeling by Neven Kresic (2007)
- Hydrogeology Field Manual by Willis D. Weight (1998)
- Hydrogeology: An Introduction by W. Gregory and Ray R. Lewis (1995)
- Hydrogeology: Field Evaluation of Aquifers in Karst Terranes by Aley K. Ehrlich and William B. White (1999)
- Hydrogeology: Objectives, Methods, Applications by R. H. Kirsch (2013)
- Hydrogeology: Principles and Practice by E. P. F. Rose and W. G. Darling (2002)
- Hydrogeology: Principles and Practice by Kevin M. Hiscock (2005)
- Hydrogeology: Problems with Solutions by A.M. Piper (1998)
- Hydrogeophysics by Yoram Rubin and Shlomo P. Neuman (2010)

- Introduction to Hydrogeology by Warren Viessman Jr. and Gary L. Lewis (2003)
- Introduction to Hydrogeophysics by Harry Vereecken and Andrew Binley (2015)
- Introduction to Physical Hydrology by Martin Hendriks (2010)
- Karst Hydrogeology and Geomorphology by Derek Ford and Paul Williams (2007)
- Practical Hydrogeology by David George and Malcolm Reeves (2016)
- Principles of Environmental and Engineering Geophysics by G. P. Mahar (1995)
- Principles of Hydrogeology by Paul F. Hudak and Michael J. Starn (2018)
- Quantitative Hydrogeology: Groundwater Hydrology for Engineers by L. P. Bear and Jacob Bear (1996)

INDEX

www.ingramcontent.com/pod-product-compliance
Lightning Source LLC
LaVergne TN
LVHW062114200926
848484LV00007B/1668